ANDEAN MAGMATISM
CHEMICAL AND
ISOTOPIC CONSTRAINTS

ANDEAN MAGMATISM
CHEMICAL AND ISOTOPIC CONSTRAINTS

R.S. Harmon and B.A. Barreiro (Editors)
Southern Methodist University, Dallas and Dartmouth College, Hannover

Shiva Publishing Limited

SHIVA PUBLISHING LIMITED
4 Church Lane, Nantwich, Cheshire CW5 5RQ, UK

Available in North America from:
BIRKHAUSER BOSTON, INC.
P.O. Box 2007, Cambridge, MA 02139, USA

British Library Cataloguing in Publication Data

Andean magmatism.—(Shiva geology series)
1. Magmatism—Andes
I. Harmon, R.S.—Barreiro, B.A.
552′.1 QE461

ISBN 0-906812-60-7
ISBN 0-906812-61-5 Pbk

Typeset by Gilbert Composing Services, Leighton Buzzard
and printed by J.W. Arrowsmith Ltd., Bristol.

Contents

Contributors

M.P. Atherton
Department of Geology, University of Liverpool, P.O. Box 147, Liverpool L69 3BX, UK

B.A. Barreiro
Department of Earth Sciences, Dartmouth College, Hannover, New Hampshire 03755, USA

D.S. Bartholomew
Department of Geology, University of Leicester, Leicester LE1 7RH, UK

M. Boily
Departement de Geologie, Universite de Montreal, CP 6128, Succ. 'A', Montreal, Quebec H3C 3J7, Canada

C. Brooks
Departement de Geologie, Universite de Montreal, CP 6128, Succ. 'A', Montreal, Quebec H3C 3J7, Canada

R. Charrier
Departamento de Geologia y Geofisica, Universidad de Chile, Casilla 13518, Santiago, Chile

A.H. Clark
Department of Geological Sciences, Queen's University, Kingston, Ontario K7L 3N6, Canada

F.M. Dobbs
Departamento de Minas, Universidad de Santiago, Casilla 10233, Santiago, Chile

E. Farrar
Department of Geological Sciences, Queen's University, Kingston, Ontario K7L 3N6, Canada

P.W. Francis
Lunar and Planetary Institute, 3303 NASA Road 1, Houston, Texas 77058, USA

F.A. Frey
Department of Earth, Atmospheric and Planetary Sciences, Massachusetts Institute of Technology, Cambridge, Massachusetts 02139, USA

L.A. Fultz
Museum of Natural History, University of Nevada, Las Vegas, Reno, Nevada 89503, USA

K. Futa
USGS, MS963 Denver Federal Center, Denver, Colorado 80225, USA

D.C. Gerlach
Department of Earth, Atmospheric and Planetary Sciences, Massachusetts Institute of Technology, Cambridge, Massachusetts 02139, USA

E. Godoy
Departamento de Geologia y Geofisica, Universidad de Chile, Casilla 13518, Santiago, Chile

N.K. Grant
Department of Geology, Miami University, Oxford, Ohio 45056, USA

M. Hammill
Lunar and Planetary Institute, 3303 NASA Road 1, Houston, Texas 77058, USA; *Now at* The Open University, Milton Keynes, UK

R.S. Harmon
Department of Geological Sciences, Southern Methodist University, Dallas, Texas 75275, USA

R.L. Hickey
Department of Earth, Atmospheric and Planetary Sciences, Massachusetts Institute of Technology, Cambridge, Massachusetts 02139, USA

J. Hoefs
Geochemisches Institut der Universitat Gottingen, D-3400 Gottingen, West Germany

D.E. James
Carnegie Institution of Washington, Department of Terrestrial Magnetism, 5241 Broad Branch Road, NW Washington DC 20015, USA

D.J. Kontak
Department of Geological Sciences, Queen's University, Kingston, Ontario K7L 3N6, Canada

L. López-Escobar
Departamento de Geologia y Geofisica, Universidad de Chile, Casilla 13518, Santiago, Chile

W.F. McDonough
Lunar and Planetary Institute, 3303 NASA Road 1, Houston, Texas 77058, USA; *Now at* The Open University, Milton Keynes, UK

K. Muehlenbachs
Department of Geology, University of Alberta, Edmonton, Alberta T6G 2E3, Canada

S.B. Mukasa
Department of Geological Sciences, University of California, Santa Barbara, California 93106, USA

J. Muñoz
Servicio Nacional de Geologia y Mineria, Casilla 10465, Santiago, Chile

L.J. O'Callaghan
Lunar and Planetary Institute, 3303 NASA Road 1, Houston, Texas 77058, USA; *Now at* The Open University, Milton Keynes, UK

W.S. Pitcher
The Jane Herdman Laboratories of Geology, University of Liverpool, Brownlow Street, P.O. Box 147, Liverpool L69 3BX, UK

W.I. Rose, Jr
Department of Geology and Geological Engineering, Michigan Technological University, Houghton, Michigan 49931, USA

C.R. Stern
Department of Geological Sciences, University of Colorado, Boulder, Colorado 80309, USA

J. Tarney
Department of Geology, University of Leicester, Leicester LE1 7RH, UK

R.S. Thorpe
Department of Earth Sciences, The Open University, Walton Hall, Milton Keynes MK7 6AA, UK

G.R. Tilton
Department of Geological Sciences, University of California, Santa Barbara, California 93106, USA

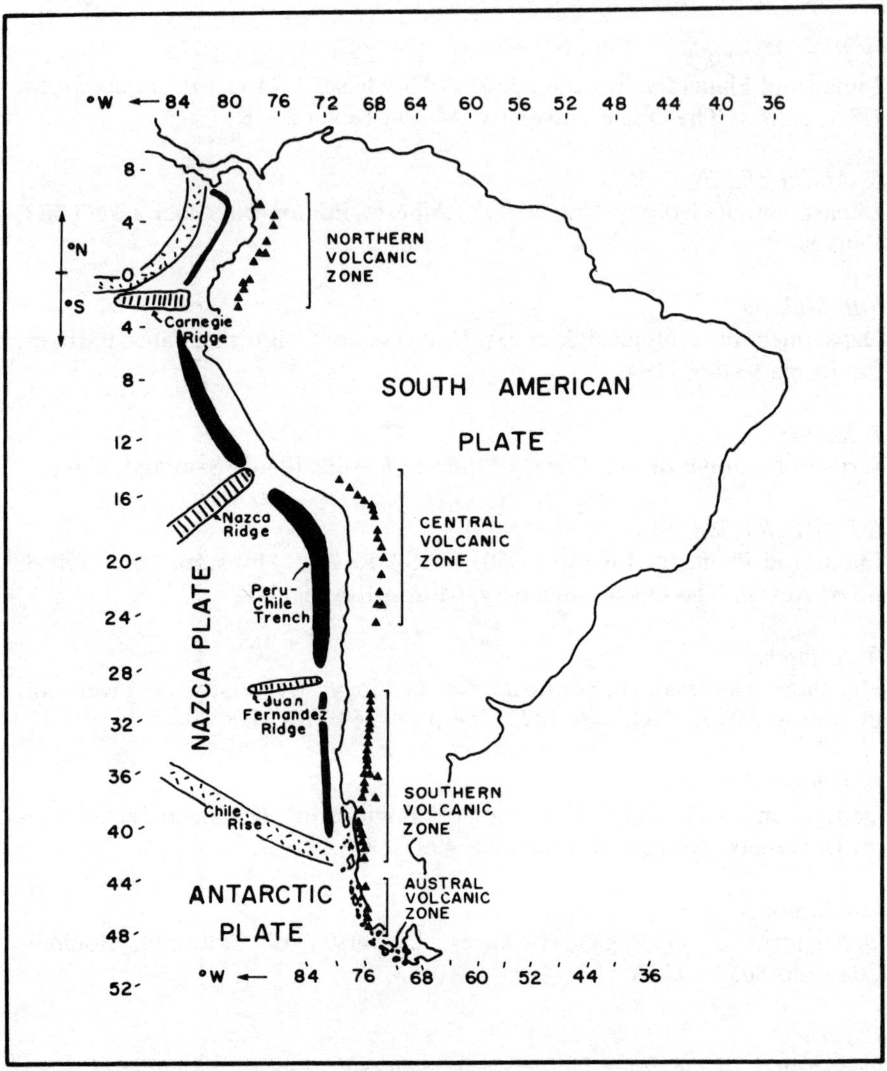

Outline map of South America showing the major tectonic features associated with orogenic magmatism in the Andean Cordillera, the boundaries of the volcanic provinces, and the locations of Late Cenozoic volcanic centres (triangles). Prepared by L. Lòpez-Escobar

Preface

Destructive plate margin magmagenesis is one of the most intensely studied and widely debated topics in the earth sciences at present. Calc-alkaline volcanic and plutonic rocks in orogenic settings exhibit such a diversity of composition and character that the subducted oceanic lithosphere, the overlying 'enriched' lithospheric mantle and the lower continental crust all have been advocated recently as the primary source region for island-arc and continental-margin basaltic to andesitic magmas. The role of the upper continental crust is also a matter of continuing controversy. It is clear that crustal contamination is a common, but not universal, feature of destructive plate margin magmatism. Whether this contamination is introduced at source by subduction-related processes or occurs during magma transit by bulk anatexis and magma mixing, selective contamination or coupled fractional crystallization-assimilation mechanisms is central to most current discussions of andesite petrogenesis.

This book presents a series of papers which directly address these and other important geological and geochemical problems within the context of the Mesozoic-Cenozoic calc-alkaline magmatism characteristic of the Andean Cordillera of western South America. Although it is aimed primarily at postgraduate students and researchers familiar with the Andes, it is also a useful general reference for workers in other fields who wish to gain an insight into current thoughts, ideas and speculations on 'andesitic' magmatism at destructive plate margins. The 16 papers contained in the book were presented as part of a special symposium of the American Geophysical Union held in Baltimore, Maryland (USA) in June 1983. The contributors include geologists, petrologists and geochemists of various specializations who each bring their own varied perspectives to bear on a problem of common interest—Andean calc-alkaline magmatism.

The individual papers in the volume are divided into two groups, those dealing with aspects of the Late Cenozoic 'andesitic' volcanism (of which the Andes is the type locality) and those considering primarily Mesozoic-Tertiary plutonism. Within each group the ordering of papers represents an evolution in scale from regional reconnaissance studies to detailed case studies of individual volcanic centres or intrusive complexes. The last paper in the volcanic section provides a useful comparison between Late Cenozoic 'andesitic' volcanism in central America with that in the Andes. The geochemical and isotopic data presented in the book do indeed provide some constraints on the origin and pre-eruption evolutionary history of Andean magmas. It is important to note at the outset, however, that the papers contained in the book do not arrive at a concensus regarding the questions addressed. The reader will find real differences in interpretation and opinion as well as a wide range of hypotheses advocated to explain the different aspects of Andean magmatism addressed by the various authors.

The debate is by no means over and many of the problems discussed here will be a focus of continuing controversy for some time to come. Nevertheless, we hope that this volume points out the utility and importance of multi-isotope and combined chemical-isotope approaches when undertaken within a framework of detailed field and petrographic studies. Our thanks are extended to the authors who contributed to the book. Their enthusiasm and timely preparation of the manuscripts after the Baltimore meeting made it possible to get the book completed without significant delay. Also, we would like to thank the symposium participants who did not contribute to the book for providing a stimulating forum for discussion and, in particular, P. J. Patchett who was instrumental in planting the seed of an idea that eventually blossomed into the symposium. We are very grateful to J. L. Atkinson for her assistance with reading the proofs. Finally our collective thanks are due to the staff of Shiva Publishing.

R.S. HARMON, Southern Methodist University, Dallas
B.A. BARREIRO, Dartmouth College, Hannover

November 1983

Section 1: Volcanic Andes

The Tectonic Setting of Active Andean Volcanism

R. S. THORPE

Abstract

The Andean plate margin of South America is characterized by the seismicity, tectonism and magmatism associated with the descent of the oceanic Nazca plate below the South American plate. The active volcanism occurs within segments in south Colombia—north Ecuador (Northern Volcanic Zone, NVZ: *ca.* 5°N–2°S), south Peru—north Chile (Central Volcanic Zone, CVZ: *ca.* 16–28°S) and in south Chile (Southern Volcanic Zone, SVZ: south of *ca.* 33°S). The active volcanic zones are underlain by more steeply-dipping oceanic lithosphere (*ca.* 30°) than that below the intervening non-volcanic zones (*ca.* 5–10°). This correlation indicates that the processes responsible for Andean magmatism are initiated within the mantle. The petrogenesis of Andean magmas involves formation of basaltic melts, followed by assimilation and fractional crystallization (AFC) during crustal ascent. Within this framework the basaltic-andesite-rhyolite association of the SVZ evolved largely by fractional crystallization of basaltic magma, while the basaltic andesites and andesites of the NVZ may have evolved by AFC of basaltic magma. The more evolved andesite-dacite lavas of the CVZ result from fractional crystallization of basaltic magma in association with assimilation (AFC) or crustal melting. Crustal growth within the central Andes has been accompanied by an increase in the role of crustal material as a result of AFC, and crustal melting processes relative to fractional crystallization of mantle-derived basaltic magmas. There are hence important links between lithospheric structure, tectonic history and magmatism within the Andes.

The descent of the oceanic Nazca plate below western South America is marked by the seismicity, tectonism and magmatism considered characteristic of an Andean plate margin. The Andean margin comprises a magmatic (intrusive/volcanic) arc flanked on the west by an oceanic trench (the Peru-Chile trench) and on the east by a foreland thrust belt and basin. Although the Andes form a morphologically continuous mountain chain from *ca.* 5–45°S (*ca.* 4000 km), distinct broad-scale tectonic segments are located above segments of similar scale in the subducted Nazca plate, indicated by major localized variations in the dip and morphology of the Benioff zone (Barazangi and Isacks, 1976, 1979). The tectonic segments have distinctive tectonic and magmatic characteristics, and some boundaries coincide with ancient crustal features. The mechanism of segmentation may therefore involve interaction of features in the descending plate with structures in the upper lithospheric plate (Jordan *et al.*, 1983).

The major tectonic segments of the Andes in relation to the depth of the Benioff zone are shown in Figure 1. Based upon the seismic characteristics of the descending Nazca plate the Andes may be divided into five segments. In Peru (*ca.* 2–15°S) and central Chile (*ca.* 27–33°S), the Benioff zone dips at only 5–10° towards the east, and upper plate seismicity is relatively abundant and shows localized clustering. By contrast, in south Colombia-north Ecuador (*ca.* 5°N–2°S), south Peru—north Chile (*ca.* 16–28°S) and south Chile (south of *ca.* 33°S), the Benioff zone is inclined at *ca.* 30° towards the east.

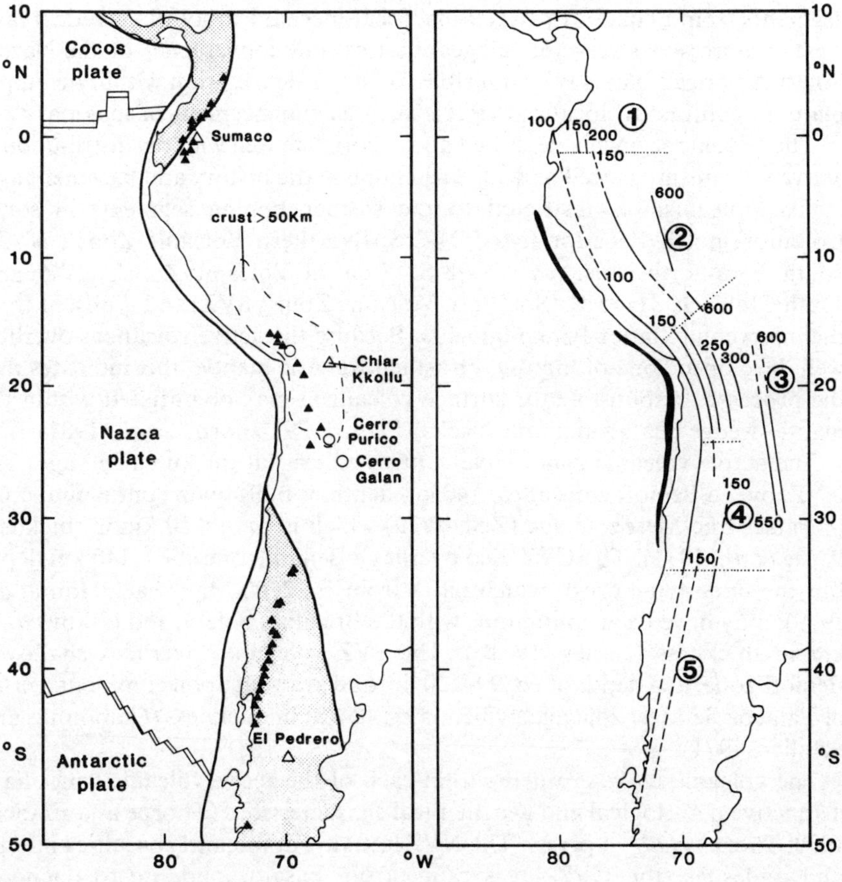

Figure 1 (a) Map of western South America showing the plate tectonic framework in relation to active volcanism, crustal age and thickness. ▲—active calc-alkaline basalt-andesite, dacite volcanoes (MacDonald, 1972); △—some alkaline volcanoes; ○—volcanic areas referred to in the text; stippled areas represent continental areas underlain by crust of Palaeozoic or younger age, and other areas may be underlain by Precambrian crust. The location of crust exceeding 50 km in thickness is shown (from Thorpe *et al.*, in press). (b) Contours to the top of the Benioff zone below the Andes (km). Dashed lines indicate that the contours are based on fewer data than shown by solid lines. Solid ornament indicates oceanic trenches exceeding 5000 m depth. The numbers indicate the location of the major tectonic segments of the Andes as discussed in the text (from Barazangi and Isacks, 1976)

The down-slope lengths of the two shallow-dipping plates are *ca.* 750 km from the trench and reach a depth of *ca.* 160 km in comparison with *ca.* 650–750 km for the more steeply dipping segments which reach a depth of *ca.* 300 km. The southernmost segment (south of *ca.* 33°S) reaches only 500 km from the trench axis at a depth of *ca.* 160 km. These lengths of subducted lithosphere are consistent with subduction during the time period since the Miocene (Wortel and Vlaar, 1978). The crust above the more steeply-dipping segments of the subducted slab is characterized by less seismicity than that above the relatively flat/subducted segments. For upper plate earthquakes within *all* segments from Ecuador to Argentina, focal-mechanism solutions reflect east-west compressive stress and suggest that overall convergence of the Nazca-South American plates is responsible for stress distribution within the upper plate, in contrast to localized effects such as emplacement of magma.

The tectonic segments described above correlate well with the distribution of active volcanism, as well as with earlier magmatic history and basement age. Active volcanism is restricted to the steeper-dipping segments in south Colombia-north Ecuador (*ca.* 5°N–2°S; 'Northern Volcanic Zone', NVZ), south Peru-north chile, (*ca.* 16–28°S; 'Central Volcanic Zone', CVZ) and south Chile (*ca.* 31–55°S; 'Southern Volcanic Zone', SVZ) and is absent from the intervening flatter-lying segments. Because the active volcanoes overlie a well-developed zone of high-Q, or asthenospheric mantle, this indicates that the processes responsible for surface volcanism may be initiated within the mantle wedge (Barazangi and Isacks, 1976, 1979; Thorpe *et al.*, 1981).

The active volcanic zones have contrasted crustal thickness and age. The NVZ, over a Benioff zone at *ca.* 140 km depth, is built upon continental crust of Palaeozoic-Mesozoic age (Zeil, 1979) which is *ca.* 40–50 km in thickness (Case *et al.*, 1971). The CVZ also overlies a Benioff zone at *ca.* 140 km depth but the underlying crust, which dates from *ca.* 2000 Ma (Shackleton *et al.*, 1979), may have been contiguous with the Brazilian Shield, and is now *ca.* 70 km in thickness (James, 1971b). The SVZ volcanoes overlie a shallower Benioff zone, at a depth of *ca.* 90–120 km and are built upon continental crust of Palaeozoic-Mesozoic age, which is *ca.* 35 km in thickness (Cummings and Schiller, 1971).

The volcanic rocks erupted within each of the active volcanic zones have distinctive petrological and geochemical characteristics (Thorpe and Francis, 1979; Thorpe *et al.*, in press). The NVZ lavas are dominantly basaltic andesite and andesites, the CVZ lavas range from basaltic andesite to dacite in composition but are dominantly andesite and dacite, while the SVZ lavas are dominantly basalt and basaltic andesite in composition. The geochemical characteristics of the SVZ lavas correspond to island-arc calc-alkaline volcanic rocks while those of the CVZ lavas are enriched in K, Rb, Th and U in comparison with such lavas and approach the overall composition of the continental crust. The NVZ lavas are intermediate between the CVZ and SVZ lavas in these characteristics (see Thorpe *et al.*, in press; Harmon *et al.*, in press). The lavas of the active volcanic zones also show distinctive Sr-, O-, Nd- and Pb-isotopic characteristics. The NVZ and SVZ lavas generally show narrow, but slightly distinct ranges in Sr-, O-, Nd- and Pb-isotope composition, while the CVZ lavas have higher (more radiogenic) Sr- and Pb-isotope ratios, lower (less radiogenic) Nd isotope ratios and higher $\delta^{18}O$ values, accompanied by a much greater *range* in isotope composition in comparison with the NVZ and SVZ lavas (see discussions which follow).

The petrogenesis of the Andean lavas may be explained on the basis of a model involving dehydration of the subducted Nazca plate, causing addition of 'subduction zone' components into the overlying mantle wedge. Below areas underlain by steeply-dipping lithosphere, dehydration takes place over a more restricted lateral extent in comparison with areas underlain by shallow-dipping lithosphere. In the latter segments, the 'subduction-zone' mantle does not immediately yield magma, but may be capable of extensive regional magmatism as a result of a later change in tectonic situation. Basaltic magmas derived from such mantle will be underplated at the base of the crust and may experience assimilation and fractional crystallization during subsequent uprise through the crust. The chemical and isotopic characteristics of the NVZ basaltic andesites and andesites, and the SVZ basalts, basaltic andesites and andesites are consistent with derivation by fractional crystallization of basaltic parent magmas formed by partial melting of the asthenospheric mantle wedge containing components from subducted oceanic lithosphere. However the higher Sr-, Pb- and O-isotope ratios of the NVZ lavas and their more evolved compositions in comparison with the SVZ lavas may indicate some assimilation of Palaeozoic or younger continental crust during fractional crystallization (Harmon *et al.*, in press; James, in press).

A variety of petrogenetic schemes have been proposed to account for the characteristics of the CVZ lavas (Thorpe *et al.*, in press). The CVZ basaltic andesites may be derived from mantle containing subduction zone components possibly accompanied by minor assimilation-fractional crystallization (AFC) during upward transit through the crust. The more evolved andesites and dacites may result from more extensive AFC processes (cf. Thorpe *et al.*, in press) involving local Precambrian crust, or from anatexis of lower crustal material (Hawkesworth *et al.*, 1982). In addition some CVZ lavas and pyroclastic rocks show petrological and geochemical evidence for magma mixing and the petrogenesis of the CVZ lavas may therefore be a complex process in which mantle-derived magmas experience assimilation, fractional crystallization and magma mixing during uprise through the continental crust.

These models may be used to comment on Andean magmagenesis from the Mesozoic to the present time. Within the CVZ there is a general age progression from Mesozoic (or older) igneous rocks near to the coast through Cenozoic rocks at intermediate elevation to the active volcanoes of the Andes. Transects at *ca.* 17–20°S and *ca.* 26–29°S reveal a broad correlation between magma composition and the physiographic development of the Andes, and the time-space evolution of the magmatic arc (Clark and McNutt, 1982). Within 26–29°S, rapid middle Miocene (14–11 Ma) arc broadening towards the continent (from *ca.* 40 to 350 km) coincided with a late event in episodic

Cenozoic uplift. At 17–20°S, late Oligocene-early Miocene uplift (*ca.* 26 Ma) uplift was also coeval with contemporaneous arc broadening which was more extensive than in the northern transect. In both areas, Pliocene topographic uplift (*ca.* 10 Ma) was accompanied by arc contraction and was followed by Pliocene-Recent uplift and volcanism (since *ca.* 4 Ma). The greater crustal thickness in the north (*ca.* 70 km) in comparison with that in the south (*ca.* 50 km) implies more rapid thickening in the north, but the timing and reason for localization of such thickening are still uncertain.

Chemical studies of Mesozoic-Recent igneous rocks from the southern transect (McNutt *et al.*, 1975; Longstaffe *et al.*, 1983) show a gradual increase in initial $^{87}Sr/^{86}Sr$ from *ca.* 80 Ma to the present, of 0.703 to 0.707, without an increase in $\delta^{18}O$ (within the range +6.2 to +8.3‰). This may be interpreted in terms of a shallowing of the site of magma generation due to rise of isotherms within a heterogeneous mantle, or the increasing assimilation of continental crustal material with time. By contrast, within the northern transect, the initiation of Oligocene-early Miocene uplift was accompanied by an abrupt increase in initial $^{87}Sr/^{86}Sr$ in volcanic rocks, such that Pliocene-Recent andesite and dacite lavas have initial $^{87}Sr/^{86}Sr$ values in the range of 0.706–0.710 and $\delta^{18}O$ values of +8 to +10‰ (Harmon *et al.*, 1981; Hawkesworth *et al.*, 1982). This indicates that crustal thickening within the area was accompanied by a sudden increase in assimilation associated with fractional crystallization processes (Thorpe *et al.*, in press) or a change from a mantle to a crustal magma source region (Hawkesworth *et al.*, 1982).

The data summarized above indicate that there appear to be important links between lithospheric structure, tectonism and magmatism within the Andes. These links are interpreted in terms of tectonic segmentation and there may be further correlations between surface geology and segmentation such as those identified elsewhere (e.g. Central America; Carr *et al.*, 1982). This segmentation may be used to interpret the recent geological history of the Andes and may form the basis for studies of the initiation of magmatism, the rise of magma and the subsequent interaction with the mantle and continental crust prior to eruption within an active volcanic zone.

Oxygen Isotope Ratios in Late Cenozoic Andean Volcanics

R. S. HARMON and J. HOEFS

Abstract

O-isotope ratios for 127 Cenozoic volcanic rocks from the Andean Cordillera between 5°N and 4°S range from +5.2 to +14.0‰ (SMOW). Lavas from eight volcanoes in the SVZ between 36–42°S exhibit a very narrow range of $\delta^{18}O$ values (+5.2 to +6.7‰), with O-isotope variations independent of bulk chemical composition. The highest O-isotope ratios occur in the CVZ where lavas from 16 volcanic centres between 16–26°S have a wide range of $\delta^{18}O$ values (+6.9 to +14.0‰), with $^{18}O/^{16}O$ ratios well correlated with both bulk chemical composition and radiogenic isotope variations at some centres. In the NVZ, between 5°N and 2°S, lavas from nine volcanoes have intermediate O-isotope compositions ($\delta^{18}O = +6.3$ to 7.7‰), which are unrelated to chemical composition. The low $^{18}O/^{16}O$ ratios for the lavas of the SVZ are most easily explained by derivation from a mantle source, with the parent magmas rising largely unmodified through the thin (<35 km) continental crust and differentiation occurring by fractional crystallization, which did not significantly modify their O-isotope composition. The high $^{18}O/^{16}O$ ratios and large range of $\delta^{18}O$ values documented for the lavas of the CVZ requires a large crustal component in all lavas. This was probably introduced by the extensive interaction of mantle-derived melts with lower and/or upper crust during transit through abnormally thick (>60 km) continental crust of the region. The intermediate O-isotope character for the lavas of the NVZ suggests that both mantle and crustal sources were important in the petrogenesis of lavas in the region.

Introduction

The O-isotope composition of any unaltered igneous rock is determined by four basic factors: (a) the $^{18}O/^{16}O$ ratio of the source region in which the original magma was generated, (b) the temperature of magma generation and crystallization, (c) its mineralogical composition, and (d) the evolutionary history of the magma prior to its eruption or emplacement and final crystallization. Provided an igneous rock has not been affected by post-crystallization, subsolidus isotope exchange or hydrothermal alteration, its O-isotope composition should be a sensitive petrogenetic indicator which will provide first-order information about both its origin and evolutionary history.

Source-liquid fractionation effects are small at the high temperatures at which magmas are typically generated (Muehlenbachs and Kushiro, 1974; Kyser *et al.*, 1981) and independent of pressure (Clayton *et al.*, 1975). Thus, a partial melt should have an O-isotope composition representative of the source region in which it was generated. In silicate minerals the highly polymerized Si–O–Si structure concentrates ^{18}O relative to the Si–O–Al structure, which in turn, concentrates ^{18}O relative to the Si–O–M structure (Taylor and Epstein, 1962). As a result, the continental crust is more ^{18}O rich than the mantle and mafic lower crust. This feature is enhanced by the fact that crustal materials, which have been through a low-temperature sedimentary cycle at the earth's surface involving weathering, chemical precipitation and diagenesis, are highly enriched in ^{18}O (Savin and Epstein, 1970).

Relatively large differences in O-isotope composition are observed between fine-grained volcanic rocks and coarse-grained plutonic rocks of equivalent composition (Taylor, 1968; Anderson *et al.*, 1971). Mineral fractionations increase with decreasing temperature (see recent summary of mineral-fluid fractionation data in Friedman and O'Neil, 1977). For example, plagioclase crystallized from a melt at 1000°C would be depleted in ^{18}O by about 2‰ relative to plagioclase crystallized from a melt of the same composition at 600°C (Matsuhisa *et al.*, 1979).

Because it is mineral structure and stoichiometry which influence the O-isotope composition of silicate minerals and not simple atomic substitution (O'Neil and Taylor, 1967; Anderson *et al.*, 1971; Matsuhisa, 1979), crystal fractionation might be expected to play an important role in determining the ultimate $^{18}O/^{16}O$ ratio of any particular igneous rock. The large-scale crystallization and removal of a sufficient quantity of a phenocryst phase significantly depleted or enriched in ^{18}O relative to a parental magma might be expected to produce a compensated shift in the $^{18}O/^{16}O$ ratio of the melt, provided there is no subsequent re-equilibration of the crystallized phase with the magma. However, crystal-melt fractionations are sufficiently small at volcanic temperatures ($\gtrsim$ 1000°C) that magmatic differentiation should proceed without causing a large shift in the O-isotope composition of the magma. This is substantiated by a variety of data from oceanic-island (Garlick, 1966), island-arc (Matsuhisa *et al.*, 1973) and continental-margin (Taylor, 1968; Deruelle *et al.*, 1983) volcanics which show that closed-system crystal fractionation can only produce small (0.5–0.8‰) variations in $^{18}O/^{16}O$ ratios within a cogenetic magma series without very radical changes in chemical composition.

The mantle and the various crustal reservoirs which may be important in the petrogenesis of andesitic lavas can be reasonably well discriminated on the basis of their O-isotope composition. Fresh, unaltered mid-ocean ridge tholeiitic basalts have $\delta^{18}O$ values* of +5.7 ± 0.5‰ (Muehlenbachs and Clayton, 1972; Pineau *et al.*, 1976; Kyser *et al.*, 1982). Oceanic-island tholeiitic basalts are similar in O-isotope composition to mid-ocean ridge tholeiites, whereas oceanic-island alkali basalts are slightly more ^{18}O-rich with $\delta^{18}O$ values of 6.2 ± 0.5‰ (Kyser *et al.*, 1982). Most island-arc and continental

* $\delta^{18}O = \left[\frac{(^{18}O/^{16}O)_{sample} - (^{18}O/^{16}O)_{reference}}{(^{18}O/^{16}O)_{reference}}\right] \times 10^3$ where the reference is Standard Mean Ocean Water (SMOW)

basalts have $^{18}O/^{16}O$ ratios similar to those of their oceanic counterparts, but $\delta^{18}O$ values in excess of 7.5‰ have been recorded from a variety of localities (Garlick, 1966; Taylor, 1968; Taylor *et al.*, 1979; Hoefs *et al.*, 1980). Calc-alkaline andesitic lavas, potassic lavas and carbonatite lavas from oceanic and continental environments typically have $\delta^{18}O$ values in the range +5.5 to +8.5‰, with dacites and rhyolites slightly more ^{18}O rich with $\delta^{18}O$ values in the range +6.5 to +10.5‰ (Garlick, 1966; Taylor, 1968; Pineau and Javoy, 1969; Matsuhisa *et al.*, 1973; Suwa *et al.*, 1975; Turi and Taylor, 1976; Magaritz *et al.*, 1978; Matsuhisa, 1979).

Hydrothermally altered oceanic crust and continental crust of intermediate composition tend to have quite variable $^{18}O/^{16}O$ ratios, typically within the range of $\delta^{18}O$ = +7 to +10‰ (Muehlenbachs and Clayton, 1972; Taylor, 1980). As previously noted, sedimentary rocks and their metamorphic equivalents tend to be particularly ^{18}O rich and generally have $\delta^{18}O$ values in the range of +10 to +30‰. Thus, it is to be expected that magmas derived predominantly in such ^{18}O-rich continental crust will themselves have high $^{18}O/^{16}O$ ratios. Turi and Taylor (1976) have observed that $\delta^{18}O$ values for calc-alkaline volcanic rocks of the Tuscan Magmatic Province range from +11.2 to 13.4‰, and argued that these lavas were produced by melting of the large-scale assimilation of ^{18}O-rich argillaceous sedimentary upper crust. It should also be pointed out here that there is no process of magmatic differentiation that can produce lavas with $\delta^{18}O$ values less than 5.0 to 5.5‰. Such rocks are only thought to form by interaction with ^{18}O-depleted meteoric waters in the magmatic state (Friedman *et al.*, 1974), during later subsolidus exchange (Taylor and Forester, 1971) or by the melting of, or subsequent exchange with, crustal rocks previously depleted in ^{18}O by such processes. For example, basalts in Iceland with $\delta^{18}O$ values as low as 2.3‰ are thought to be the result of contamination by low-^{18}O meteoric waters (Muehlenbachs *et al.*, 1974).

Oxygen isotope composition of Andean lavas

A total of 127 O-isotope analyses of Andean Late Cenozoic lavas from 33 volcanic centres are available from this study and the data of Magaritz *et al.* (1978), Harmon *et al.* (1981), James (1982a, b), and Deruelle *et al.* (1983). These data are summarized in Table 1 and their distribution as a function of latitude and composition is illustrated in Figure 1. Although a few centres, such as the Cerro Galan complex in northwestern Argentina (Harmon *et al.*, 1981), the Arequipa volcanics in south Peru (James, 1982a), and the Galeras volcano in southwestern Columbia (James, 1982b), have been examined in some detail, most were sampled on a limited basis as part of regional reconnaissance studies. Here we use this relatively large data set to place some basic constraints on the source region of Andean lavas, to define the role of continental crust in their origin and evolution and to compare the petrogenesis of 'andesitic' lavas occurring in the different tectonic settings of the Andean Cordillera.

Table 1 **Summary of O-isotope composition of Late Cenozoic Andean lavas**

Locality	Latitude	Composition*	Number of samples	$\delta^{18}O$ (‰ SMOW)	References
NVZ					
Southwest Columbia					
Galeras	1°18′N	A	14	6.5–7.8	James (1982b)
Purace	2°18′N	A	3	7.1–7.5	James (1982b)
Ruiz	4°52′N	A	6	6.8–7.4	James (1982b)
Ecuador					
Sincholahua	0°32′S	A	2	6.6–6.7	This study
Cotopaxi	0°40′S	BA–A	5	6.5–7.7	Harmon *et al.* (1981); this study
Punalica	0°41′S	A	2	6.6–7.7	This study
Chimborazo	1°27′S	A	1	6.9	This study
Calpi	1°37′S	A	2	6.8–7.5	Harmon *et al.* (1981); this study
Tungurahua	1°47′S	BA–A	4	6.3–7.2	This study
CVZ					
South Peru					
Arequipa volcanics	16°20′S–16°25′S	A	16	6.9–7.3	Magaritz *et al.* (1978); James (1982a)
Barroso volcanics	16°42′S–17°30′S	A–D	7	6.9–7.4	Magaritz *et al.* (1978); James (1982a)
Southwest Bolivia					
Sierra de Lipez	21°52′S	A–R	6	8.0–9.8	Deruelle *et al.* (1983)
Tocorpuri	22°26′S	A–R	5	7.3–8.8	Deruelle *et al.* (1983)
Sairecabur	22°43′S	A	1	9.0	This study

Locality	Latitude	Composition*	Number of samples	$\delta^{18}O$ (‰ SMOW)	References
North Chile					
San Pablo-San Pedro	21°52′S	A–D	4	7.3–10.6	Harmon *et al.* (1981)
Lascar	22°23′S	A–D	5	7.3–8.4	Deruelle *et al.* (1983)
Miscanti	23°41′S	A	2	8.3–8.6	This study
Cordon Puntas Negras	23°35′S–23°43′S	A	1	7.8	This study
Volcan Puntas Negras	23°45′S	A	1	8.0	This study
Laco	23°47′S	A	2	8.6–11.5	This study
El Negrillar	24°05′S–24°30′S	BA–A	2	7.6–9.1	Deruelle *et al.* (1983)
Northwest Argentina					
Cerro Chorrillos	24°14′S	BA	2	8.0–8.1	Deruelle *et al.* (1983)
San Jeronimo	24°14′S	A	1	8.0	Deruelle *et al.* (1983)
Socampa	24°25′S	BA–D	3	7.5–7.9	This study
Cerro Galan	26°00′S	B–D	14	8.2–14.0	Harmon *et al.* (1981)
SVZ					
South-Central Chile					
Nevados de Chillan	36°50′S	D	1	5.6	Deruelle *et al.* (1983)
Antuco	37°25′S	B–A	2	6.2–6.4	Deruelle *et al.* (1983)
Lonquimay	38°24′S	BA–D	3	5.2–6.0	Deruelle *et al.* (1983)
Llaima	38°43′S	BA–A	2	5.5	Deruelle *et al.* (1983)
Villarrica	39°25′S	B–R	3	5.8–6.3	Deruelle *et al.* (1983)
Choshuenco	39°56′S	BA–A	2	5.4–6.1	Deruelle *et al.* (1983)
Osorno	41°07′S	B–BA	2	6.8	Deruelle *et al.* (1983)
Calbuco	41°20′S	B–A	2	6.4–6.7	Deruelle *et al.* (1983)

*B—basalt; BA—basaltic andesite; A—andesite; D—dacite; R—rhyolite.

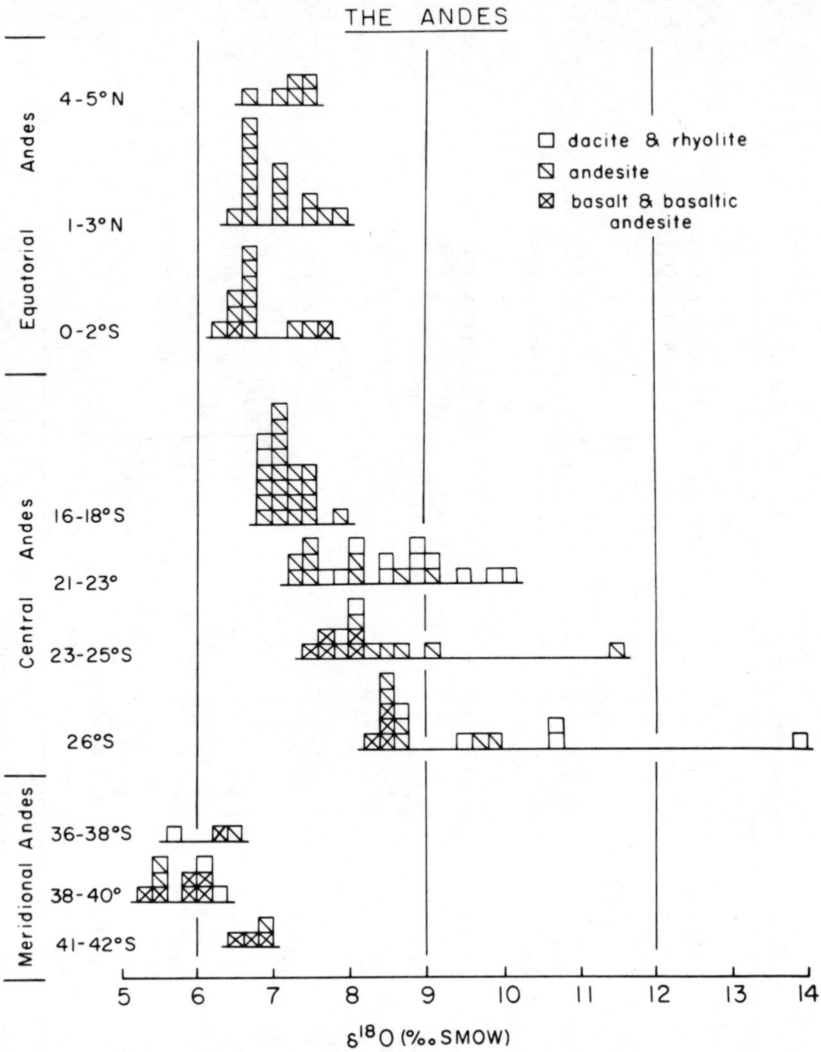

Figure 1 Histograms of whole-rock O-isotope compositions of Late Cenozoic Andean calc-alkaline lavas shown as a function of geographic distribution (latitude) and composition. The Meridional Andes, Central Andes, and Equatorial Andes groupings in the figure correspond to the SVZ, CVZ and NVZ respectively. Data from Magaritz *et al.* (1978), Harmon *et al.* (1981), James (1982a, b), Deruelle *et al.* (1983), and this study

Overall, $^{18}O/^{16}O$ ratios for Late Cenozoic Andean lavas range from +5.2 to +14.0‰ (Table 1). However, $\delta^{18}O$ ranges within the three individual volcanic provinces are significantly smaller: NVZ (5°N–2°S) = +6.5 to +7.8‰; CVZ (16–28°S) = +6.9 to +14.0‰; and SVZ (33–46°S) = +5.2 to +6.8‰ (Table 2). In both the NVZ, where the lavas are basaltic andesites and andesites, and in the SVZ, where the lavas range in composition from basalts

Table 2 **O-isotope ratios for Late Cenozoic volcanic rocks in the Andean Cordillera (reported as 'δ' values in ‰ relative to SMOW)***

	SVZ (36°S–42°S)	CVZ (16°S–28°S)	NVZ (5°N–2°S)
Basalts and basaltic andesites	5.2–6.8	7.9–8.5	6.6–7.7
Andesites	5.4–6.4	6.9–11.5	6.3–7.8
Dacites and rhyolites	6.0–6.3	7.0–14.0	—

*From Magaritz *et al.* (1978), Harmon *et al.* (1981), James (1982a, b), Deruelle *et al.* (1983), and this study.

to rhyolites, O-isotope variations are independent of chemical composition. For example, lavas at the Villarrica centre at 39°S range from basalt with 51% SiO_2 and a $\delta^{18}O$ value of +5.8‰ to rhyolite with 69% SiO_2 and a $\delta^{18}O$ value of +6.3‰. Andesites at Villarrica are equivalent to basalts in their O-isotope composition. Within the CVZ, there is a general trend for $^{18}O/^{16}O$ ratios to be well correlated with compositional variations, with the more silicic lavas being the more ^{18}O-rich. This trend is most pronounced in the Cerro Galan centre at 26°S where $\delta^{18}O$ values range from +8.2 to +8.5‰ for basalts and basaltic andesites, from +8.5 to +9.9‰ for andesites, and from +8.6 to +14.0‰ for dacites (Harmon *et al.*, 1981). Throughout the remainder of the CVZ dacites generally tend to have higher $\delta^{18}O$ values than basaltic andesites, with andesites intermediate in O-isotope composition, although this trend is less well developed for most other volcanic centres than it is for Cerro Galan (see Figure 1). It is clear from Table 1 that $\delta^{18}O$ values tend to be higher for a given SiO_2 content in the CVZ than in the NVZ and higher in the NVZ than in the SVZ. Also, it can be seen in Figure 1 that there is a general tendency for $\delta^{18}O$ values to increase from north to south along the CVZ. The former observation undoubtedly is either a direct or indirect manifestation of the very thick (> 60 km) crustal keel of the CVZ; the latter feature is attributed to the presence of Precambrian granulitic basement which is considered to be largely absent to the south in southwestern Bolivia, northwestern Argentina, and north-central Chile where Late Cenozoic volcanoes are developed on igneous and metasedimentary upper crust of Palaeozoic-Mesozoic age.

Discussion

The large range of O-isotope ratios (>8‰) documented for the Late Cenozoic lavas of the Andean Cordillera (see Table 1) cannot be explained simply in terms of derivation from a single source region. Rather, the data require the involvement of both mantle and continental crust in magmagenesis.

Overall, $\delta^{18}O$ values for the lavas of the SVZ range from +5.2 to +6.8‰ (see Table 1). Basalts and basaltic andesites from six of the eight volcanoes sampled by Deruelle *et al.* (1983) have $\delta^{18}O$ values within the $+5.7 \pm 0.5$‰ range considered representative of uncontaminated basaltic magmas derived from a primitive mantle source region (Pineau *et al.*, 1976; Kyser *et al.*, 1982). Low-pressure fractional crystallization can produce small (<1‰) degrees of ^{18}O enrichment within a differentiating magma (Matsuhisa *et.al.*, 1973). Thus, the minor intravolcano $\delta^{18}O$ variations of 0.2 to 0.8‰ observed within the five volcanic centres for which different lava compositions were analysed are attributed to the removal of olivine and pyroxene phenocrysts as modelled by Lopez-Escobar *et al.* (1981) for the Antuco volcano.

By contrast, neither the lavas of the NVZ nor those of the CVZ have such low $\delta^{18}O$ values (see Table 1). Basaltic andesites and andesites in Columbia (James, 1982b) and Ecuador (Harmon *et al.*, 1981; this study) have $\delta^{18}O$ values which range from +6.3 to +7.8‰ with no apparent correlation of O-isotope and chemical variations. In the CVZ andesites and dacites from south Peru (Magaritz *et al.*, 1978; James, 1982a) are similar in O-isotope to the lavas of the NVZ, whereas those of all compositions to the south in southwest Bolivia, north Chile, and northwestern Argentina are relatively ^{18}O-rich by comparison (see Figure 1). Also, intravolcano $\delta^{18}O$ variations within the NVZ (0.4 to 1.3‰) and the CVZ (0.3 to 5.8‰) are generally larger than in the SVZ (0.2 to 0.8‰), and basalts and basaltic andesites from some volcanic centres such as Cerro Galan and El Negrillar have $^{18}O/^{16}O$ ratios which are significantly higher than those for most andesites and some dacites from the other volcanoes sampled in the CVZ. These features—the ^{18}O-enriched character and large internal variation in O-isotope compositions—require a substantial crustal component to the CVZ lavas.

The O-isotope data for the Andes is compared with that for different volcanic settings in Figure 2. Unaltered mid-ocean ridge and oceanic-island and tholeiitic basalts have $\delta^{18}O$ values largely between +5.5 and +6.0‰ (Garlick, 1966; Taylor, 1968; Muehlenbachs and Clayton, 1972; Pineau *et al.*, 1976; Kyser *et al.*, 1982). Oceanic-island alkali basalts tend to be slightly enriched in ^{18}O with ^{18}O values between +6.0 to +7.0‰ (Taylor, 1968; Kyser *et al.*, 1982), suggesting that the mantle may be slightly heterogeneous in terms of its O-isotope composition. Island-arc and continental-margin basalts and basaltic andesites from destructive plate margin settings exhibit a much larger variation in O-isotope composition with $\delta^{18}O$ values ranging between +5.0 to +8.0‰, essentially equivalent to the +5.2 to +8.5‰ range observed for the basalts and basaltic andesites of the Andes. Overall, the +5.2 to +14.0‰ range in $^{18}O/^{16}O$ ratios documented for compositional spectrum of Andean lavas compares well with the +5.0 to +13.8‰ range for destructive plate margin lavas in general.

The O-isotope variations observed for the Late Cenozoic lavas of the Andes between 5°N and 42°S are best explained by the model schematically

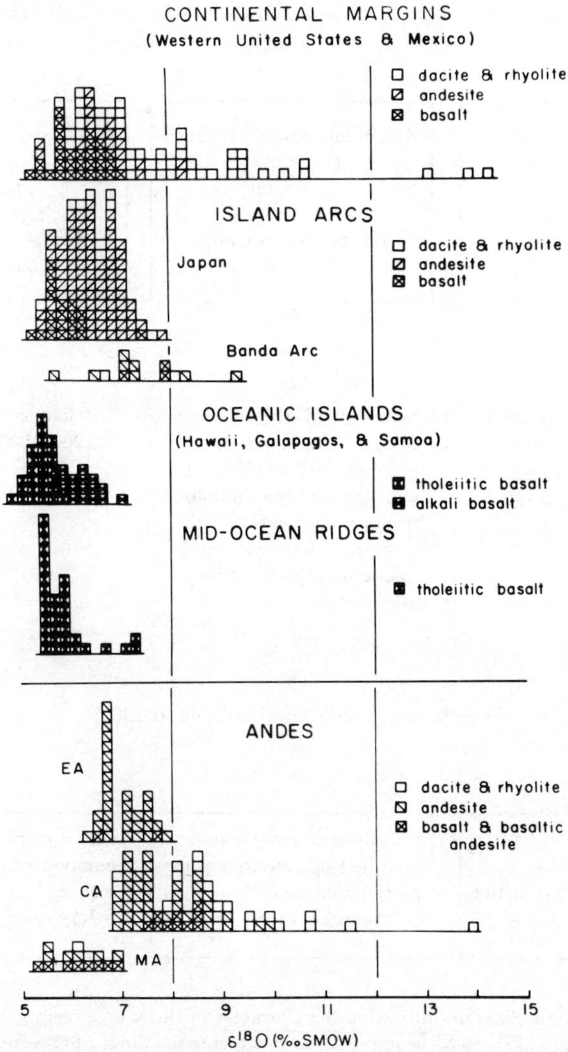

Figure 2 Histograms comparing the whole-rock O-isotope composition of calc-alkaline Late Cenozoic Andean lavas with tholeiitic and calc-alkaline lavas from other tectonic environments. The abbreviations EA, CA and MA refer to the Equatorial Andes (NVZ), Central Andes (CVZ) and Meridional Andes (SVZ) respectively. Data from the references in Figure 1 and Garlick (1966), Taylor (1968), Muehlenbachs and Clayton (1972), Matsuhisa *et al.* (1973), Pineau *et al.* (1976), Matsuhisa (1979) and Kyser *et al.* (1982)

illustrated in Figure 3. Calc-alkaline magmatism throughout the length of the Andean Cordillera is envisaged to be derived directly from a lithospheric mantle source region enriched in large-ion-lithophile elements and volatiles as a result of dehydration of the underlying oceanic lithosphere during subduction. The low $\delta^{18}O$ values (~5.5‰) observed within the SVZ suggests

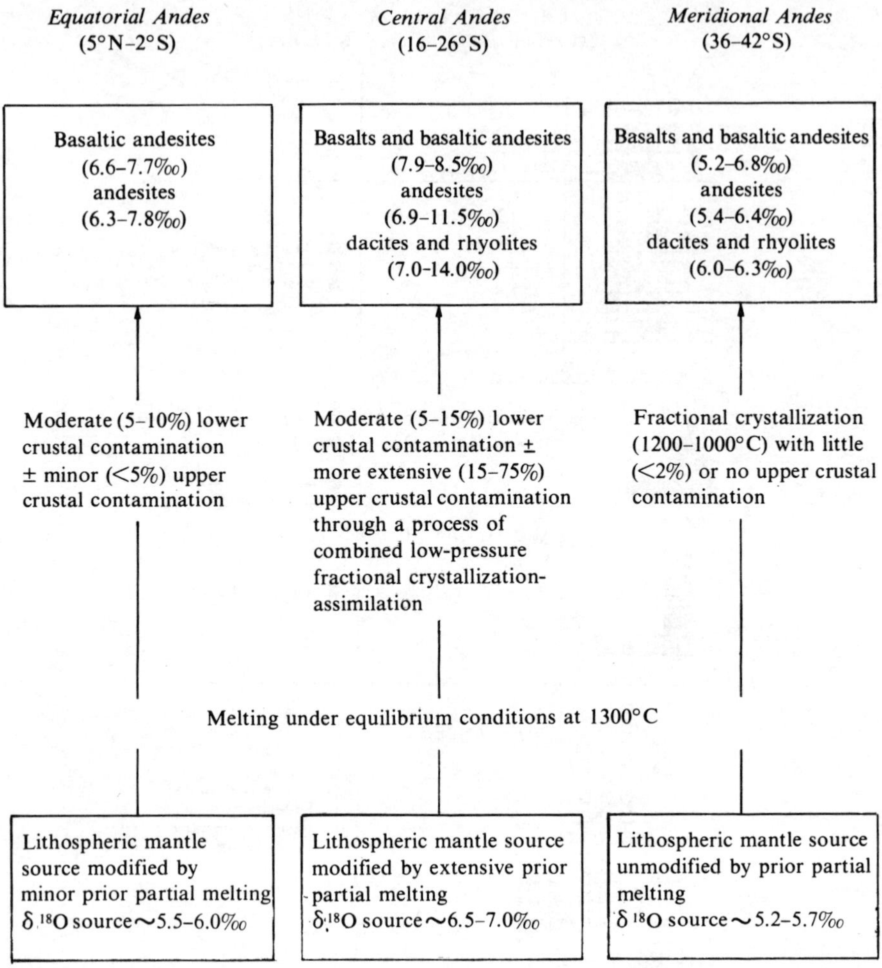

Figure 3 Schematic diagram indicating the character of the source regions and post-melting processes which are likely to be responsible for producing the range of O-isotope compositions observed in Andean lavas

derivation of parental basaltic magmas through a relatively large degree (10–15%) of partial melting of this mantle source region (Lopez-Escobar *et al.*, 1977) which had not been subject to extensive prior partial melting. Subsequently, the high-Al basaltic magmas evolved by fractional crystallization which occurred largely free of concomitant contamination by ^{18}O-rich continental crust. The highly variable O-isotope composition ($\delta^{18}O$ = −3 to +10‰) of hydrothermally altered oceanic lithosphere (Muehlenbachs and Clayton, 1972; Gregory and Taylor, 1981) and the extremely ^{18}O-rich character ($\delta^{18}O$ = +20 to 30‰) of entrained pelagic sediment (Savin and Epstein, 1970) rule out significant (> 1%) participation of the subducted slab in the generation of these and other Andean magmas.

In the CVZ the higher $^{18}O/^{16}O$ ratios ($\delta^{18}O$ = +6.9 to +8.5‰) of the chemically least evolved lavas are attributed to a combination of low degrees (5–10%) of partial melting of a mantle source region, whose isotopic composition had been modified prior to partial melting by the large-scale extraction of basaltic melts since Mesozoic time, and the subsequent extensive (> 20%) contamination of these partial melts by the ^{18}O-rich continental crust of the CVZ, which was more than doubled in thickness as a result of extensive Mesozoic and younger plutonism (James, 1971a). Although plagioclase is a ubiquitous phenocryst phase in the lavas of the CVZ, only four volcanic centres exhibit well-correlated O-isotope and compositional variations indicative of upper-crustal contamination by low-pressure fractional crystallization-assimilation processes. In the other 21 centres $\delta^{18}O$ variations are largely independent of composition suggesting that contamination was dominated by bulk assimilation (perhaps magma mixing) processes. Large-scale crustal anatexis, such as that recently advocated by Hawkesworth *et al.* (1982), is considered unlikely to be the sole process involved in the generation of the 'andesitic' volcanic centres in the CVZ on the basis of both experimental evidence (Wyllie, 1977) and geological arguments (e.g. see Francis *et al.*, 1977, 1980, this volume; Coulon and Thorpe, 1981).

The situation in the NVZ is considered to be intermediate between that of the SVZ and the CVZ. Here the mantle source region may have been somewhat modified by earlier basalt extraction so that parental magmas have slightly higher $\delta^{18}O$ values (+6.0 to +6.5‰) than do those in the SVZ ($\delta^{18}O \sim$ +5.5‰). The higher $^{18}O/^{16}O$ ratios of the more evolved andesitic lavas ($\delta^{18}O$ = +6.5 to +7.8‰) and the lack of a strong correlation between O-isotope and compositional variations indicates that deep crustal contamination is also an important process in the evolution of the lavas of the NVZ. The fact that crustal xenoliths are largely absent in these andesitic lavas suggests that the contamination occurred at depth in the lower crust through a magma mixing process rather than by partial fusion of assimilated basement rocks. Bulk contamination of parental basaltic magmas with $\delta^{18}O$ = +6.0 to 6.5‰ by anatectic partial melts of igneous lower crust with $\delta^{18}O$ = +7.5 to 8.5‰ could produce the slight ^{18}O-enrichment observed in the lavas of the NVZ.

Summary

The O-isotope data for Late Cenozoic Andean volcanic centres discussed here indicate a mantle source for the low-^{18}O lavas of the SVZ and document that the isotopic composition of the lavas of the CVZ is dominated by ^{18}O-rich continental crust. The intermediate isotopic character of the lavas of the NVZ suggests that both mantle and crustal sources were important in their petrogenesis. The actual mechanisms of magma generation and crustal interaction remain poorly understood. Experimental evidence precludes the derivation of large volumes of andesitic magma from either subducted oceanic lithosphere (Stern and Wyllie, 1978) or the continental crust (Wyllie, 1977). The low $\delta^{18}O$ values and very small range of O-isotope variations observed in the volcanic centres of the SVZ imply a mantle source for parental basaltic magmas with the generation of more silicic, but less voluminous, lavas through crystal-liquid fractionation processes. Similar low $\delta^{18}O$ values for basalts and andesites in the northernmost portion of the SVZ at 33–34°S coupled with substantially higher $\delta^{18}O$ values for rhyolitic lavas at these volcanic centres (Stern, Futa, Muehlenbachs, Dobbs, Muñoz, Godoy and Charrier, this volume) implies that calc-alkaline magmas across the Andean Cordillera in the CVZ and NVZ are produced in the same source region and subsequently modified relative to those of the SVZ by crustal interaction. This observation, together with the fact that the deepest part of the Chile-Peru trench of the CVZ, where the evidence for a crustal component to the 'andesitic' lavas is greatest, is virtually sediment free (Thorpe *et al.*, 1981), suggests that the crustal component present in the Andean lavas of the CVZ and NVZ is not introduced at source from the melting of subducted sediment. The direct correlation between ^{18}O-enrichment and the abnormally thick continental crust and high elevations of volcanoes in the CVZ is unlikely, therefore, to be simply a fortuitous coincidence. We infer a cause and effect relationship.

Available trace element and isotopic evidence suggests that crustal contamination occurs after magma generation in the lithospheric mantle. It may be that there are fundamental differences in rates of mantle convection and crustal thermal regime, magmatic plumbing systems and hydrodynamics and/or regional structure between the CVZ (where the Benioff zone is deep (~140 km), the crust abnormally thick (>60 km) and elevations high) and the SVZ (where the Benioff zone is shallower (~90 km), the crust of normal thickness (~35–40 km), and elevations are lower) which promote extensive partial melting and magma mixing in the lower crust and/or assimilation in the upper crust in the former. Detailed studies combining trace element petrology with stable and radiogenic isotope geochemistry on several different individual volcanic centres in each region of the Andes are required to resolve this question.

Lead Isotopes and Andean Magmagenesis

B. A. BARREIRO

Abstract

Late Cenozoic volcanics from the northern and southern Andes have restricted Pb isotope compositions which appear unrelated to rock type or basement age. Although regional differences in $^{206}Pb/^{204}Pb$ exist between the two areas, the uniform enrichment in ^{207}Pb and ^{208}Pb in the lavas is likely to be due to Pb leached from sediments and the altered igneous rocks of the subducted slab, which have with time been well mixed into the mantle wedge above the slab.

In the central Andes, igneous rocks of Mesozoic and Cenozoic age have Pb isotope compositions which indicate varying degrees of interaction with both U- and Th-rich upper crustal rocks and with Th-rich but U-poor high grade gneisses exposed in the Arequipa massif.

Introduction

The volcanic and plutonic rocks of the South American Andes have long been viewed as typical examples of igneous rocks generated along continental margins. Other papers in this volume discuss some of the techniques used to determine how these rocks were formed. The study of Pb isotope ratios has been one of these methods. Isotopes of Pb occasionally can provide petrogenetic information unavailable through petrographic, geochemical, trace element or other isotope studies. Most often, the combination of all these techniques can shed light on the processes involved in the generation of a related suite of rocks.

Radioactive decay of ^{238}U, ^{235}U and ^{232}Th ultimately produces the three isotopes ^{206}Pb, ^{207}Pb and ^{208}Pb. The equations describing this decay are given in Table 1. The isotopic composition of lead is generally reported as the ratio of a radiogenic isotope to the non-radiogenic ^{204}Pb, i.e. the three ratios $^{208}Pb/^{204}Pb$, $^{207}Pb/^{204}Pb$ and $^{206}Pb/^{204}Pb$. As is the case for all isotope systems, the ratio of a daughter isotope to a non-radiogenic isotope depends on the initial isotopic composition, the parent/daughter ratio and the length of time involved in the radioactive decay of parent to daughter. In the case of the Pb system, the evolution of ^{206}Pb and ^{207}Pb are intimately coupled because natural chemical processes do not fractionate ^{238}U and ^{235}U to any measurable degree. Thus, $^{206}Pb/^{204}Pb$ and $^{207}Pb/^{204}Pb$ ratios cannot only distinguish systems with histories of high U/Pb from low U/Pb, but can also provide information about the timing of U-Pb fractionation, without any measurement of U and Pb concentrations. Slopes of straight lines in

Table 1 **The equations of radioactive decay**

$$\left(\frac{^{206}Pb}{^{204}Pb}\right) = \left(\frac{^{206}Pb}{^{204}Pb}\right)_o + \left(\frac{^{238}U}{^{204}Pb}\right)\left(e^{\lambda_1 t}-1\right) \quad \lambda_1 = 1.155125 \times 10^{-10}a^{-1}$$

$$\left(\frac{^{207}Pb}{^{204}Pb}\right) = \left(\frac{^{207}Pb}{^{204}Pb}\right)_o + \left(\frac{^{238}U}{^{204}Pb}\right)\left(e^{\lambda_2 t}-1\right) \quad \lambda_2 = 9.8485 \times 10^{-10}a^{-1}$$

$$\left(\frac{^{208}Pb}{^{204}Pb}\right) = \left(\frac{^{208}Pb}{^{204}Pb}\right)_o + \left(\frac{^{232}Th}{^{204}Pb}\right)\left(e^{\lambda_3 t}-1\right) \quad \lambda_3 = 4.948 \times 10^{-11}a^{-1}$$

It follows that:

$$\left(\frac{^{207}Pb}{^{206}Pb}\right)_{radiogenic} = \left(\frac{^{235}U}{^{238}U}\right)\left(\frac{e^{\lambda_2 t}-1}{e^{\lambda_1 t}-1}\right)$$

$^{206}Pb/^{204}Pb$ versus $^{207}Pb/^{204}Pb$ diagrams correspond to model ages which may have geological meaning; alternatively, straight lines are also the result of two-component mixing. Choosing between these two alternatives must be based on geological reasoning.

Mantle heterogeneity

Isotope data from mid-ocean ridge basalts and oceanic-island volcanics, unquestionably derived from suboceanic mantle, show a large range of Pb isotope composition (Figure 1). Mid-ocean ridge basalt Pb data generally define a line in $^{207}Pb/^{204}Pb$ versus $^{206}Pb/^{204}Pb$ diagrams, the slope of which corresponds to an age of ~1.7 Ga (Tatsumoto, 1978; Dupré and Allègre, 1980; Cohen and O'Nions, 1982). Many oceanic-island volcanics have Pb compositions which fall close to the MORB array or its extension to higher or lower $^{206}Pb/^{204}Pb$ values. The MORB array may represent mixing of two mantle sources. Several oceanic islands, however, have higher $^{207}Pb/^{204}Pb$ and $^{208}Pb/^{204}Pb$ than data on the MORB trend. Zindler *et al.* (1982) and Tera (1982) reason that there are at least three sources of Pb in the suboceanic mantle. The higher $^{207}Pb/^{204}Pb$ and $^{208}Pb/^{204}Pb$ in the third source must be the result of an old enrichment in U and Th relative to the depleted MORB sources.

Continental volcanic rocks and mantle nodules record a complex history for the subcontinental mantle. Tilton (1983) summarizes the evidence that depleted mantle with low $^{207}Pb/^{204}Pb$ and $^{208}Pb/^{204}Pb$ is present beneath the continents as well as the oceans. Some continental flood basalts, however, have high $^{207}Pb/^{204}Pb$ and $^{208}Pb/^{204}Pb$, but are depleted in $^{206}Pb/^{204}Pb$. Their

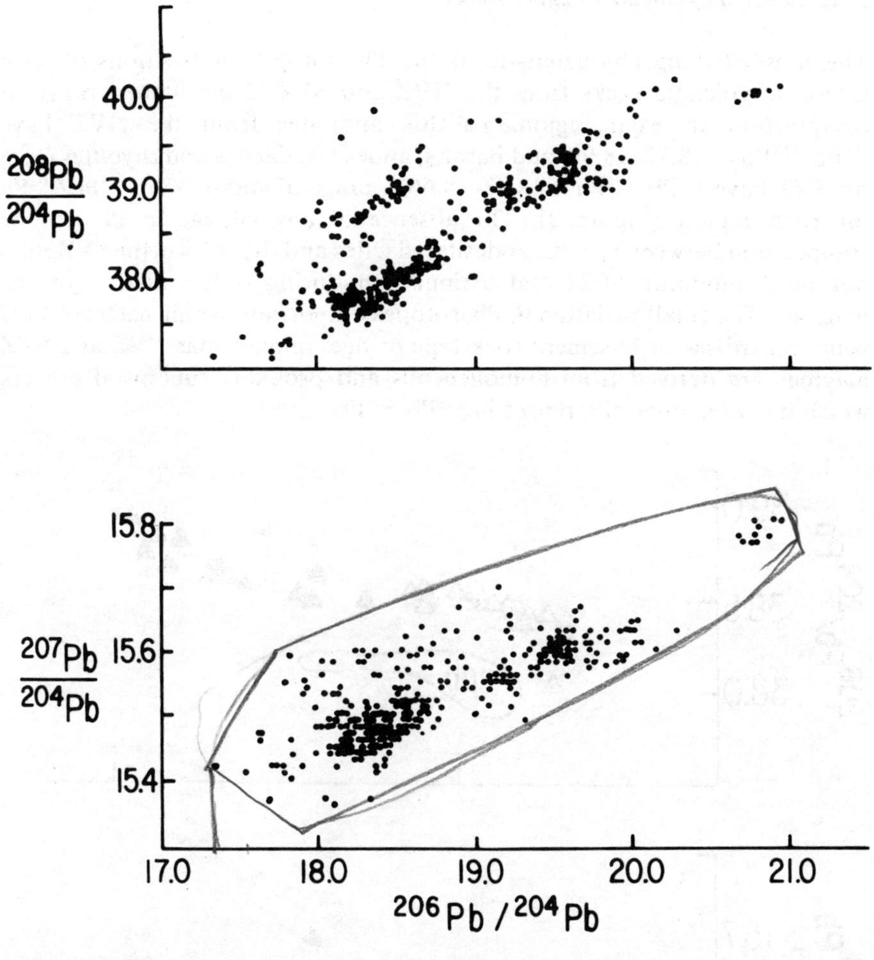

Figure 1 Lead isotope compositions of MORB and ocean-island volcanics. Most data fall along linear trends corresponding to an age of ~1.7 Ga and Th/U = 3.8, but some South Atlantic and Indian Ocean basalts have higher $^{207}Pb/^{204}Pb$ and $^{208}Pb/^{204}Pb$ than the main trend. Lead isotope data from Church and Tatsumoto (1975), Cohen *et al.* (1980), Cohen and O'Nions (1982), Cumming (1976), Dosso *et al.* (1979), Dupré and Allègre (1980, 1983), Dupré *et al.* (1982), Harris *et al.* (1982), Kay *et al.* (1978), Oversby (1971, 1972), Richardson *et al.* (1982), Sun and Jahn (1975), Sun *et al.* (1975), Sun (1980), Tatsumoto (1978), Unruh and Tatsumoto (1976), White (1979)

Pb isotope compositions may form isochrons which occasionally mimic the age of the basement rocks. These isotopic characteristics may be the manifestation of old enriched subcontinental lithosphere (Doe *et al.*, 1982). Most Andean Pb isotope data have little in common with either depleted, MORB-type Pb or with the Pb characteristic of basalts from old enriched subcontinental lithosphere.

Late Cenozoic Andean volcanic rocks

The most striking characteristic of the Pb isotope compositions of Late Cenozoic volcanic rocks from the NVZ and SVZ is the limited range of composition in each region. Basaltic andesites from the NVZ have $^{206}Pb/^{204}Pb$ = 18.72–18.99, and basalts, andesites, dacites and rhyolites from the SVZ have $^{206}Pb/^{204}Pb$ = 18.48–18.66, a range of about 1% in $^{206}Pb/^{204}Pb$ for each region (Figure 2). The absence of any change in Pb isotope composition between basalts, andesites, dacites and rhyolites in the SVZ rules out large amounts of crustal assimilation during differentiation of the magmas. The small variation in Pb isotope composition within each volcanic zone, regardless of basement rock type or age, implies that NVZ and SVZ magmas are derived from homogeneous and probably subcrustal sources which have regional differences in $^{206}Pb/^{204}Pb$.

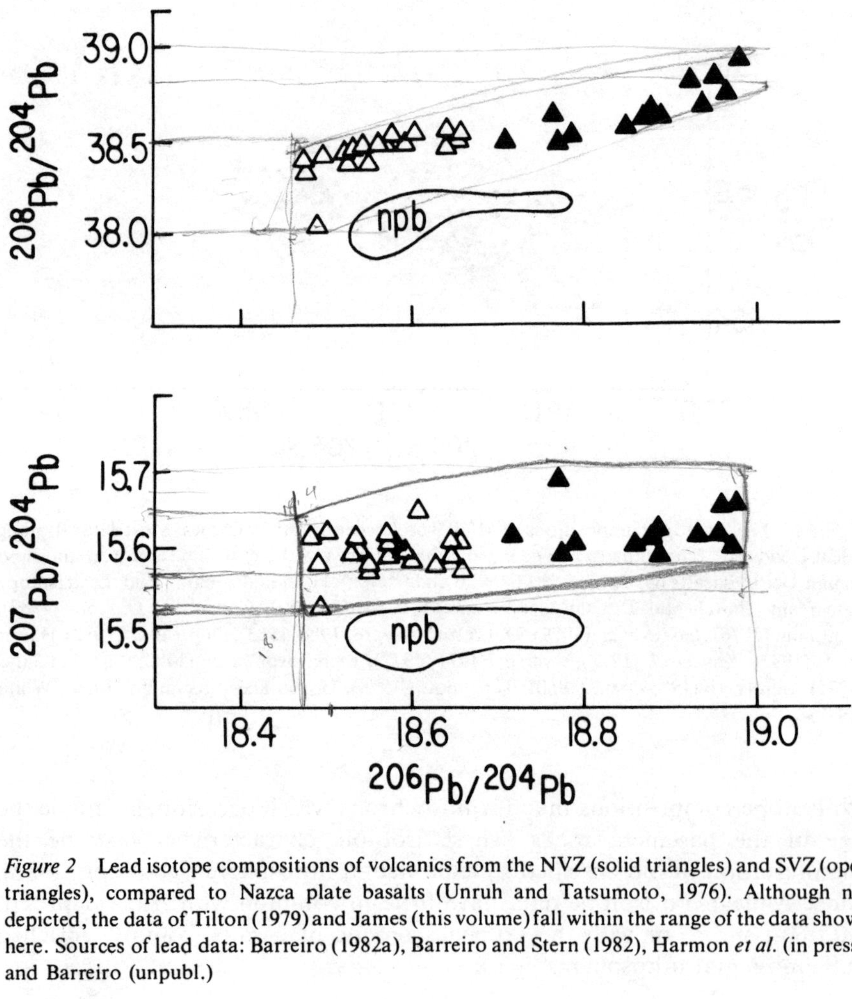

Figure 2 Lead isotope compositions of volcanics from the NVZ (solid triangles) and SVZ (open triangles), compared to Nazca plate basalts (Unruh and Tatsumoto, 1976). Although not depicted, the data of Tilton (1979) and James (this volume) fall within the range of the data shown here. Sources of lead data: Barreiro (1982a), Barreiro and Stern (1982), Harmon *et al.* (in press), and Barreiro (unpubl.)

Although rocks from the northern Andes have higher $^{206}Pb/^{204}Pb$ and $^{208}Pb/^{204}Pb$ than do rocks from the southern Andes, the $^{207}Pb/^{204}Pb$ of rocks from both zones are indistinguishable (usually 15.58–15.65). The constant $^{207}Pb/^{204}Pb$ places constraints on the relative age difference between northern and southern sources. Because ancient boosts in U/Pb ratios would result in increased $^{207}Pb/^{204}Pb$ as well as $^{206}Pb/^{204}Pb$ ratios, any increase in U/Pb ratio in a northern source relative to the southern source must have occurred within the past few hundred million years or less.

Nearly all of the volcanic rocks from the NVZ and SVZ have higher values of $^{207}Pb/^{204}Pb$ and $^{208}Pb/^{204}Pb$ than do MORB and most oceanic-island lavas with similar $^{206}Pb/^{204}Pb$ ratios (Figure 2). Although it is conceivable that these values indicate old enriched mantle similar to that hypothesized beneath some ocean islands and in the old subcontinental lithosphere, the homogeneity of Pb isotope compositions over the large areas involved in Andean volcanism implies a dynamic source that is too well mixed to preserve small scale mantle heterogeneities.

Because extensive (>20%) crustal contamination of the magmas has been ruled out by the arguments above, and by Sr, Nd and O isotope data (see other papers this volume and references therein), the observed enrichment in ^{207}Pb and ^{208}Pb might be attributed to a small component of recycled continental material in the form of subducted sediments, as in the case of purely oceanic-island arcs. The Pb isotope compositions of southern Andean volcanic rocks are within the range of values measured in Nazca plate metalliferous and pelagic sediments (Chow and Patterson, 1962; Dasch, 1981), in accord with this hypothesis. However, northern Andean volcanic Pb has higher $^{206}Pb/^{204}Pb$ than the majority of Nazca plate sediments. Intraoceanic-island arcs commonly show a rough correspondence between sediment and volcanic Pb isotope compositions, and often form subvertical mixing arrays between suitable mantle and sediment components in Pb versus Pb diagrams. Because these features are not apparent in the Andean Pb isotope data, incorporation of subducted continental Pb into Andean magmas is clearly more complex than the two-component mixing proposed for intraoceanic-island arcs. Fluids derived from the subducted slab probably leach Pb from peridotite, gabbro, basalt and sediments, with the sediment Pb by far the largest component. If this Pb-rich fluid percolates through a mantle wedge for a long period of time, on the order of the length of the Andean orogeny, small-scale variations in Pb isotope compositions within the wedge are likely to be averaged out, and little trace would remain of an original depleted mantle signature.

Large-scale variations in $^{206}Pb/^{204}Pb$ ratios between the NVZ and SVZ are unexplained at present. Long-term differences in the provenance of subducted sediments may be the controlling factor. Alternatively, a small amount of enriched mantle often deduced from intraplate volcanoes might be responsible for the more radiogenic $^{206}Pb/^{204}Pb$ ratios in the NVZ. The extremely anomalous radiogenic Pb measured in an otherwise normal andesite from Colombia (James, this volume) can be attributed to this effect.

A generalization one can make of NVZ and SVZ Pb data is that only one

Andean volcanic rock analysed to date, a magnesian andesite from the southernmost Andean volcano, Cook Island, has Pb with an isotopic composition similar to that of MORB (see the next paper by Stern *et al.*, this volume, for a more detailed description of this unusual rock). All the other NVZ and SVZ volcanic rocks contain Pb with a distinct imprint of a moderately high ^{207}Pb and ^{208}Pb source. The ultimate cause for this distinctive Pb isotope signature is recycled continental material, most likely in the form of subducted sediments and slab-derived fluids which enrich the mantle wedge.

Volcanic rocks from the CVZ have a far greater range of Pb isotope composition than both the NVZ and SVZ combined (^{206}Pb/^{204}Pb = 17.35–19.01, ^{207}Pb/^{204}Pb = 15.57–15.69, ^{208}Pb/^{204}Pb = 38.47–39.14; Figures 3 and 4). Many CVZ andesites and dacites have ^{207}Pb/^{204}Pb within the same range (15.57–15.62) as most northern and southern Andean samples. Other volcanics in the central Andes, particularly the ignimbrites, have somewhat

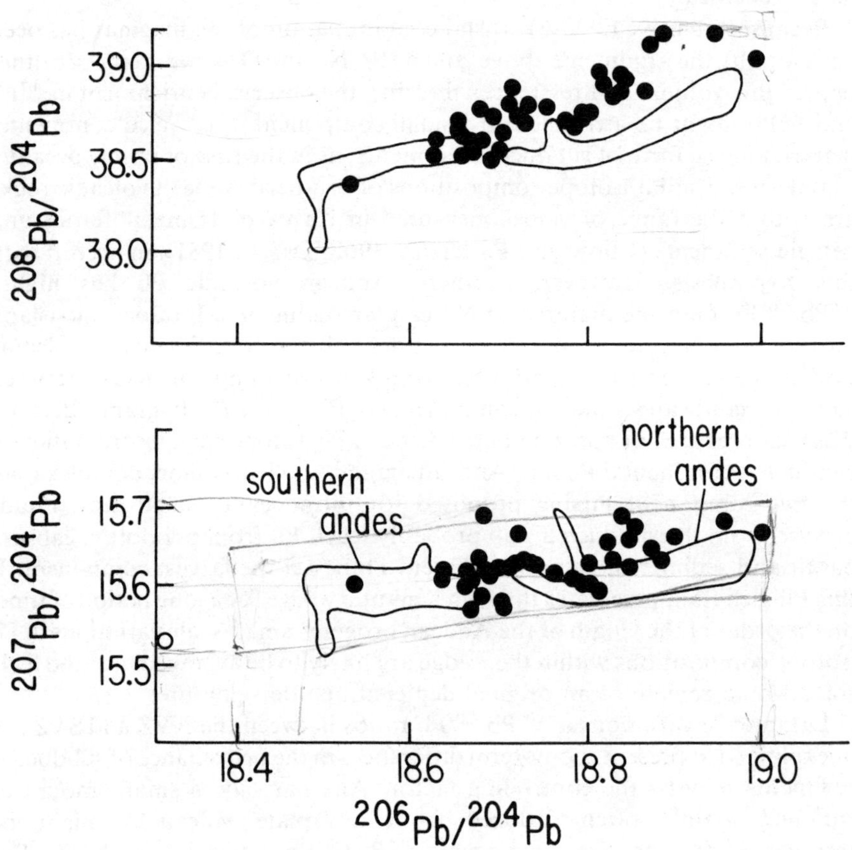

Figure 3 Lead isotope compositions of Late Cenozoic Central Andean volcanics from Chile, Argentina and Bolivia (solid circles) compared to data from the northern and southern Andes. Data: Barreiro (1982a) and Harmon *et al.* (in press)

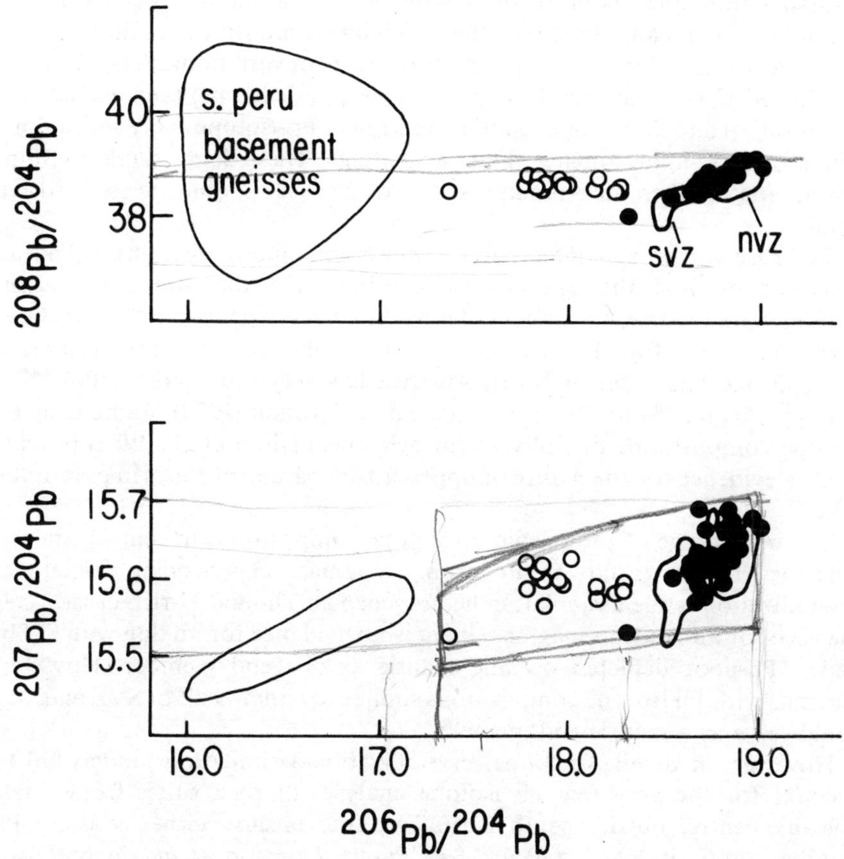

Figure 4 Lead isotope compositions of Late Cenozoic Central Andean volcanics, including those from southern Peru, (open circles, Tilton and Barreiro, 1980, Barreiro, 1982a). While volcanics of southern Peru appear to have assimilated material from the granulitic basement gneisses of the Arequipa massif, other Central Andean volcanics have interacted with U- and Th-rich crust with radiogenic Pb isotopic compositions

higher $^{207}Pb/^{204}Pb$ (>15.62). In addition, nearly all CVZ volcanics have higher $^{208}Pb/^{204}Pb$ ratios than NVZ and SVZ rocks with equal $^{206}Pb/^{204}Pb$ ratios. Some rock suites may even exhibit an inverse correlation of $^{206}Pb/^{204}Pb$ and $^{208}Pb/^{204}Pb$ (Tilton, 1979). Thus, magmagenesis in the central Andes involves a component with higher Th/Pb and Th/U than any involved in the northern or southern Andes. The tendency towards higher $^{207}Pb/^{204}Pb$ ratios in the CVZ suggests that this component also has average-to-high time-integrated U/Pb ($^{238}U/^{204}Pb$, μ) ratios. Clearly a reasonable candidate for this component is the continental crust through which the magmas ascend.

The Arequipa and Barroso volcanics of south Peru have distinctly lower $^{206}Pb/^{204}Pb$ ratios than other Andean volcanics, but similar $^{207}Pb/^{204}Pb$ and $^{208}Pb/^{204}Pb$ ratios. James (this volume) shows how combined fractional crystallization and assimilation of the ~2 Ga granulite gneisses of the Arequipa massif can account for the Pb isotope compositions of the volcanics as well as their Sr, Nd, O isotope and trace element variations. Six rocks from the El Misti volcano exhibit a gross correspondence between Pb isotope composition and SiO_2 content and form a linear Pb-isotope array with a slope which roughly approximates the basement age. These detailed relationships are eliminated when the Arequipa and Barroso volcanic data are grouped as a whole.

The case for crustal interaction is more ambiguous for central Andean volcanics erupted through basement other than the Arequipa massif. Although no basement samples from north Chile, Argentina or Bolivia have been analysed for Pb isotope composition, similar non-cratonized Precambrian basement in North America has very radiogenic $^{206}Pb/^{204}Pb$, $^{207}Pb/^{204}Pb$ and $^{208}Pb/^{204}Pb$ ratios (Doe and Zartman, 1979). Radiogenic Pb isotope compositions of Bolivian tin belt ores (Tilton *et al.,* 1981) provide further evidence for the nature of upper crustal basement rocks in part of the CVZ.

The wide range of radiogenic Pb isotope compositions in central Andean volcanic rocks could result from magmas experiencing fractional crystallization while assimilating heterogeneous, Th- and U-rich crust. As is the case for all Andean magmas, there is no evidence for mixing with ^{207}Pb- and ^{208}Pb-poor depleted oceanic mantle as one end member. However, magmas with Pb isotope compositions similar to either NVZ or SVZ magmas could serve as a mantle end member.

However, in detail, fractional crystallization-assimilation models fail to account for the very few Pb isotope analyses of rock suites from single volcanic centres outside south Peru. Two studies show either constant Pb isotope composition (San Pedro-San Pablo, Harmon *et al.,* in press) or seemingly random variation of Pb isotope composition with differentiation from basaltic andesite to rhyolite (Cerro Galan, Barreiro, in prep.), contrary to the patterns predicted from AFC models. Thus the Pb isotope compositions of some CVZ lavas permit the involvement of crustal melts in their genesis, a point reiterated by Hawkesworth *et al.* (1982).

Lead and magmagenesis through time

A few Pb isotope studies of pre-Quaternary igneous rocks of the CVZ permit speculation upon the changing patterns of Andean magmagenesis through time. In southernmost Peru, Andean plutonic and volcanic rocks of Jurassic to Eocene age have measured Pb isotope compositions which are typical of other Andean igneous rocks, with $^{206}Pb/^{204}Pb$ ratios > 18.5. Miocene to Recent volcanics, in contrast, have non-radiogenic Pb ($^{206}Pb/^{204}Pb < 18.3$) which implies significant interaction of the magmas with the granulite gneisses of the Arequipa massif (Figure 5). Barreiro and Clark (in press) correlate this abrupt change in the style of magmagenesis with the inception of uplift of the

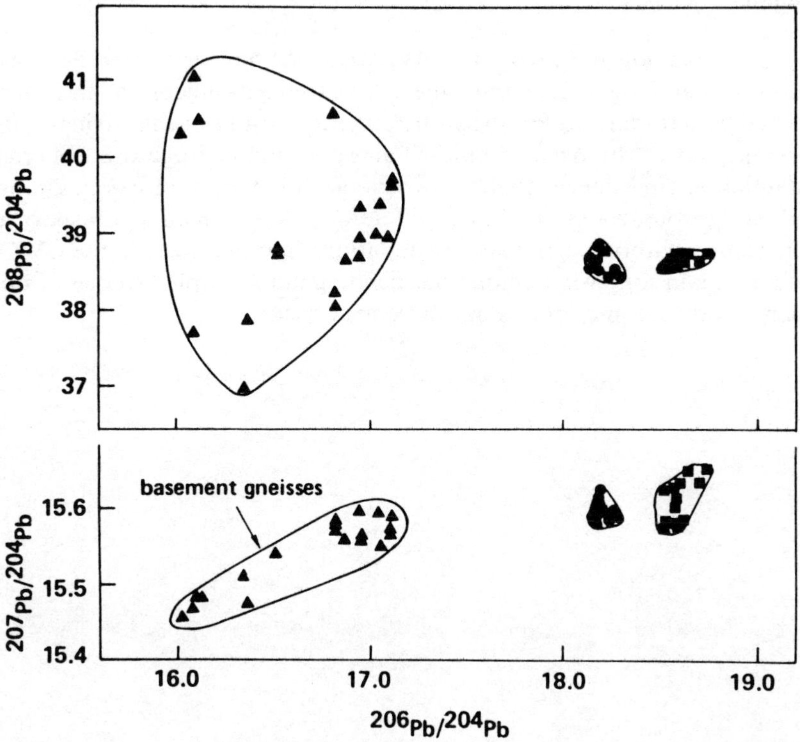

Figure 5 Lead isotope composition of Andean igneous rocks near Toquepala, south Peru. Squares (right, circled field) represent data from Jurassic to Eocene plutonic and volcanic rocks. Circles (centre) are data from Miocene to Recent volcanics which show evidence of crustal interaction with the basement gneisses. Only one Palaeogene high-silica ash flow tuff falls within the centre field of 'more contaminated' rocks

Andes mountains in this region and the inferred crustal over-thickening. In their view, the exceptionally thick crust beneath southernmost Peru encouraged protracted fractional crystallization and increased the proportion of crustal material incorporated into the magmas. These simple relationships do not hold up in other parts of south Peru, where the history of crustal interaction with granulites appears to be more complex (see Mukasa and Tilton, this volume).

In north Chile, McNutt *et al.* (1979) have recognized a trend of increasing $^{87}Sr/^{86}Sr$ ratios through time in a series of plutonic and volcanic rocks. However, they found no pattern at all in their Pb isotope compositions. This inconsistency is probably due to the heterogeneous Pb isotope composition of the crust in this area which obscures any trends related to greater amounts of crustal interaction.

Summary

Late Cenozoic volcanics from the NVZ and SVZ have restricted Pb isotope compositional ranges probably due to well-mixed subcontinental mantle enriched by slab-derived Pb. In south Peru, the non-radiogenic Pb in the high-grade gneisses of the Arequipa massif acts as a unique fingerprint of crustal contamination in Miocene to Recent volcanics, and the absence of this imprint in older igneous rocks implies a different style of magma transport and fractionation without significant assimilation. In other areas in the CVZ the Pb data remain somewhat ambiguous pending more detailed studies of single volcanic and plutonic centres and basement rocks.

Isotope and Trace Element Data for Orogenic Andesites from the Austral Andes

C. R. STERN, K. FUTA and K. MUEHLENBACHS

Abstract

The Austral Volcanic Zone of the Andes (AVZ, 49–55°S) consists of six Pleistocene to Recent volcanic centres resulting from the subduction of the Antarctic plate beneath southernmost South America. Andesites from the southernmost volcano in the AVZ, the Cook Island volcanic complex, have very distinctive geochemical characteristics, including high MgO, CaO, Sr, K/Rb and LREE/HREE, low FeO/MgO, K, Rb, Ba and HREE, and Sr, Nd, O and Pb isotope compositions similar to mid-ocean ridge basalts. These high-MgO andesites are interpreted to have formed by small (<5%) degrees of partial melting of subducted oceanic lithosphere (MORB), followed by rapid upward migration of these partial melts so that near-surface fractional crystallization and crustal contamination were minimal. Andesites and dacites from Mt. Burney, the next volcanic centre to the north, and from Aguilera, Nunatak and Lautaro, the three most northerly volcanoes in the AVZ, have progressively higher SiO_2, K, Rb, Ba, $^{87}Sr/^{86}Sr$ and $\delta^{18}O$, and lower MgO, CaO, Sr, Ni, Cr, K/Rb and $^{143}Nd/^{144}Nd$. These rocks are interpreted to have formed by increasing degrees of near-surface fractional crystallization, combined with small amounts of crustal contamination of a parental composition similar to the high-MgO andesites from Cook Island, the most primitive magma type identified in the AVZ. These south-to-north changes in the geochemistry and petrogenesis of the volcanic centres of the AVZ may result from variations in the stress pattern within the continental lithosphere due to the change from direct convergence to strike-slip plate motion which occurs along the southernmost margin of South America.

Introduction

Three separate linear zones of recently active orogenic volcanoes have been distinguished in the Andes by Thorpe and Francis (1979); the northern volcanic zone (NVZ, 5°N–2°S), the central volcanic zone (CVZ, 16–28°S) and the southern volcanic zone (SVZ, 33–52°S). The southern volcanic zone actually consists of two distinct sections (Figure 1). Volcanoes in central-south Chile from 33°S to 46°S result from the subduction of the Nazca plate beneath the South American continent. South of 46°S the Antarctic plate is converging against the South American plate resulting in at least six Pleistocene to Recent Andean volcanoes in southernmost Chile between 49–55°S (Stern *et al.*, 1976; Stern, 1979). As described in this paper, these volcanoes in the southernmost Andes are quite different geochemically from the volcanoes north of the Chile Rise-Trench triple junction, as well as being

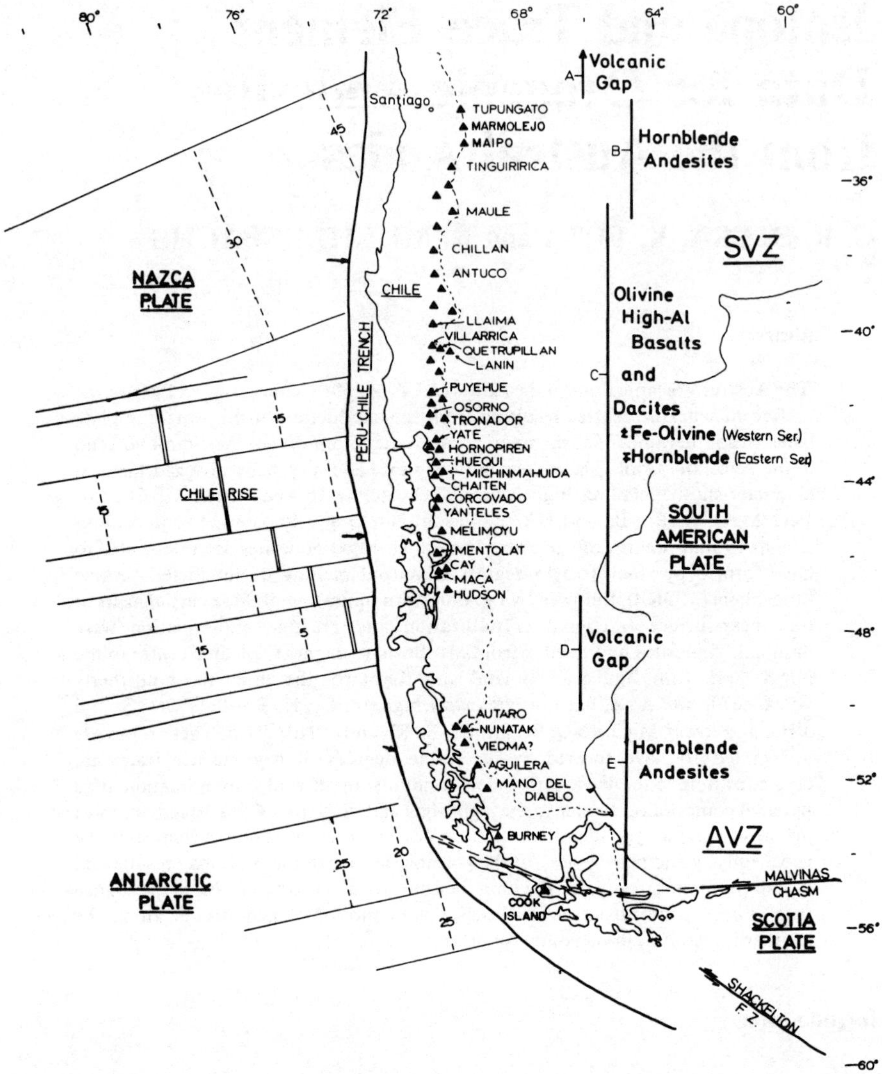

Figure 1 Map of southernmost South America showing the location of the volcanic centres of both the Austral and Southern Volcanic Zones of the Andes, and the major plates and plate boundaries in the region

separated from these more northerly volcanoes by a gap in active orogenic volcanism between 46–49°S (Figure 1). The southernmost Andean volcanoes are thus considered to form a fourth distinct linear volcanic zone in the Andes, herein referred to as the 'Austral Volcanic Zone' (AVZ, 49–55°S). The SVZ is considered to extend only as far south as the Chile Rise-Trench triple junction at 46°S.

Seismicity is negligible south of the Chile Rise-Trench triple junction, but active subduction of the Antarctic plate beneath the South American plate is suggested by (a) the small but finite rate of convergence between these two plates (2 cma^{-1}, Minster *et al.*, 1974); (b) the negative gravity anomaly associated with the Chile Trench which is a well-defined structural feature as far south as 56°S even though south of 46°S it is morphologically masked by sediments (Hayes, 1966); and (c) the active orogenic volcanism in the region. The lack of seismicity prevents the angle of subduction or the depth beneath the active volcanoes of the subducted slab from being known. The low seismicity south of 46°S may result in part from the slow rate of convergence between the Antarctic and South American plates, and in part from the young age of the ocean crust being subducted (Forsyth, 1975).

Forsyth (1975) has suggested that a small plate, the Scotia plate, occurs between the South American and the Antarctic plates southeast of South America (Figure 1). The Scotia plate is separated from the Antarctic plate on the west by the Shackelton Fracture Zone, and from the South American plate on the north by a strike-slip fault zone within the Malvinas Chasm. These fault zones extend westward and northwestward into the Magellanes fault zone and ultimately the Chile Trench (Fuenzalida, 1974; Herron *et al.*, 1977). With this plate configuration, the southeastern tip of South America and the southernmost Andean volcano, the Cook Island volcanic complex, actually lie within the Scotia plate (Figure 1).

As discussed in this paper, progressive north to south changes in the geochemistry of the volcanoes of the AVZ may be explained in terms of the effects of changes in stress within the continental lithosphere associated with the change from direct convergence to strike-slip motion along the southernmost tip of South America.

Geology and petrology of the AVZ

Only the summits are accessible and only a few samples have been collected from the northernmost four volcanoes in the AVZ, Lautaro, Nunatak, Aguilera and Mano del Diablo, which all occur within the area of the Patagonian ice cap. The Mt. Burney volcanic complex consists of three volcanic units: a group of glaciated domes that form a ring of low topography approximately six kilometres in diameter, Mt. Burney itself which is a Pleistocene-Holocene stratovolcano on the western flank of the older ring of domes and a mid-Holocene pyroclastic flow unit which has both filled up the depression within and flowed out beyond the ring formed by the pre-Holocene domes (Stern, 1982). The Cook Island volcanic complex consists of a group of coalesced Holocene domes sitting unconformably upon glaciated terrain on one of the many islands that form the non-submerged part of the Andean Cordillera at this latitude (Stern *et al.*, 1979).

The only documented historic eruption in the AVZ was a minor eruption of Mt. Burney in 1911 (Quensel, 1911), but widespread Holocene ash layers in southernmost Patagonia and Tierra del Feugo attest to sporadic post-

Pleistocene activity of significant magnitude throughout the region (Auer, 1974).

The available samples from the different volcanoes in the AVZ are all pyroxene + hornblende ± biotite andesites and dacites which are very distinct petrologically from the olivine basalts which dominate the volcanic centres in the SVZ north of the Chile Rise-Trench triple junction (see Figure 1). Cook Island consists of clinopyroxene + hornblende andesites without phenocrysts of plagioclase, which only occurs in the groundmass. All three volcanic units of the Mt. Burney volcanic complex are two-pyroxene + hornblende andesites and dacites with orthopyroxene dominant over both clinopyroxene and hornblende, and significant amounts of modal plagioclase both as phenocrysts and in the groundmass. Samples from Aguilera, Nunatak and Lautaro are all orthopyroxene + hornblende + biotite dacites, petrologically similar to dacites from the CVZ (Deruelle, 1982). Samples collected from ash layers exposed in the pampas of Patagonia east of each of these volcanoes have a mineralogy which is similar to that observed in the lavas sampled from the individual centres indicating that the available sampling is representative.

Geochemistry of the AVZ

Each individual volcanic centre of the AVZ is characterized by a very limited range of compositional variation, even where detailed mapping and extensive collecting have been accomplished such as for Mt. Burney and Cook Island. The average compositions presented here are therefore considered representative.

With respect to major element chemistry (Table 1 and Figure 2), one of the significant features of the samples collected from throughout the entire AVZ is the limited range of SiO_2 content, which varies only from 60 to 67 wt. %; neither basalts nor rhyolites have been encountered in the volcanic centres of the AVZ. In this respect, these volcanic centres differ markedly both from those centres in the SVZ north of the Chile Rise-Trench triple junction which are formed predominantly of basalt, and from the volcanic centres in the CVZ in which rhyolite is common. As a group, the volcanic rocks of the AVZ have chemical compositions which are similar to typical orogenic calc-alkaline andesites and dacites, having high Al_2O_3 and low TiO_2 contents.

In contrast to these overall similarities, the volcanic centres of the AVZ also exhibit significant intravolcano geochemical differences that show systematic north-to-south regional variations over the few degrees of latitude within which these centres occur. The range and average value of SiO_2 increases from the southernmost volcano, Cook Island, to the more northerly ones, despite the small range of SiO_2 for the entire AVZ noted above (Table 1 and Figure 2). Greater systematic south-to-north variations of regional extent occur for other major elements, particularly K_2O which varies from 0.6 wt. % for Cook Island andesites to an average value of 2.1 wt. % for the more northerly volcanoes in the AVZ which contain biotite phenocrysts. Besides low K_2O contents, Cook Island andesites are characterized by relatively high MgO

Table 1 **Average major element composition of samples from the volcanic centres of the AVZ, andesites from the CVZ and high-Mg andesites from the Aleutian Islands**

	AVZ			CVZ[1]	Aleutians[2]
	Cook Is.	Mt. Burney	A,N,L		
Samples	3	4	6	16	2
SiO_2 range	59.5–61.5	63.0–64.5	64.0–67.0	58.0–66.2	55.5–58.5
SiO_2	60.8	64.1	65.9	62.4	57.0
TiO_2	0.8	0.4	0.6	0.7	0.9
Al_2O_3	18.2	17.8	16.9	16.4	15.4
FeO*	3.6	4.0	3.8	4.6	4.9
MgO	3.8	2.4	2.1	2.5	5.3
CaO	8.1	5.6	4.5	4.8	8.6
Na_2O	4.4	3.7	3.2	4.2	3.5
K_2O	0.6	1.0	2.1	2.7	1.8
P_2O_5	0.3	0.2	0.2	0.2	0.4
Mg#	0.45	0.32	0.30	0.30	0.46

FeO*—total Fe as FeO;[1]Thorpe *et al.* (1976), SiO_2> 58 wt. %;[2]Kay (1978); A—Aguilera; N—Nunatak; L—Lautaro

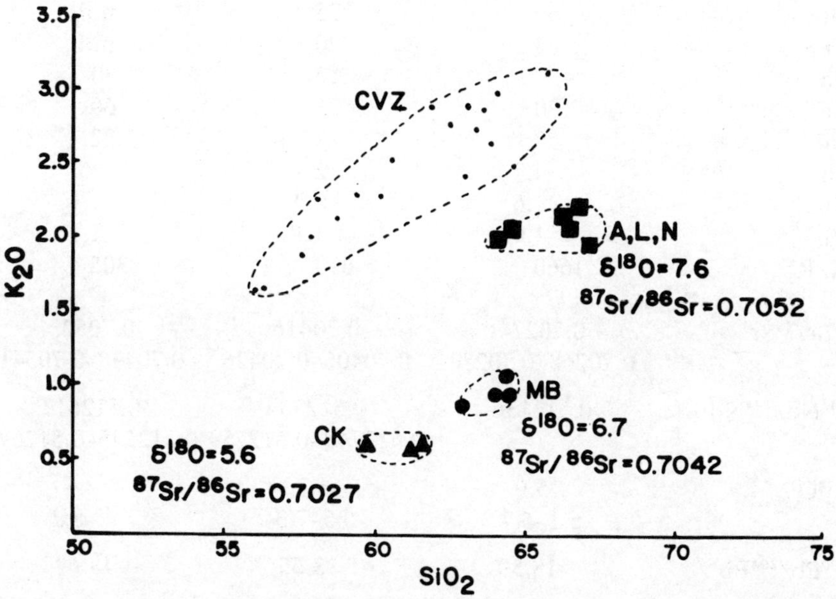

Figure 2 A plot of SiO_2 versus K_2O for samples from Cook Island (CK; solid triangles), Mt. Burney (MB; solid circles), Aguilera, Nunatak and Lautaro (A,L,N; solid squares) and intermediate rocks from the central Andes (CVZ; dots, Thorpe *et al.*, 1976). The diagram illustrates the limited compositional range of each individual centre within the AVZ and the systematic south-to-north increase in SiO_2 and K_2O contents, $^{87}Sr/^{86}Sr$ ratios, and $\delta^{18}O$ values

contents and low FeO/MgO ratios, as well as high CaO and Na_2O contents compared to more typical orogenic andesites (Table 1). In these respects they are more comparable to high-Mg andesites from other volcanic arcs (Table 1). Samples from Aguilera, Nunatak and Lautaro volcanoes, which are similar petrochemically and thus grouped together in the tables, have lower MgO, CaO and Na_2O contents and are more similar to typical orogenic andesites and dacites, such as those from the central Andes, although they do not have as high a K_2O content as the andesites from the CVZ (Table 1 and Figure 2).

Table 2 **Average selected trace element and isotopic composition of samples from the Austral Volcanic Zone of the Andes**

Volcano	Cook Is.	Mt. Burney	A,N,L
Rb	3	14	57
Ba	134	304	533
Cs	n.d.	0.22	1.9
Sr	2003	602	533
Zr	125	81	135
Th	4.4	1.5	11.9
U	1.3	0.66	1.7
Pb	1.5	5.37	13.4
Ni	60	25	n.d.
Cr	64	20	n.d.
La	31	12	30
Ce	70	25	60
Nd	28	12	22
Sm	4.7	2.4	3:7
Eu	1.6	0.9	1.1
Yb	1.0	1.1	1.1
K/Rb	1660	610	305
$^{87}Sr/^{86}Sr$[1]	0.70274 0.70268–0.70280	0.70416 0.70406–0.70426	0.70523 0.70495–0.70541
$^{143}Nd/^{144}Nd$[1]	0.513136	0.512754 0.512748–0.512759	0.512617 0.512545–0.512693
$\delta^{18}O$[1]	5.6 5.1–5.7	6.7 6.5–7.3	7.6 5.7–8.3
$^{206}Pb/^{204}Pb$	18.50	18.65	18.62
$^{207}Pb/^{204}Pb$	15.53	15.60	15.59
$^{208}Pb/^{204}Pb$	38.04	38.48	38.52

n.d.—not detectable;[1] both average and range of more than one sample

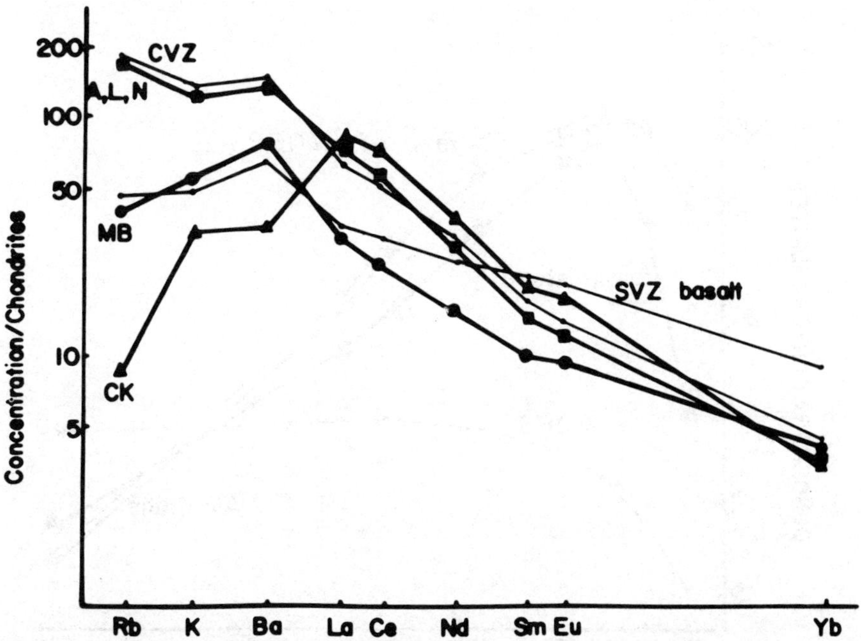

Figure 3 Large-ion-lithophile and rare-earth element compositions normalized to chondritic abundances, of samples from Cook Island (CK; solid triangle), Mt. Burney (MB; solid circle), Aguilera, Nunatak and Lautaro (A,L,N; solid square), the average of intermediate rocks from the central Andes (CVZ; dots; Thorpe *et al.*, 1976) and a representative olivine bearing high-Al basalt just north of the Chile Rise-Trench triple junction (SVZ; dots, Stern *et al.*, 1976). The diagram shows that the andesites and dacites from both the AVZ and the CVZ have more fractionated REE patterns with lower HREE than the basalt from the SVZ

Mt. Burney is intermediate both compositionally and spatially although it has noticeably lower TiO_2 contents than the other AVZ centres.

Also, the volcanic centres of the AVZ are characterized by little trace element variation within each centre, but marked systematic south-to-north intravolcano variations. Andesites from Cook Island have very low concentrations of the highly incompatible large-ion-lithophile elements such as Cs, Rb and Ba (but not La) which are chemically similar to K (Table 2). For this volcanic centre, the larger the ionic radius and thus the more incompatible an element of this group of elements is, the lower is its concentration relative to chondritic meteorites (Figure 3), similar to mid-ocean ridge basalts as well as the high-Mg andesites from the Aleutians (Figure 4). As a result the Cook Island andesites have very high K/Rb ratios. Mt. Burney has higher concentrations of these LIL elements, and the most northerly group of three volcanoes in the AVZ have LIL element abundances which approach those of typical Andean andesites from the CVZ. In these latter centres, the more incompatible an element, the higher its relative concentration (Figure 3), so that the K/Rb ratio is very much lower than it is for the Cook Island andesites.

Given the marked differences in incompatible element concentrations, Cook Island and the three most northerly volcanoes in the AVZ have

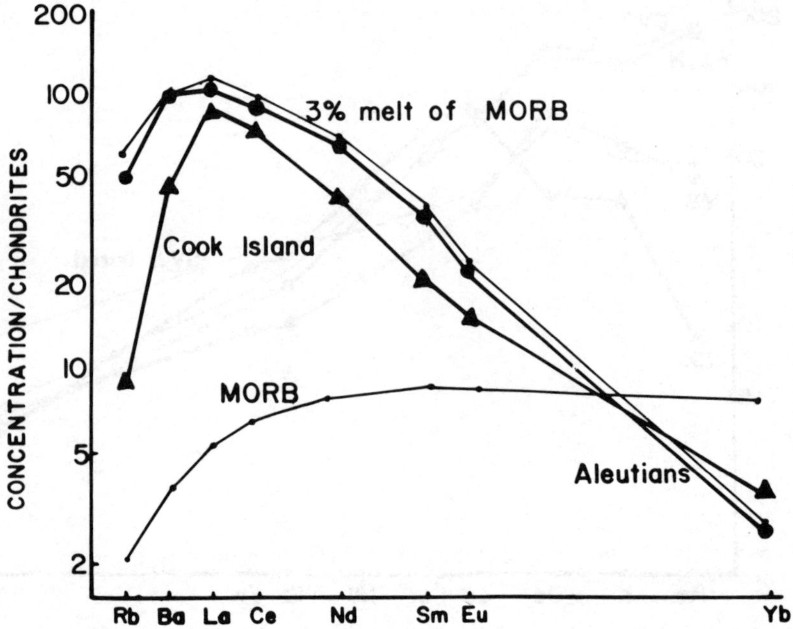

Figure 4 LIL and REE concentrations normalized to chondritic abundance for high-Mg andesites from Cook Island (solid triangle) and from the Aleutians (solid circles, Kay, 1978), and for a MORB and a liquid derived by 3% partial melting of that basalt with eclogite mineralogy (dots, Kay, 1978)

remarkably similar rare-earth element (REE) compositions (Table 2 and Figure 3). The light REE are enriched relative to the heavy REE, which are depleted compared to both olivine basalts from the orogenic volcanoes of the SVZ (Figure 3) and mid-ocean ridge basalts (Figure 4). In this respect they are similar to the typical Andean andesites from the CVZ (Figure 3). Although Cook Island has higher La concentration than the more northerly volcanic centres of the AVZ, it has a lower La/Ce ratio, a feature noted by Kay (1978) for high-Mg andesites from the Aleutians. Mt. Burney has similar HREE but lower LREE and consequently lower LREE/HREE ratio than the other volcanic centres of the AVZ. Cook Island andesites have a small positive Eu anomaly, but none of the samples from the other centres in the AVZ have detectable Eu anomalies.

The abundance of Sr decreases drastically from values over 2000 ppm in samples from Cook Island, to more typical average values of 600 ppm for Mt. Burney and 500 ppm for the three more northerly AVZ centres. Contents of Ni and Cr also decrease and Pb concentration increases in the same regional pattern; Zr, Th and U abundances are noticeably low in samples from Mt. Burney.

Sr, Nd and O isotope ratios show a wide range of correlated and regionally systematic south-to-north variations within the AVZ (Table 2, Figures 2 and

5; Stern, 1981; Meuhlenbachs and Stern, 1980; Stern and Futa, 1982a). Andesites from Cook Island are similar isotopically to mid-ocean ridge basalts with respect to these isotopes as well as isotopes of Pb (Stern and Futa, 1982b; Barreiro and Stern, 1982). Samples from Mt. Burney have Sr and Nd isotope ratios similar to basalts, andesites and dacites from central-south Chile (36–46°S), although the average $\delta^{18}O$ value is higher in the Mt. Burney samples (see the second paper by Stern *et al.,* this volume). Samples from the three most northerly volcanoes in the AVZ have higher $^{87}Sr/^{86}Sr$ and $^{18}O/^{16}O$ ratios, and lower $^{143}Nd/^{144}Nd$ ratios than Mt. Burney. However, their Pb isotope compositions are comparable to basalts, andesites, dacites and rhyolites in the SVZ (Barreiro and Stern, 1982), but distinct from the Pb

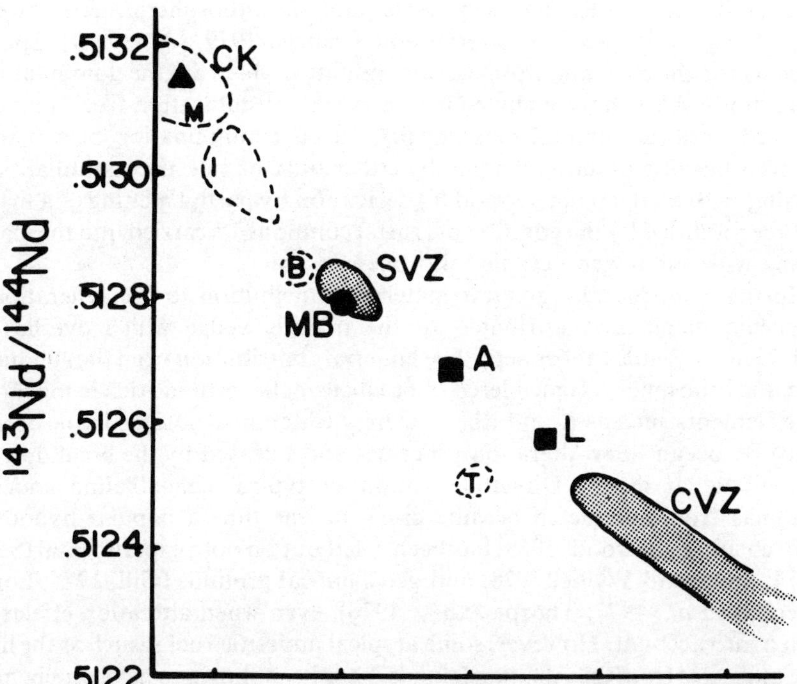

Figure 5 $^{87}Sr/^{86}Sr$ versus $^{143}Nd/^{144}Nd$ for andesites and dacites from Cook Island (CK; solid triangle), Mt. Burney (MB; solid circle), Aguilera and Lautaro (A,L; solid square) and the central Andes (CVZ, Hawkesworth *et al.*, 1979), and basalts, andesites, dacites and rhyolites from central-south Chile south of 36°S (SVZ, Stern and Futa, 1982a) compared to the mantle array as defined by samples from the Atlantic including MORB, Iceland (I), Bouvet Is. (B) and Tristan de Cuhna (T). The diagram shows that Cook Island andesites are similar isotopically to MORB and that the other samples from the AVZ become progressively more radiogenic northward, but not as radiogenic as similar rocks from the CVZ

isotopic composition of similar rocks from the CVZ (Barreiro, this volume).

When the samples from the AVZ are considered together as a coherent group of spatially and genetically related rocks, they exhibit correlated geochemical variations such that as SiO_2 increases MgO, CaO, Al_2O_3, Na_2O, Sr, Ni, Cr, K/Rb and $^{143}Nd/^{144}Nd$ decrease whereas K_2O, Rb, Ba, Cs, $^{87}Sr/^{86}Sr$ and $\delta^{18}O$ increase. Changes in total REE, LREE/HREE and Pb isotope compositions are not correlated with changes in SiO_2.

Discussion

The generation of Andean orogenic magmas may involve components derived from the continental crust, subducted oceanic lithosphere, including altered and unaltered MORB and sediments, and the lithospheric mantle wedge separating these two (see Thorpe and Francis, 1979, for review). Specific models for the derivation of Andean andesites, which are the dominant rock type in the AVZ, have included (a) fractional crystallization from a mantle-derived basaltic parental magma; (b) crustal contamination of a mantle-derived basaltic parental magma by either bulk or selective assimilation or mixing with anatectic melts; and (c) derivation by partial melting of a mantle source modified by the addition of crustal components carried into the mantle along with subducted oceanic lithosphere.

In these models the greatest material contribution to the generation of orogenic magmas is attributed to the mantle wedge which overlies the subducted oceanic lithosphere. The material contribution from the subducted oceanic lithosphere is considered to be silicic melts or fluids rich in incompatible elements such as K and Rb, or merely water incorporated in the oceanic crust by ocean floor alteration processes and released by the breakdown of amphibole at depth. Direct derivation of typical calc-alkaline andesitic magmas from subducted oceanic crust, at one time a popular hypothesis (Green and Ringwood, 1968) has been ruled out on both experimental (Stern, 1974; Stern and Wyllie, 1978) and geochemical grounds (Gill, 1974; Lopez-Escobar *et al.*, 1977; Thorpe *et al.*, 1976), even when alteration effects are taken into account. However, some atypical andesitic rocks, such as the high-Mg andesites from the Aleutian Islands, have been shown to have geochemical characteristics compatible with direct derivation from subducted oceanic crust (Kay, 1978). The high-Mg andesites from Cook Island have similar geochemical characteristics, and their petrogenesis is discussed independently from the petrogenesis of the more northerly volcanic centres in the AVZ. Near-surface fractional crystallization and crustal contamination are shown to be important processes in the petrogenesis of the more typical andesites from these latter volcanic centres. It is suggested that intermediate magmas derived directly from subducted oceanic crust, such as reach the surface in Cook Island, may be parental to the generation of these and possibly other typical Andean andesites.

PETROGENESIS OF COOK ISLAND ANDESITES

The geochemical data for the andesites from Cook Island places some strong constraints on models for their petrogenesis. The high MgO and Sr content and the low FeO/MgO, K, Rb and Ba content of these rocks preclude much near-surface crystal-liquid fractionation, involving low-pressure mafic phases and plagioclase, in the formation of these rocks. In particular, no plausible low-pressure fractional crystallization scheme can be conceived that will produce the Cook Island andesites from the high-Al basalts common in central-south Chile, which are frequently suggested as the mantle-derived liquids parental to other more silicic Andean magmas (see e.g. Lopez-Escobar *et al.,* 1977). Not only are these basalts isotopically distinct from the Cook Island andesites (see Figure 5), but they have higher Rb and lower Sr contents which would be increased and decreased respectively by near-surface crystal-liquid fractionation.

The isotopic composition of the Cook Island andesites rules out any significant component of continental crust, subducted sediments or highly altered oceanic crust in their genesis. Taken together, the isotope and trace element data indicate that the Cook Island andesites are derived, without significant near-surface fractionation or crustal contamination, from subducted mid-oceanic ridge basalt or from subcontinental mantle with MORB source-type geochemical characteristics.

The extremely fractionated REE pattern and the low concentration of HREE suggest a derivation of the Cook Island andesites by a low degree of partial melting in the presence of garnet. Kay (1978) has calculated partial melting models and concluded that either less than 5% melting of subducted MORB in the eclogite (garnet + clinopyroxene) facies or 1% melting of a garnet-lherzolite mantle could produce the REE concentration observed in a high-Mg andesite from the Aleutians (see Figure 4). The Cook Island andesite has a slightly less fractionated REE pattern and lower total LIL and LREE content than the Aleutian high-Mg andesites suggesting a slightly greater degree of melting, but still less than 5% for subducted MORB. The low relative abundances of K, Rb, Ba and Cs, and the low La/Ce and high K/Rb ratios in these high-Mg andesites would reflect in part the LIL-depleted nature of the source material. In addition, the low K/La, Rb/La and Ba/La ratios suggest the presence in the source residue of a mineral such as amphibole, phlogopite, or sanidine that fractionates K, Rb and Ba from the REE.

LIL-depleted basalts, similar to MORB, were formed in this region of southernmost South America during an episode of back-arc spreading in Cretaceous time (Stern, 1980), supporting the possibility of the local existence of sub-continental MORB source-type mantle. However, the low Ni and Cr content of the Cook Island andesites compared with mantle-derived mafic magmas suggests that these andesites may have formed by remelting of subducted MORB, rather than by direct partial melting of a MORB-source type mantle, as does the very low percentage of partial melting required to generate magmas of this composition directly from the mantle.

Experimental studies have indicated that melting of subducted oceanic basalts in the eclogite facies, either dry or with small amounts of water present, initially would produce silicic liquids and then liquids of more intermediate composition as the degree of melting increased. However, these intermediate liquids would have high CaO/(FeO + MgO) ratios than typical calc-alkaline andesites (Stern, 1974; Stern and Wyllie, 1978). The high CaO content is one of the distinctive features of the Cook Island andesites. Kay (1978) has suggested that the high-Mg andesites from the Aleutians may have originated as dacitic melts derived from partial melting of subducted oceanic crust, which subsequently equilibrated to some degree with the overlying mantle as they migrated towards the surface, as indicated by reverse zoning in pyroxene phenocrysts. Phenocrysts of clinopyroxene and amphibole in the Cook Island andesites are normally zoned with MgO-rich cores and FeO-rich edges. Taken as a whole, the petrochemical characteristics of the Cook Island andesites are consistent with direct derivation of these intermediate rocks by melting of subducted MORB without significant interaction with the overlying crust or mantle.

PETROGENESIS OF THE MT. BURNEY, AGUILERA, NUNATAK AND LAUTARO VOLCANOES

Andesites and dacites from the more northerly volcanic centres in the AVZ have petrologic features (such as extensively zoned phenocrysts) and geochemical features (such as higher SiO_2, K_2O, Rb, Ba and Cs contents and FeO/MgO ratios and lower MgO, CaO, Al_2O_3, Sr, Ni and Cr contents and K/Rb ratios) that are consistent with their derivation in part by fractional crystallization from a more primitive parental magma. The high $\delta^{18}O$ values, radiogenic Sr and non-radiogenic Nd composition of the dacites from the Aguilera, Lautaro and Nunatak volcanoes suggest that contamination of a more primitive parental magma by crustal material also played a part in the generation of these rocks. The greater the degree of contamination as indicated by increased $^{87}Sr/^{86}Sr$ ratio, the higher the $\delta^{18}O$ value of the resultant magma (see Figure 2). This stable-radiogenic isotope correlation is indicative of near-surface contamination processes rather than source-region contamination in which O-isotope composition is buffered by a large mantle reservoir of oxygen (James, 1981).

Despite the spatial separation of the AVZ volcanoes, their coherent geochemical character, in particular the positive correlation of increasing SiO_2, K_2O, Rb, $\delta^{18}O$, $^{87}Sr/^{86}Sr$, and decreasing MgO, Sr, $^{143}Nd/^{144}Nd$ and K/Rb, are qualitatively consistent with a model of near-surface fractional crystallization combined with a small but significant amount of crustal contamination, by either bulk or selective assimilation, of a parental magma similar in composition to the high-Mg andesite from Cook Island, the most primitive magma type observed in the AVZ. Figure 6 shows that samples from the AVZ define a linear array on a plot of $^{87}Sr/^{86}Sr$ versus 1/Sr. This is

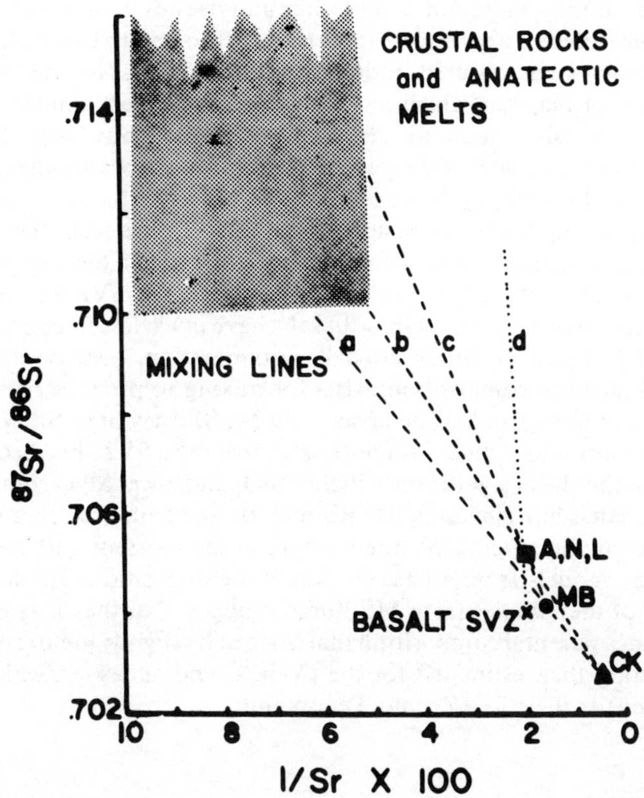

Figure 6 Plot of 1/Sr versus $^{87}Sr/^{86}Sr$ for samples from Cook Island (CK; solid triangle), Mt. Burney (MB; solid circle) and Aguilera, Nunatak and Lautaro (A,N,L; solid square). These latter samples fall along mixing lines (dashed line; a,b,c) between the high-Mg andesite from Cook Island and either crustal materials or anatectic melts from the southern Andes (shaded area; Sr = 100–200 ppm, $^{87}Sr/^{86}Sr$ = 0.71–0.73; Nelson and Elthon, in press; Halpern, 1973). No realistic crustal end-member exists in the southern Andes for a mixing model (dotted line; d) to produce the dacites found in the three most northerly volcanoes in the AVZ (A,N,L) from a typical basalt from central-south Chile (SVZ; x)

indicative of two-component mixing between a high-Sr and low-$^{87}Sr/^{86}Sr$ composition, in this case the Cook Island andesite, and a low-Sr and high-$^{87}Sr/^{86}Sr$ end-member, such as average crustal material. Nelson and Elthon (in press) estimate the average Sr content of the continental basement rocks of southernmost South America to be 115 ppm, and the average Sr content of silicic melts derived from these basement lithologies to be even lower. As noted in Figure 6, the whole-rock $^{87}Sr/^{86}Sr$ ratios of Palaeozoic metamorphic basement and Mesozoic-Cenozoic plutonic rocks from southernmost South America fall in the range 0.71 to 0.73 (Halpern, 1973). These values are lower than the Sr isotope ratios of the Precambrian basement rocks from the CVZ which range up to 0.78 (Francis *et al.,* 1980). Although crustal contamination is important in the generation of andesites and dacites in both the AVZ and the CVZ, the different ages and isotopic composition of the basement rocks in these two regions has resulted in distinctly different Sr, Nd and Pb isotope compositions of magmas which in other respects are very similar.

High-Mg andesite, such as reaches the surface on Cook Island, is considered to be the appropriate parental composition for the other andesites and dacites in the AVZ, both because of the absence of basalts in this region and because of geochemical considerations. The dacites from the Aguilera, Nunatak and Lautaro volcanoes could not form from the high-Al basalt magmas that characterize the volcanic centres in the SVZ by either low-pressure fractionation, because these basalts have much less fractionated REE patterns (see Figure 3), or by crustal contamination, because no realistic crustal end-member composition exists for mixing to produce these dacites from the basalts (Figure 6). The lavas from Mt. Burney have Sr, Nd and Pb isotope compositions similar to the basalts from the SVZ, but because Mt. Burney andesites have significantly higher SiO_2 and lower MgO contents than the SVZ basalts, but similar K_2O, Rb and Ba contents (see Figure 3) and higher Sr contents (Figure 6), their origin is inconsistent with reasonable fractionation models from a basaltic parental composition. The low LREE abundance of the samples from Mt. Burney suggest that they may have been derived from a parental composition that formed by slightly greater degrees of partial melting than estimated for the Cook Island andesites, which would also account for their low Zr and Ti contents.

Conclusions

The petrogenetic models outlined above for the volcanic centres of the Austral Volcanic Zone suggest that the andesites erupted at Cook Island, the southernmost volcanic centre of the AVZ, underwent very little near-surface fractional crystallization or crustal contamination. Northward along the AVZ, progressively greater amounts of near-surface fractionation and contamination occurred. These south-to-north changes in the petrogenesis of the orogenic volcanic rocks in this region of the Andes suggests systematic regional changes in the integrated rate at which magmas move from their source to the surface. These south-to-north changes in the magmatic plumbing in the AVZ probably reflect variations in the stress pattern in the continental lithosphere related to the change along the southernmost margin of South America from strike-slip plate motion to direct convergence (see Figure 1). At Cook Island the rapid rise of the high-Mg andesitic magmas could have taken place along structural pathways associated with strike-slip faulting within the Straits of Magellan and the Beagle Canal. Mt. Burney occurs along the northern limit of these strike-slip fault zones, in a region in which some component of direct convergence is occurring so that the migration of magmas at Mt. Burney from their source to the surface was not so rapid that they escaped all near-surface fractionation and contamination effects. The more northerly centres in the AVZ occur in a region dominated by direct convergence and magmatic evolution in these centres involved significant near-surface fractionation and associated crustal contamination.

Dacites from the Aguilera, Nunatak and Lautaro volcanoes are very similar petrochemically to dacites from the 'Andean' calc-alkaline suite of the CVZ. A significant amount of crustal contamination plays an important role in the origin of intermediate and silicic rocks from the CVZ (Francis *et al.,* 1980; Harmon *et al.,* 1981; Deruelle *et al.,* 1983). The dacites from the AVZ are less radiogenic with respect to Sr and Pb, but more radiogenic with respect to Nd isotope composition than similar rocks from the CVZ. This reflects the age and isotopic character of the different sections of continental crust through which the magmas in each region passed and interacted, not the degree of contamination that they underwent.

The abnormally thick continental crust and the high elevation at which the volcanoes in the CVZ occur have been cited as the factors resulting in crustal contamination of magmas in this region. The continental crust beneath southernmost South America is of normal ($\leqslant$40 km) thickness and the base of the volcanoes in the AVZ is at sea level. The conclusions reached concerning the petrogenesis of the volcanoes in the AVZ suggest that the effects of stress within the continental lithosphere on magmatic plumbing in regions of normal continental thickness and elevation can also cause crustal contamination of magmas resulting in the formation of typical 'Andean' calc-alkaline volcanics. However, no rhyolites occur in the AVZ, and the generation of voluminous rhyolites within convergent plate boundary arc systems may depend on such factors as crustal thickness or elevation (Grunder, 1983).

The petrogenetic models for the generation of the high-Mg andesites of Cook Island have important implications for models of continental growth and the evolution of the subducted oceanic lithosphere as it descends into the mantle. Similar rocks have been described from the Aleutians, the Cascades and Japan, but just as in the Andes they are not as volumetrically important in these convergent plate boundary volcanic arcs as more typical orogenic andesites with lower MgO contents, higher LIL abundances and more radiogenic isotope character. This may only reflect the need for special structural pathways for these deeply-derived magmas to reach the surface. The extent to which such liquids are important in the evolution of more typical orogenic andesites is speculative, but their well-documented occurrence in even a few locations does indicate that the evolution of subducted oceanic crust may involve not only dehydration and release of volatiles with dissolved solids, but melting and the generation of intermediate liquids as well.

ACKNOWLEDGEMENTS

Alexandra Skewes and Eric Leonard assisted with the collection of some of the samples, and E. Godoy, M. Suarez, and L. Dickenson supplied others. Sr and Nd isotope data were obtained at the USGS in Denver with the permission and assistance of C. Hedge, B. Doe, R. Zartman, and Z. Peterman. B. Barreiro determined the Pb isotope data referred to in the text. This study was supported by NSF grant EAR 80-20791.

Petrology and Chemistry of Volcanic Rocks of the Southern Andes

L. LÓPEZ-ESCOBAR

Abstract

Pliocene to Quaternary volcanic rocks of andesitic to dacitic composition predominate in the northern (33–37°S; Province I), and southern (48–56°S; Province III) sectors of the SVZ, whereas rocks of basaltic composition predominate in the central sector (37–46°S; Province II). Many stratovolcanoes in the latter sector have precaldera lavas of basaltic composition and postcaldera lavas of dacitic composition (e.g. Quetrupillán, Lanín and Puyehue volcanoes). Hornblende is a common phenocryst phase in lavas of Provinces I and III (e.g. Marmolejo–San José volcanic group and Mt. Burney volcano), but are uncommon in Recent lavas from Province II, particularly between 37–42°S (e.g. Antuco, Callaqui, Lonquimay, Villarrica, Puyehue, Osorno, Puntiagudo and Calbuco volcanoes). Nevados de Chillán (36°50′S) located near the boundary of Provinces I and II is transitional, with lavas of andesitic composition lacking hornblende. Lavas from Province II generally have relatively flat chondrite-normalized REE patterns with low La/Sm ratios and HREE contents greater than 8 × chondrites; lavas from Province I have variable La/Yb ratios with HREE ranging from about 15 × chondrites to less than 5 × chondrites. Most samples from stratovolcanoes such as Puyehue and Nevados de Chillán, have very subparallel REE patterns; other volcanic suites have REE patterns of varying slope with poor correlations of La/Sm, La/Yb and total REE with major elements, especially SiO_2 content. Miocene volcanism in the SVZ is represented by subvolcanic rocks of basaltic to dacitic composition distributed along the Longitudinal Depression (33–42°S) and High Andes Cordillera (33–38°S). Manifestations of Miocene igneous activity associated with porphyry copper deposits are also observed between 28–33°S where Quaternary volcanism is absent.

Introduction

This paper is centred on the Southern Volcanic Zone from 33–56°S and presents a general summary of the main petrologic and geochemical characteristics of calc-alkaline Cenozoic volcanism in this region. Two major periods of calc-alkaline volcanic activity are recognized in the Southern Andes: the Miocene volcanism, outcropping along the Longitudinal Depression (33–42°S) and High Andes Cordillera (33–38°S), and the Pliocene-Quaternary volcanism which is restricted to the High Andes Cordillera (Vergara and Munizaga, 1974; Moreno, 1976; López-Escobar and Frey, 1976; López-Escobar *et al.,* 1976, 1977; Stern *et al.*, 1976; Deruelle 1977, 1982; Muñoz and Niemeyer, 1983). As a matter of simplification, the Pliocene-Quaternary Southern Andes Volcanic belt between 34–42°S has been divided into the subparallel 'eastern' and 'western' volcanic chains. (Moreno, 1976; López-Escobar *et al.,* 1977). Both chains are separated by about 50 km and no longitudinal depression is located between them; on the contrary, in many places they are connected by transverse lines of volcanoes (e.g. the Villarrica-Quetrupillán-Lanin volcanic chain). South of 42°S the Southern Andes stratovolcanoes are distributed along a single line.

To the east of the Andean calc-alkaline belt the volcanism is alkaline in character. In the Central Andes this volcanism has been characterized as shoshonitic by Lefèvre (1973). South of latitude 33°S it is represented by plateau-type basalts outcropping mainly in Argentina (Niemeyer, 1978; Skewes and Stern, 1979; Charrier *et al.,* 1979).

Petrography

The Pliocene-Quaternary volcanism of the Southern Andes can be divided into three main petrographic regions. Between 33–37°S (Province I) rocks range in composition from basalt to rhyolites, but there exists a clear predominance of hornblende-, biotite- and magnetite-bearing andesites and dacites (Moreno, 1975, 1976). These petrographic characteristics are typical of the calc-alkaline series (Kuno, 1950, 1968a). Pyroxene andesites, like basalts and rhyolites, are comparatively less abundant. Between 37–46°S (Province II) basalts and basaltic andesites predominate (Moreno, 1975, 1976) and some stratovolcanoes present a precaldera stage of basaltic composition and a post caldera stage which includes rocks of dacitic composition (e.g. Quetrupillán Lanín and Puyehue volcanoes). Andesites are less abundant than dacites, and rhyolites are rare. In fact, only one rhyolitic tuff has been found at Cordillera Nevada (~ 40°30′S; Moreno, 1975). Magnetite and hydrous phases are absent in rocks from the western volcanic belt (Moreno, 1976). The predominant mafic minerals in this region are olivine, clinopyroxene and orthopyroxene with Fe-olivine a common phenocryst phase in dacites. These petrographic features are not typical of calc-alkaline series (Kuno, 1950, 1968a). Hornblende and magnetite have been found only in andesitic and dacitic lavas from the eastern volcanic chain (e.g. at Tronador volcano, 41°20′S; Larsson, 1941) and in dacites from Cay volcano (45°S; Stern *et al.,* 1976). The sector between 36–37°S may be considered as a transitional zone. The main volcanic group in this area is Nevados de Chillán (Deruelle and Deruelle, 1974). This is composed by lavas of andesitic to dacitic composition which lack hornblende. In Province III between 48–56°S, only five stratovolcanoes are known. They are andesitic in composition and hornblende is again a common phenocryst (Stern *et al.,* 1976 and the previous paper by Stern *et al.,* this volume). Province III andesites and dacites are similar petrographically to those from Province I which, in turn, are similar to andesites and dacites from the Central Andes volcanic zone. In the three provinces plagioclase is the most abundant phenocryst and groundmass phase, a feature which is typical of modern orogenic lavas (Ewart, 1976).

Province II, particularly between 37–42°S, has been the most studied among the three petrographic regions of the Southern Andes because it is relatively accessible and located close to centres of population. Unfortunately, some authors have used incorrectly the features from this province to characterize the whole of the SVZ.

The Miocene volcanism along the Longitudinal Depression is mainly andesitic in its northern part (33–37°S) and basaltic in its southern part (37–42°S). Like in the Pliocene-Quaternary volcanism of the Central Andes and Province I in the Southern Andes, two kinds of andesites are common in the northern part of the Longitudinal Depression volcanic belt: pyroxene andesites and hornblende andesites (Katsui and Vergara, 1966; Vergara and López-Escobar, 1980). Similar characteristics are observed in the Miocene volcanism which occurs along the High Andean Cordillera. The pyroxene andesites contain phenocrysts of plagioclase (mostly labradorite), clinopyroxene (augite), orthopyroxene (hypersthene), olivine and Fe-Ti oxides. The hornblende andesites contain phenocrysts of plagioclase (mostly andesine), hornblende and biotite. Pyroxenes are rare in the hornblende andesites. The basaltic rocks from the southern sector of the Miocene belt contain abundant phenocrysts of plagioclase (An_{30}–An_{50}), clinopyroxene and orthopyroxene (Vergara and Munizaga, 1974; López-Escobar *et al.,* 1976). Hornblende is sometimes present and secondary quartz is quite common.

Major element geochemistry

Province I rocks (33–37°S) are similar in major element composition to those from the CVZ with equivalent SiO_2 content and petrography. Although the volcanic rocks from this province range in composition from basalt to rhyolite, it is difficult to find a single volcanic centre containing the whole series. The Laguna del Maule volcanic centre at 36°S is one of the scarce Andean volcanic centres which contain rocks ranging in composition from basalt ($SiO_2 \sim 51$ wt. %) to rhyolites ($SiO_2 \sim 74$ wt. %), but basalts and rhyolites predominate with intermediate rock types much less abundant (Munizaga, 1978).

Rocks from Province I fall in the subalkaline field on a $Na_2O + K_2O$ versus SiO_2 diagram (Figure 1) and exhibit a typical calc-alkaline trend on an AFM diagram (Figure 2). There is a predominance of rocks with SiO_2 contents in the range 55–65 wt. %. Most rocks from this province are quartz-normative; olivine-normative rocks are rare.

Province II rocks also fall in the subalkaline field on a $Na_2O + K_2O$ versus SiO_2 diagram (Figure 1). However, in the AFM diagram (Figure 3) they exhibit a more tholeiitic character. On this basis, some authors have classified the basaltic rocks from this province as tholeiites, but on the basis of their high Al_2O_3 (>17 wt. %), high CaO (>8 wt. %), low TiO_2 (<1.5 wt. %) contents and relatively low $Mg/(Mg + \Sigma Fe)$ ratios (generally <0.60) they can be considered as typical calc-alkaline basalts associated with convergent plate margin volcanism (Perfit *et al.,* 1980). Actually, these major element features make these basalts very distinct from the ocean floor tholeiites. Because of their low Mg/Fe ratios, some authors prefer to classify the basalt from this province as basaltic andesites. However, they have basaltic SiO_2 contents (49–52 wt. %)

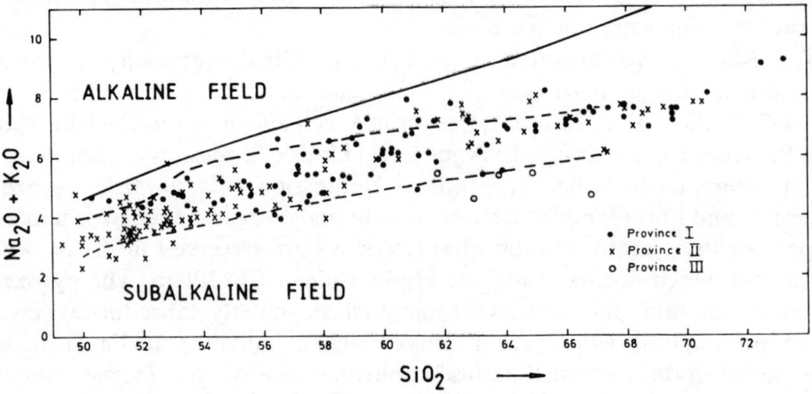

Figure 1 Total alkalis versus silica plot. Dotted lines represent boundaries of high-alumina basalt series (Kuno, 1966). Solid line is proposed boundary between alkaline-subalkaline suites (Irvine and Baragar, 1971)

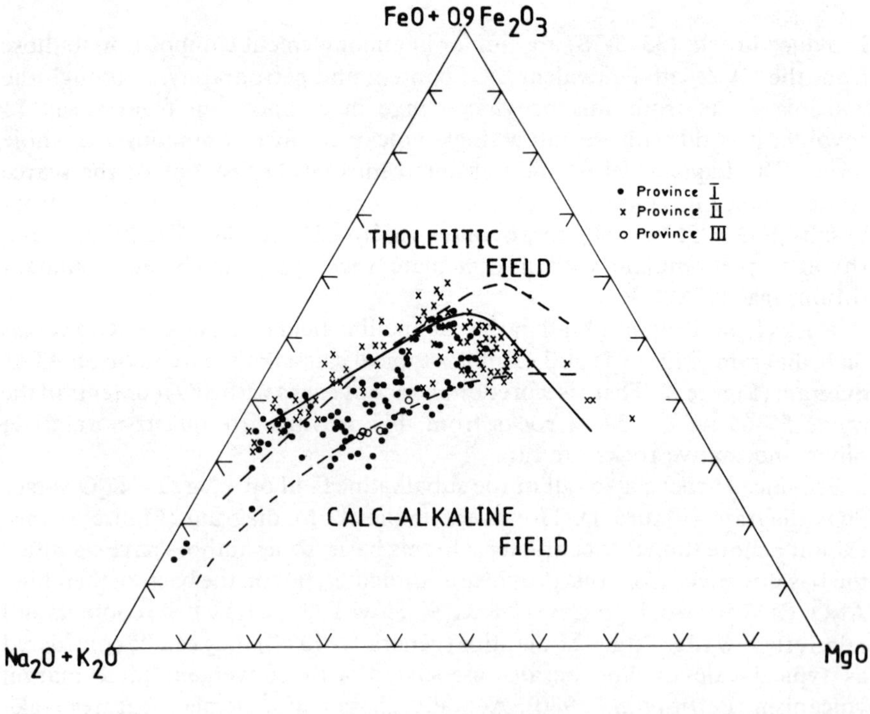

Figure 2 AFM diagram (wt. %) with total iron calculated as FeO. Solid line is proposed boundary between tholeiitic-calc-alkaline field (Irvine and Baragar, 1971). Dotted lines represent general boundaries of Japanese hypersthene (calc-alkaline) series (Kuno, 1968b)

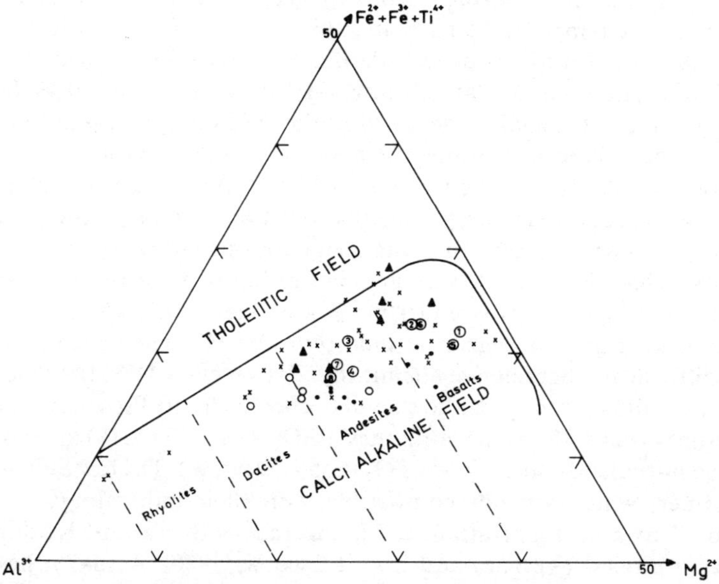

Figure 3 Jensen (1976) cation plot illustrating the division of subalkaline volcanic rocks into tholeiitic and calc-alkaline series and the distribution of the Longitudinal Depression Miocene volcanics from 33–34°S (▲ – pyroxene-bearing rocks; ○ – hornblende-bearing rocks) into the fields of calc-alkaline basalts, andesites and dacites. As a comparison, this figure also includes rocks from the Southern Andes Pliocene-Quaternary volcanic belt (×) and the Ewart (1976) averages for island-arc basalts (1), basaltic andesites (2), pyroxene andesites (3), hornblende-andesites (4), western American continental margin basalts (5), basaltic andesites (6), pyroxene andesites (7) and hornblende andesites (8)

and, on the basis of the Irvine and Baragar (1971) classification, they are basalts (Moreno, 1976; López-Escobar *et al.*, 1977). Their high Al_2O_3 contents may reflect some plagioclase accumulation, but it is possible that the magmas are inherently alumina-rich. For this reason, some authors have referred to them as high-alumina basalts (Vergara and Katsui, 1969). Similar basalts occur throughout the circumpacific orogenic region (Ewart, 1976).

For similar SiO_2 contents, the Pliocene-Quaternary volcanic rocks from this province are lower in Mg and higher in Fe and Ti than the Pliocene-Quaternary volcanic rocks of equivalent SiO_2 from the Central Andes. Deruelle (1982) attributes this feature to earlier fractionation of olivine in rocks from Province II as compared to a fractionation of Fe-Ti oxides, orthopyroxene and hornblende in the volcanics of the CVZ.

Between 34–42°S, where it is possible to distinguish two subparallel volcanic belts, it is observed that rocks from the eastern volcanic belt are richer in K than those from the western volcanic belt. Thus, the trend of increasing K with the distance from the trench, observed in island-arc settings, also appears to be characteristic of Provinces I and II in the SVZ (Vergara, 1970; Moreno, 1975; Deruelle, 1977, 1982).

Mt. Burney is the best studied stratovolcano in Province III (Stern *et al.*, 1976; the first paper by Stern *et al.*, this volume). It ranges in SiO_2 from 61.8–66.8 wt. %. Like Province I andesites and dacites, Mt. Burney lavas have high Al_2O_3 contents and relatively low Mg/(Mg + ΣFe) ratios (0.49–0.52) and follow the typical calc-alkaline trend in the AFM diagram (see Figure 2).

A common feature in some volcanoes or volcanic groups in the SVZ, particularly those located between 37–42°S in Province II, is that older rocks are more siliceous than the youngest ones. For example, Antuco volcano (37°25′S) is less siliceous than its neighbouring volcano Sierra Velluda (37°30′S) which is older. A similar relationship is observed between Sierra Nevada (38°30′S) and its younger neighbour Llaima (38°45′S).

Due to their greater degree of alteration, the Miocene volcanic rocks are more difficult to characterize chemically. Between 33–34°S, the volcanics of the Longitudinal Depression are similar chemically to Province I Pliocene-Quaternary rocks. Pyroxene 'andesites' (SiO_2 = 49 – 53 wt. %) are more basic than the hornblende 'andesites' (SiO_2 = 55 – 59 wt. %). The low SiO_2 values of the former, which are not compatible with their mineralogy, have been attributed to a remobilization of Si, together with Ca and K, during the alteration process (Vergara and López-Escobar, 1980). Actually these rocks were classified on the basis of their less mobile elements using the Jensen (1976) cation plot (see Figure 3). Although the pyroxene 'andesites' are more basic than those bearing amphibole, they exhibit higher Mg/(Mg + ΣFe) ratios suggesting that both kinds of rocks have an independent origin.

The Miocene Longitudinal Depression volcanic belt between 37–42°S is similar geochemically to calc-alkaline suites of circum-Pacific island arcs (López-Escobar *et al.*, 1976). On a AFM diagram they follow the typical calc-alkaline trend (Figure 2 of López-Escobar *et al.*, 1976). In this aspect, they are more similar to the Province I suite than to the Province II suite. However, some rocks from this region, particularly those classified as basalts, are similar to Province II volcanics of equivalent SiO_2.

Trace element geochemistry

Figure 4 shows the Sr–Ba–REE patterns of rocks of different Southern Andes volcanoes. These patterns vary along the SVZ and this variation, as well as the variation in petrography, seems to be related to the geometry of the Benioff zone. Province I rocks are generally characterized by fractionated REE patterns with relatively large La/Yb normalized ratios (17.0–5.0); the highest values are found in rocks from 33–34°S and the lowest in rocks from 36–37°S. These patterns, which are similar to those observed in CVZ andesites and dacites (Thorpe *et al.*, 1976; Dostal *et al.*, 1977a, b; Palacios and López-Escobar, 1979; James, 1981a; Deruelle, 1982), are due to a relatively low HREE abundance (generally less than 8× chondrite). By contrast, rocks from the Province II, have lower La/Yb normalized ratios (1.5–5.0) than rocks from the Province I. The only known exceptions are rocks from monogenetic cones located at 39°10′–39°30′S whose ratios fall in the range 4.1–5.7.

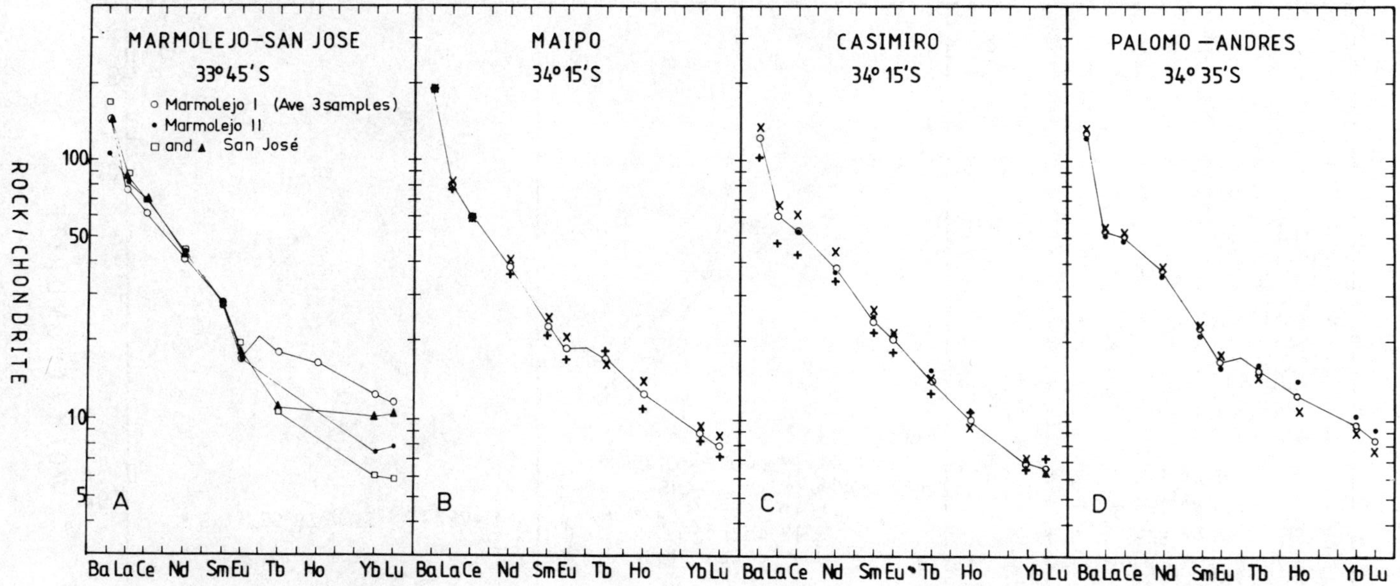

Figure 4 Rare-earth element abundances distribution patterns, normalized to chondrites, of Southern Andes calc-alkaline Cenozoic volcanic rocks. Most data belong to Frey-López-Escobar working group (MIT, Universidad de Chile). Data of Mt. Burney (52°S) were taken from Stern *et al.*, 1976

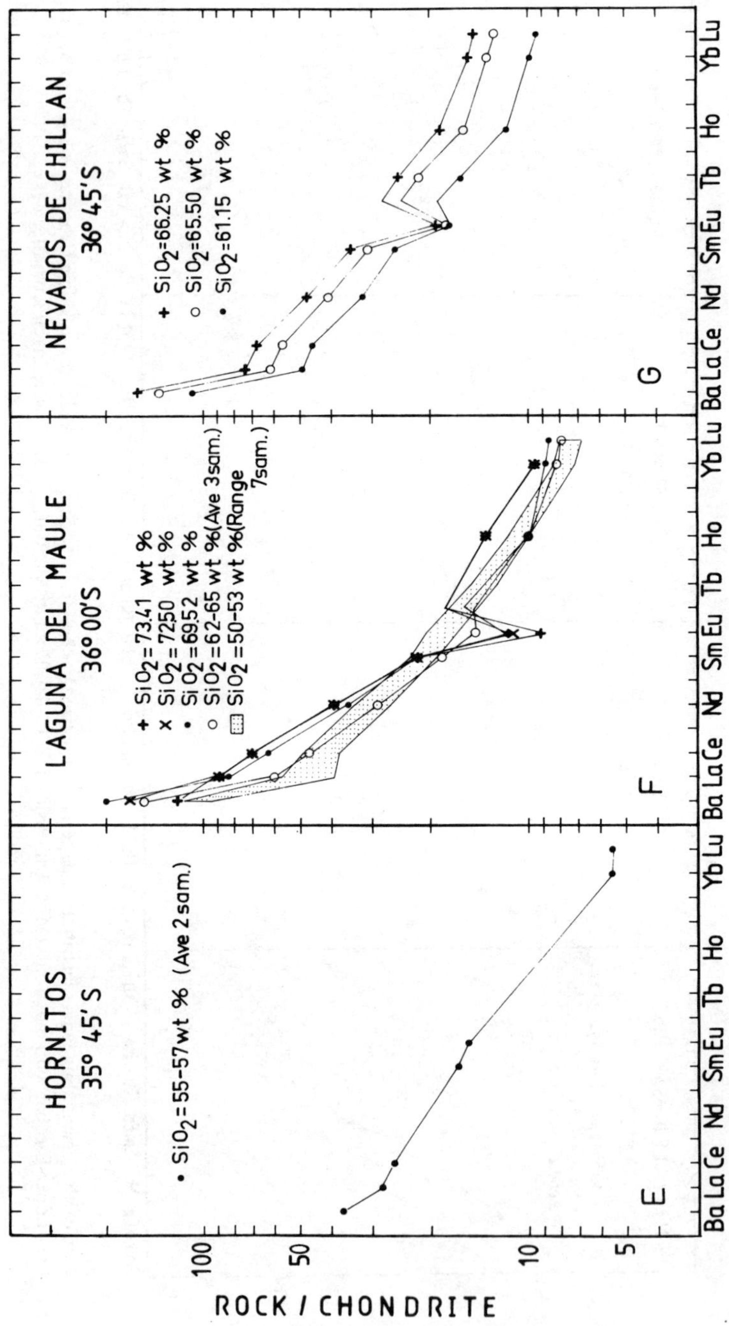

Figure 4 (contd.)

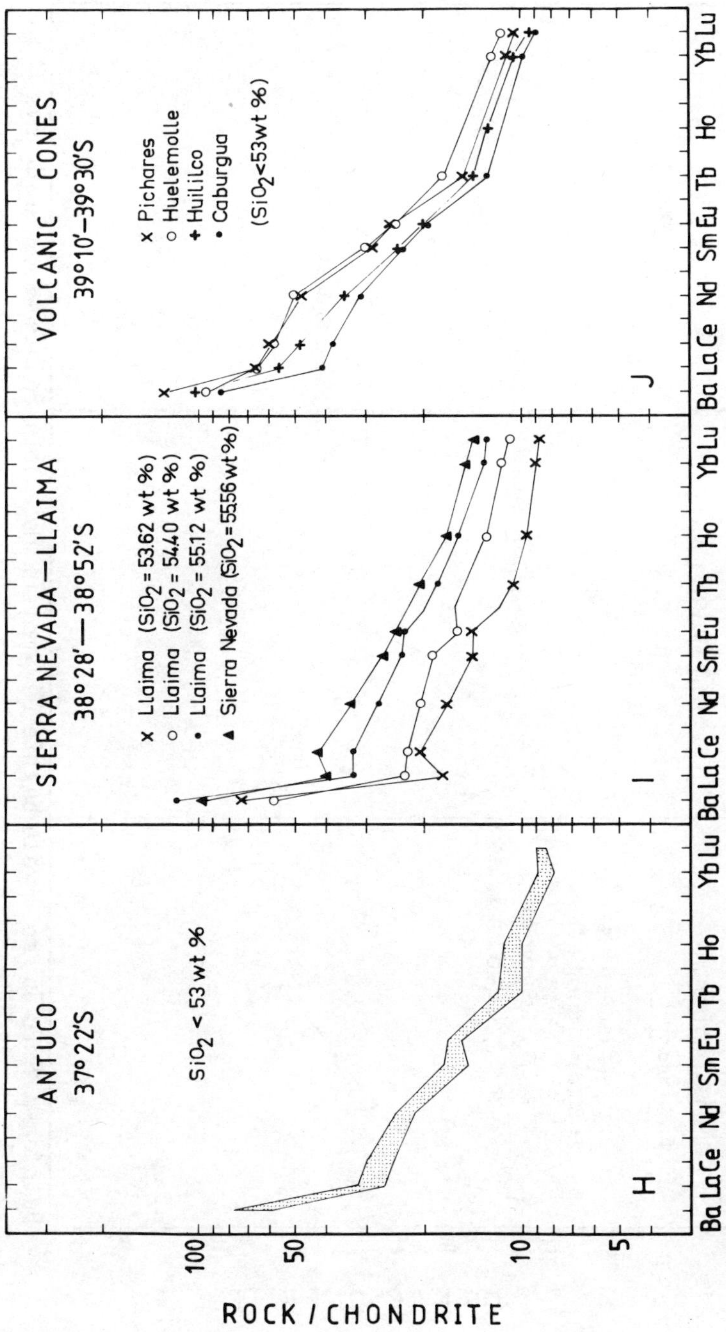

Figure 4 (contd.)

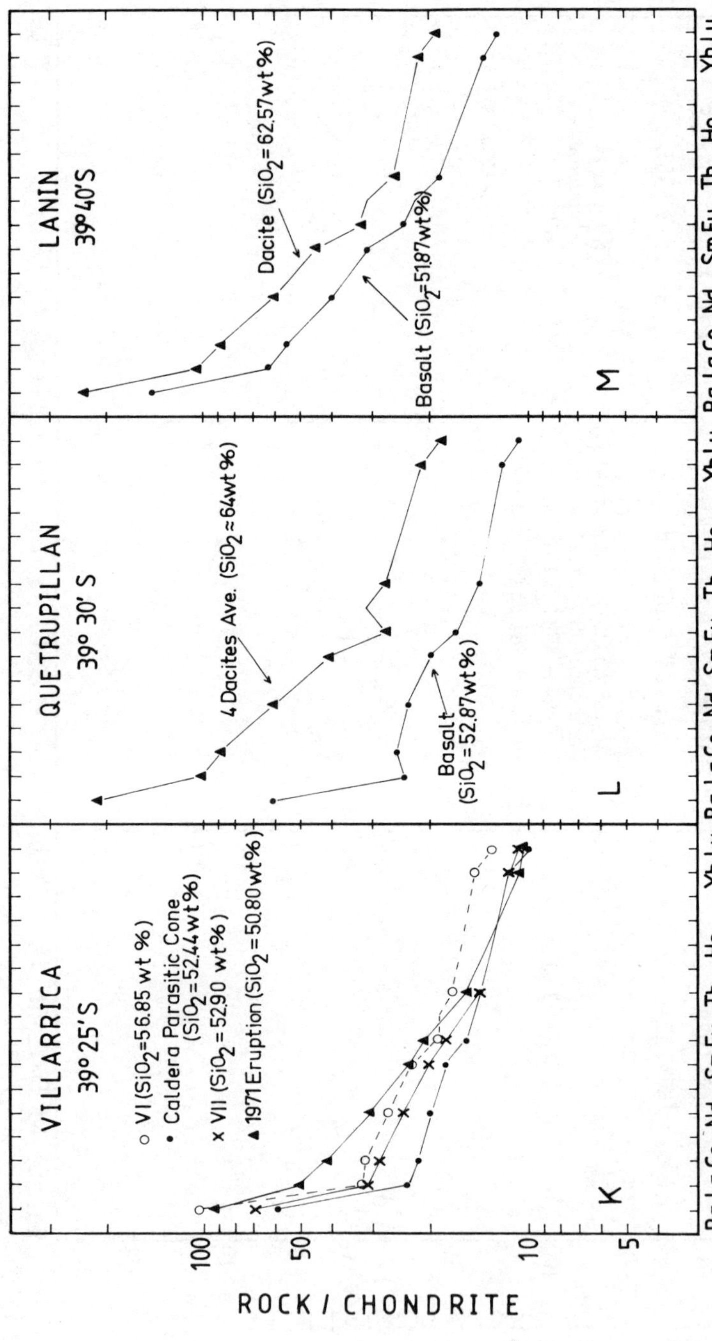

Figure 4 (contd.)

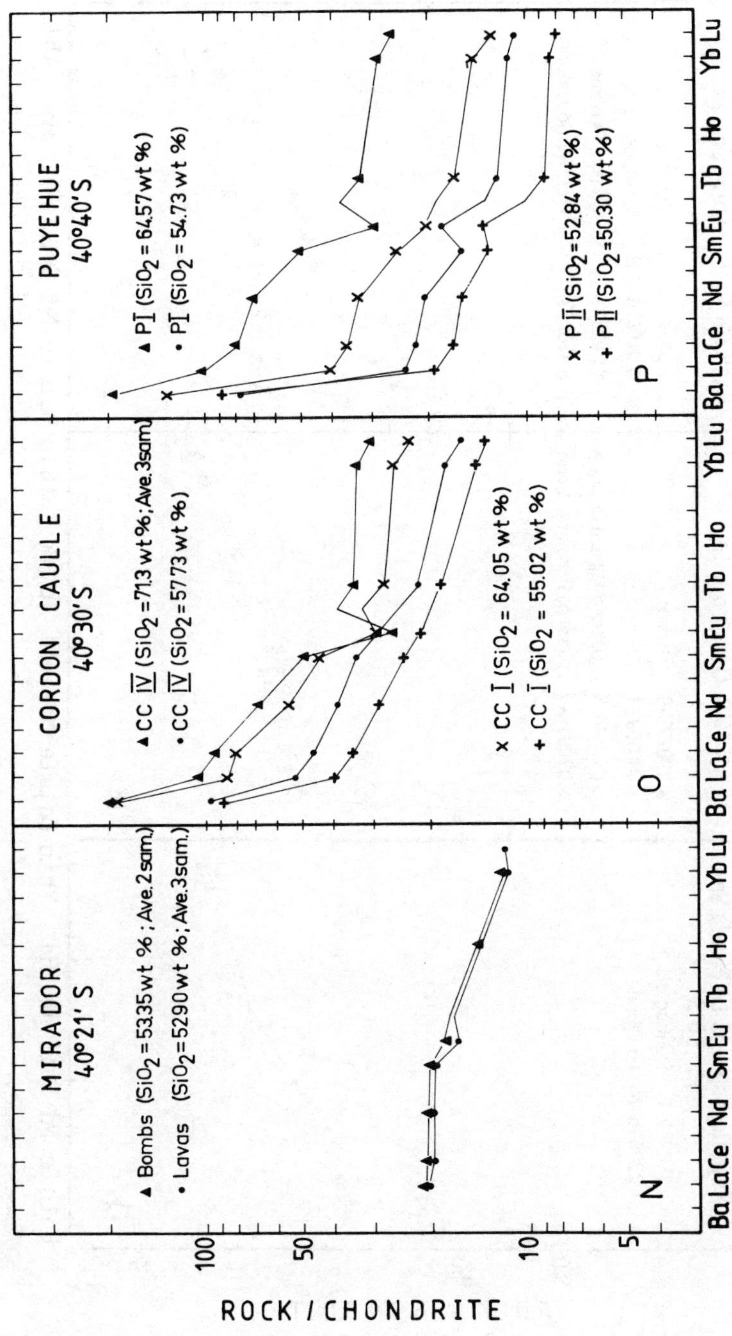

Figure 4 (*contd.*)

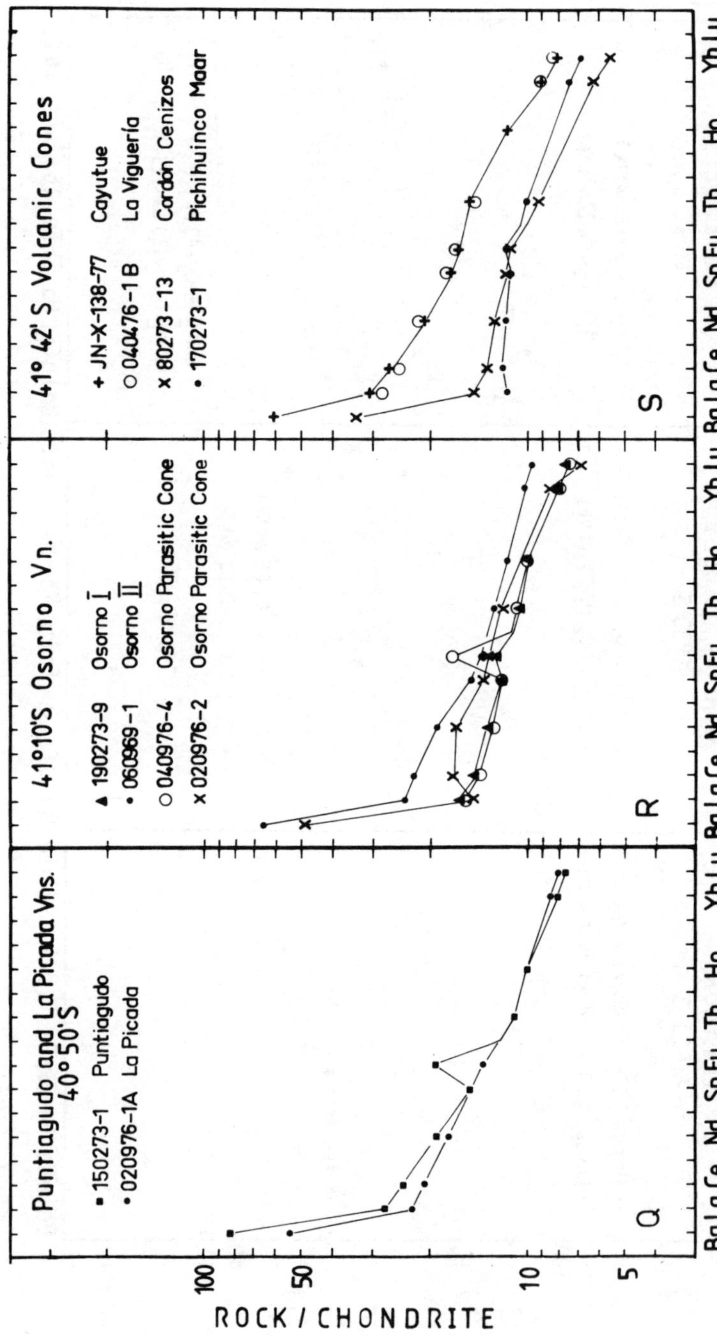

Figure 4 (contd.)

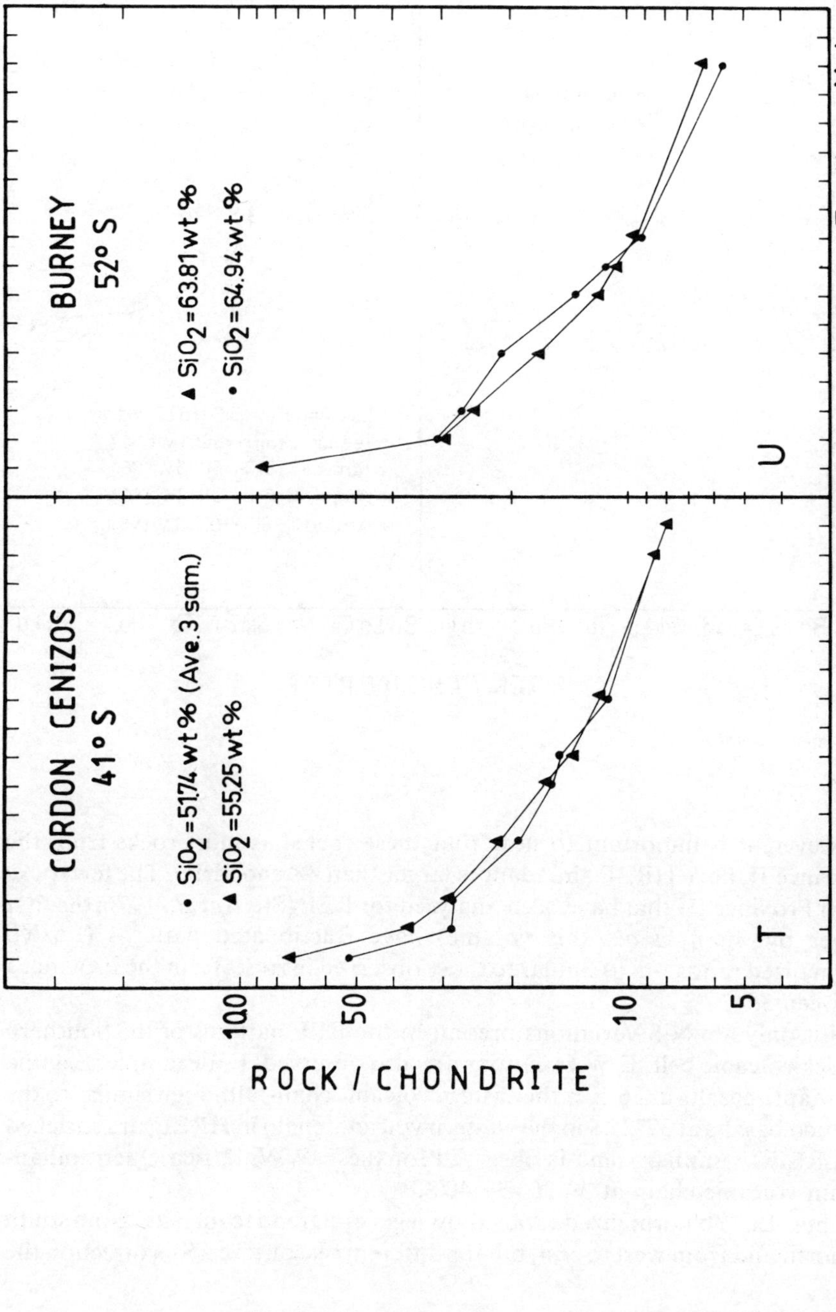

Figure 4 (contd.)

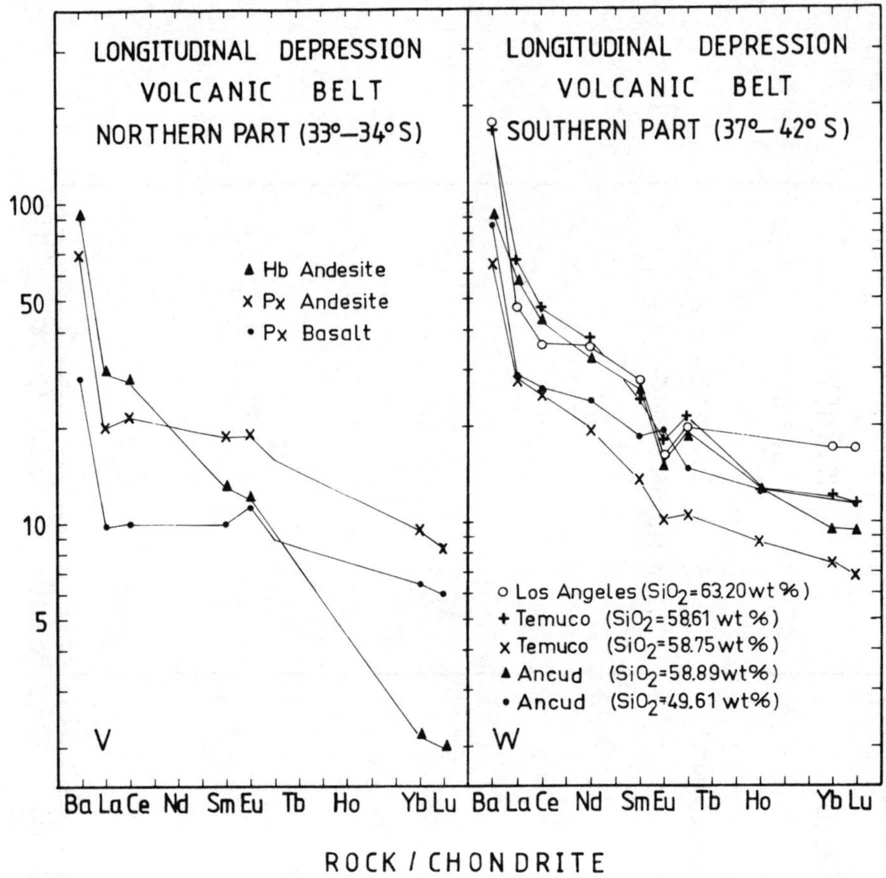

Figure 4 (contd.)

However, it is important to note that these rocks, as most rocks from the Province II, have HREE abundances larger than 8× chondrite. The few rocks from Province III that have been analysed for REE (Stern *et al.*, 1976; the first paper by Stern *et al.*, this volume) have fractionated patterns (La/Yb normalized ratios ~5.0) similar to those observed in rocks from the Province I between 35–36°S.

Not only are N-S variations present in the REE patterns of the Southern Andes volcanic belt, E-W variations are also observed. For example, Laguna del Maule basalts at 36°S in the eastern volcanic chain, although similar to the Antuco basalts at 37°25′S in the western volcanic chain in HREE, are enriched in LREE. A similar trend is observed for the E-W Villarrica-Quetrupillán-Lanin volcanic chain at 39°30′–39°40′S.

Thus, La/Yb normalized ratios show a general trend to increase from south to north and from west to east, but for different reasons. In a S-N direction the increase is due to HREE depletion, whereas in a W-E direction it is due to LREE enrichment. These trends appear to parallel the trend followed by K.

Superimposed on the regional geochemical trends, each volcano has its own individual geochemical character. Some volcanic suites exhibit significant variations in LREE but not in HREE (e.g. Villarrica volcano at 39°30′S). Others show relatively large variations in HREE but not in LREE (e.g. Marmolejo-San José volcanic group at 33°45′S). In some volcanoes the increase in REE abundance is correlated with increasing SiO_2 (e.g. Nevados de Chillán volcanic group at 36°50′S and Llaima volcano at 38°45′S); in others there is no well-defined relationship between REE abundances and SiO_2 content (e.g. Cordón Caulle at 40°30′S). Some volcanoes exhibit very homogeneous REE abundances throughout their eruptive history (e.g. Antuco at 37°25′S); others show relatively heterogeneous REE distributions (e.g. Puyehue at 40°35′S).

Province I rocks are similar in large-ion-lithophile elements such as K, Rb, Sr, Cs, Ba, Zr, Ta, Th and U to the CVZ andesites and dacites, but are enriched in these elements relative to Province II rocks equivalent in SiO_2. The Province II lavas, regardless of their age (Miocene or Pliocene-Quaternary), have trace element abundances which all similar to calc-alkaline island-arc rocks of equivalent SiO_2 content. Basalts from Province II are dissimilar to island-arc tholeiites which have low and flat or LREE depleted REE patterns.

At 33–34°S the Miocene pyroxene-bearing volcanics of the Longitudinal Depression exhibit flatter REE abundance distribution patterns (La/Yb normalized ratios about 2) than the hornblende-bearing rocks (La/Yb $\sim$15). South at 37–42°S, the Longitudinal Depression Miocene volcanics exhibit REE patterns with variable La/Yb normalized ratios (2.4–6.0) which are relatively similar to those of some Province II Pliocene-Quaternary calc-alkaline volcanic rocks.

Isotope geochemistry

The Sr isotope characteristics of Pliocene-Quaternary volcanic rocks in the SVZ are shown in Figure 5. Most data have been obtained on rocks from Province II, particularly between 37–42°S. Data from other provinces are still sparse. Despite this, Figure 5 suggests that there is a major change in the Sr-isotope composition at $\sim$37°S. From 33–37°S Sr isotope ratios decrease from the high (>0.7050) values, similar to those characteristic of the CVZ Pliocene-Quaternary volcanic rocks (Deruelle *et al.*, 1983 and references therein) to low (<0.7045) values approaching those of ocean floor basalts (0.7037). As pointed out by Deruelle *et al.* (1983), south of 37°S Sr isotope ratios show a general trend to increase reaching values close to 0.7045 at Hudson volcano (46°S). The behaviour of this parameter, including its break at 37°S, coincides with the behaviour of other parameters such as petrographic character and major and trace element geochemistry which appear to be related to such primary tectonic and geological features as subduction zone geometry and crustal thickness.

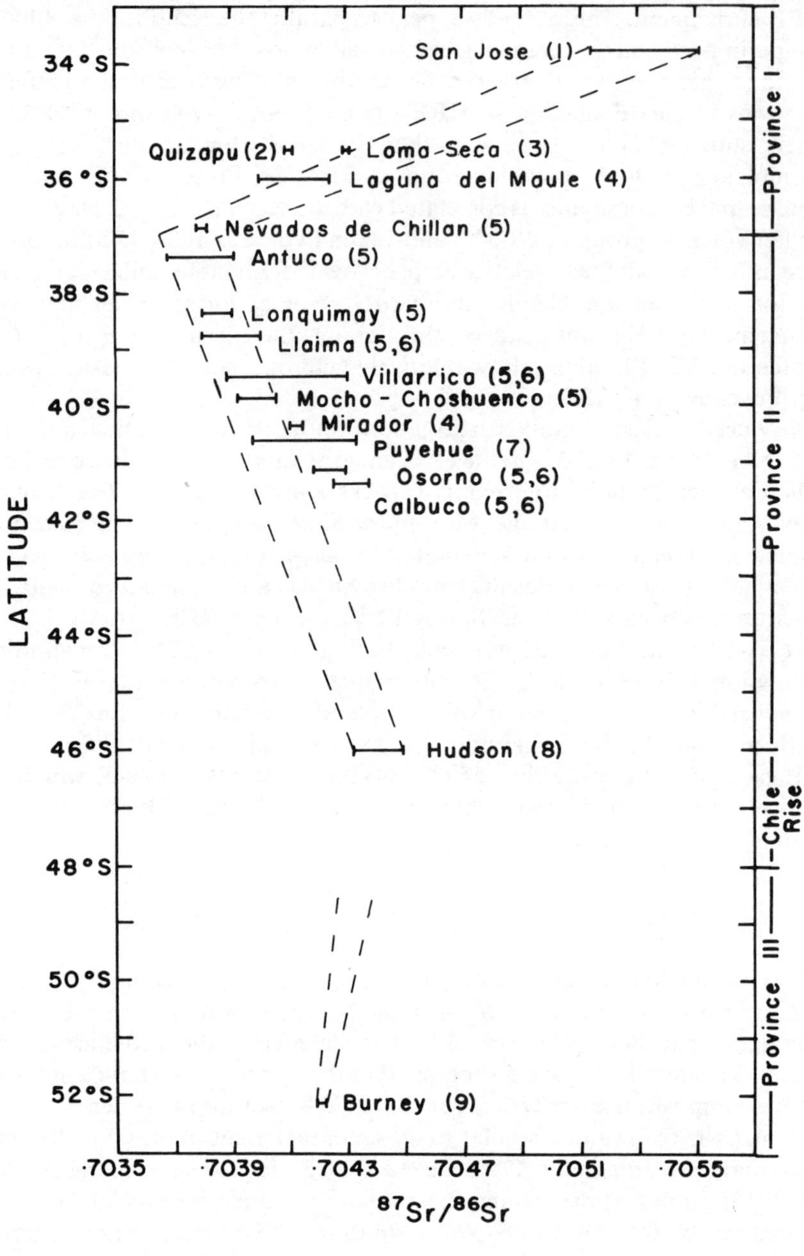

Figure 5 Southern Andes Pliocene-Quaternary volcanics Sr isotope ratios as a function of latitude. Data from: López-Escobar *et al.* (in prep.); W. Hildreth and R. Drake (pers. comm.); Hildreth *et al.* (1981); Hickey *et al.* (1982); Deruelle *et al.* (1983); Klerkx *et al.* (1977); D. Gerlach (pers. comm.); Godoy *et al.* (1981); Stern (1982). Dashed lines connect the Sr isotope ratios ranges of San José, Antuco, Hudson and Burney volcanoes

At present, Nd-isotope data are extremely limited for the Pliocene-Quaternary calc-alkaline volcanics from the Southern Andes. Most $^{143}Nd/^{144}Nd$ ratios measured have been determined on lavas from Province II. Figure 6 shows the Nd versus Sr isotope compositions of rocks from Laguna

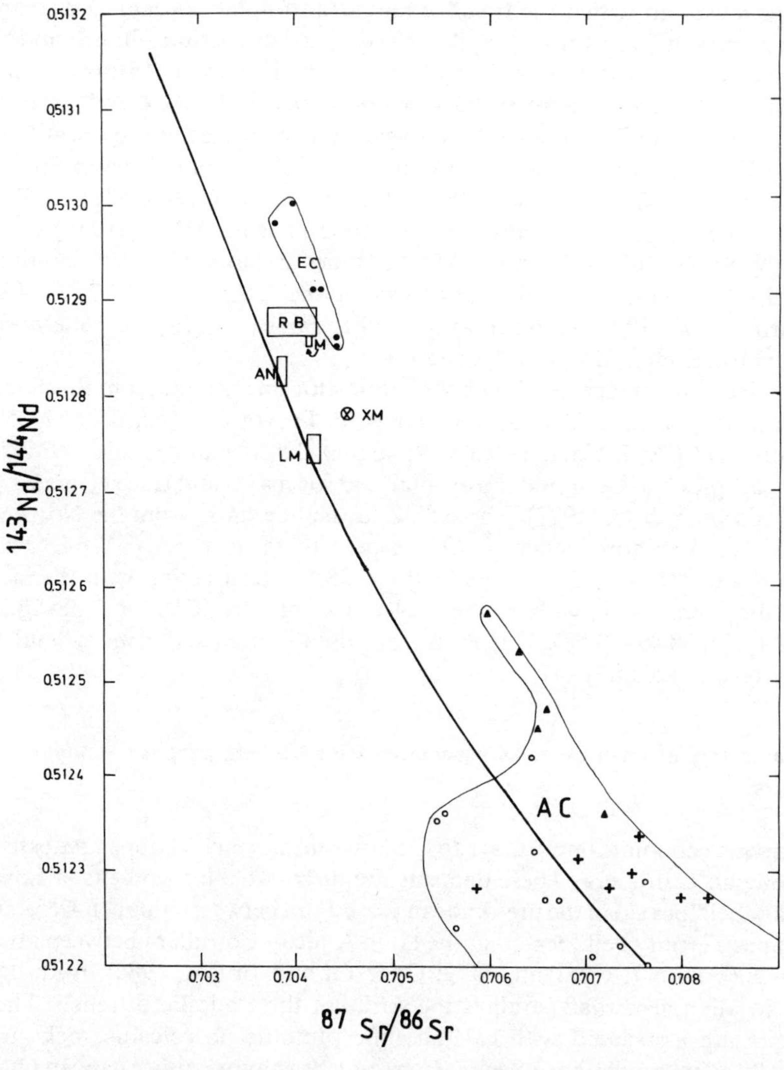

Figure 6 $^{143}Nd/^{144}Nd$ versus $^{87}Sr/^{86}Sr$ plot for Laguna del Maule (36°00′S), Antuco (37°25′S) and Mirador (40°21′S) volcanics (Hickey *et al.*, 1982). As a comparison this figure also includes data of NVZ andesites (EC – Ecuador; Francis *et al.*, 1977; Hawkesworth *et al.*, 1979b), Central Andes (AC; ▲ – data from Francis *et al.*, 1977 and Hawkesworth *et al.*, 1979b; ○ – data from James, 1982; + – data from Hawkesworth *et al.*, 1982) and Southern Andes plutonic rocks (RB – granodiorites associated with the Río Blanco-Los Bronces porphyry-copper deposit. Data from Halpern, 1979 and R. Drake, pers. comm. ● – granitic xenoliths included in Mirador volcano basalts. Data from Hickey *et al.*, 1982)

del Maule volcanic centre (36°00'S), Antuco (37°25'S) and Mirador (41°21'S) from Hickey *et al.* (1982). For comparison this diagram also includes data for Pliocene-Quaternary rocks from the NVZ and CVZ from Francis *et al.* (1977), Hawkesworth *et al.* (1979b, 1982), and James (1982a). It is interesting to note that the Nd and Sr isotope compositions of the Laguna del Maule and Antuco rocks fall within the trend of magmas from intraoceanic island arcs where continental crust is absent. A more detailed discussion of the Sr and Nd isotope relationships in the SVZ lavas is given by Hickey *et al.* (this volume).

Similar differences between the volcanic rocks of the SVZ and CVZ are observed in O and Pb isotope ratios. Basalts and andesites in Provinces I and II have $\delta^{18}O$ values which range between 5 and 7 ‰ (second paper by Stern *et al.*, this volume; Deruelle *et al.*, 1983). A similar range (5.9–7.2 ‰) has been found in rocks from the Mt. Burney volcano in Province III (Stern, 1982, first paper by Stern *et al.*, this volume). These values are notably lower than those for Pliocene-Quaternary volcanic rocks in the CVZ ($\delta^{18}O$ = 7 – 14‰; Magaritz *et al.*, 1978; Harmon *et al.*, 1981; James, 1982a; Deruelle *et al.*, 1983; Harmon and Hoefs, this volume).

Barreiro (1982c) reported Pb isotope ratios for one sample from Province I (Tupungato volcano, 33°15'S) and five from Province II (Antuco, 37°25'S; Callaqui, 37°50'S; Llaima, 38°45'S; Villarrica, 39°30'S; and Lanin, 39°40'S). These samples had been previously analysed for major and trace elements by López-Escobar *et al.* (1977). These SVZ lavas, like those from the Northern Andes, are very homogeneous with respect to Pb isotopes ($^{206}Pb/^{204}Pb$ = 18.58; $^{207}Pb/^{204}Pb$ = 15.61; $^{208}Pb/^{204}Pb$ = 38.49) contrasting with the large variability observed within the volcanics of the CVZ (17.38–18.29; 15.53–15.63; 38.49–38.78). This Pb-isotope data is discussed in more detail by Barreiro (this volume).

Geochemistry of igneous rocks associated with Chilean porphyry-copper deposits

Of extreme economic importance to Chile are the porphyry copper deposits of the Andean Cordillera. These deposits are distributed in two belts: a Lower Tertiary belt located in the pre-Andean Cordillera between about 18–28°S and an Upper Tertiary belt located in the High Andean Cordillera between about 31–34°S (Figure 7; Oyarzún, 1979; Oyarzún and Frutos, 1980). Both belts have an alignment that parallels the strike of the Chile-Peru trench. These deposits are associated with calc-alkaline plutonic or volcanic rocks that generally have modal hornblende. Cenozoic porphyry-copper deposits have not been recognized south of 37°S.

Igneous rocks associated with the El Salvador (~26°S; 41 Ma) and Río Blanco-Los Bronces (~33°S; <13 Ma) ore deposits have relatively low $^{87}Sr/^{86}Sr$ ratios (0.7037–0.7042; Gustafson, 1979; Halpern, 1979). The Río Blanco-Los Bronces rocks exhibit $^{143}Nd/^{144}Nd$ ratios in the range 0.512860–0.512890 (Drake, pers. comm.) and as shown in Figure 6, for a given $^{143}Nd/^{144}Nd$ ratio, the Río Blanco-Los Bronces granodioritic rocks are

Figure 7 Sketch map showing the geographic distribution of the main Chilean porphyry-copper deposits (after Gustafson and Hunt, 1975)

slightly enriched in radiogenic Sr relative to mantle-derived rocks. This behaviour is similar to that exhibited by basalts of the Mirador volcano (40°21′S) and NVZ andesites in Ecuador (Hawkesworth *et al.*, 1979b).

Lead isotopic composition of igneous rocks from the El Salvador porphyry copper deposit ($^{206}Pb/^{204}Pb$ = 18.51–18.53; $^{207}Pb/^{204}Pb$ = 15.59–15.62; $^{208}Pb/^{204}Pb$ = 38.44–38.52; Tilton *et al.*, 1981) agree closely with ratios from the Río Blanco-Los Bronces ore deposit ($^{206}Pb/^{204}Pb$ = 18.60–18.64; $^{207}Pb/^{204}Pb$ = 15.61–15.62; $^{208}Pb/^{204}Pb$ = 38.53–38.55; Tilton, 1979). These values are also similar to those exhibited by Pliocene-Quaternary volcanics in the SVZ (Barreiro, this volume).

Figure 8 shows that igneous rocks associated with porphyry-copper deposits, regardless of their age, are also similar in REE to Pliocene-Quaternary volcanic rocks from the CVZ and Province I lavas in the SVZ. As discussed above, these patterns are strongly fractionated and typical of hornblende-bearing andesitic rocks. The similarity in isotopic and REE composition between igneous rocks associated with Chilean porphyry-copper deposits and Pliocene-Quaternary volcanic rocks from Province I in the SVZ strongly suggests a similar origin for both.

The presence of Miocene volcanism between 28–33°S where Quaternary volcanism is absent, and the existence in this area of a major porphyry-copper deposit (Los Pelambres; 31°42′S; 9.96 Ma; Quirt *et al.*, 1971) suggests that a change occurred in subduction geometry between Upper Tertiary and Quaternary time. Nur and Ben-Avraham (1981) explain this change by the collision and consumption of the aseismic Juan Fernandez ridge under the South American plate.

Petrogenesis

Trace element and isotope data have been considered to evaluate the petrogenesis of the calc-alkaline Miocene and Pliocene-Quaternary volcanic rocks of the SVZ (López-Escobar and Frey, 1976; López-Escobar *et al.*, 1976; Stern *et al.*, 1976; Klerkx *et al.*, 1977; López-Escobar *et al.*, 1977; Deruelle, 1982). In developing petrogenetic models for these rocks two types of sources have been considered: the downgoing oceanic crust in either an amphibolitic or eclogitic mineralogy and the overlying mantle peridotite wedge.

It has been concluded that the first source cannot generate these rocks in a single-stage process for the following reasons: (a) no amphibolitic or eclogitic xenoliths have ever been found in them; (b) melts of eclogite yield strongly fractionated REE distribution patterns which are unlike the patterns exhibit by rocks from Province II and, although like some Province I rocks in HREE, they are dissimilar in LREE; (c) melts of amphibolite retain the flat or LREE depleted distribution of the Nazca plate basalts and therefore do not explain the La/Yb normalized ratios (>1) observed in the SVZ-Cenozoic volcanics; (d) melts of amphibole also do not explain the characteristically low HREE abundances in SVZ lavas relative to Nazca plate basalts; and (e) the very low Rb (0.2–1.3 ppm), Sr (92–126 ppm) and Ba (1.3–1.6 ppm) contents of Nazca

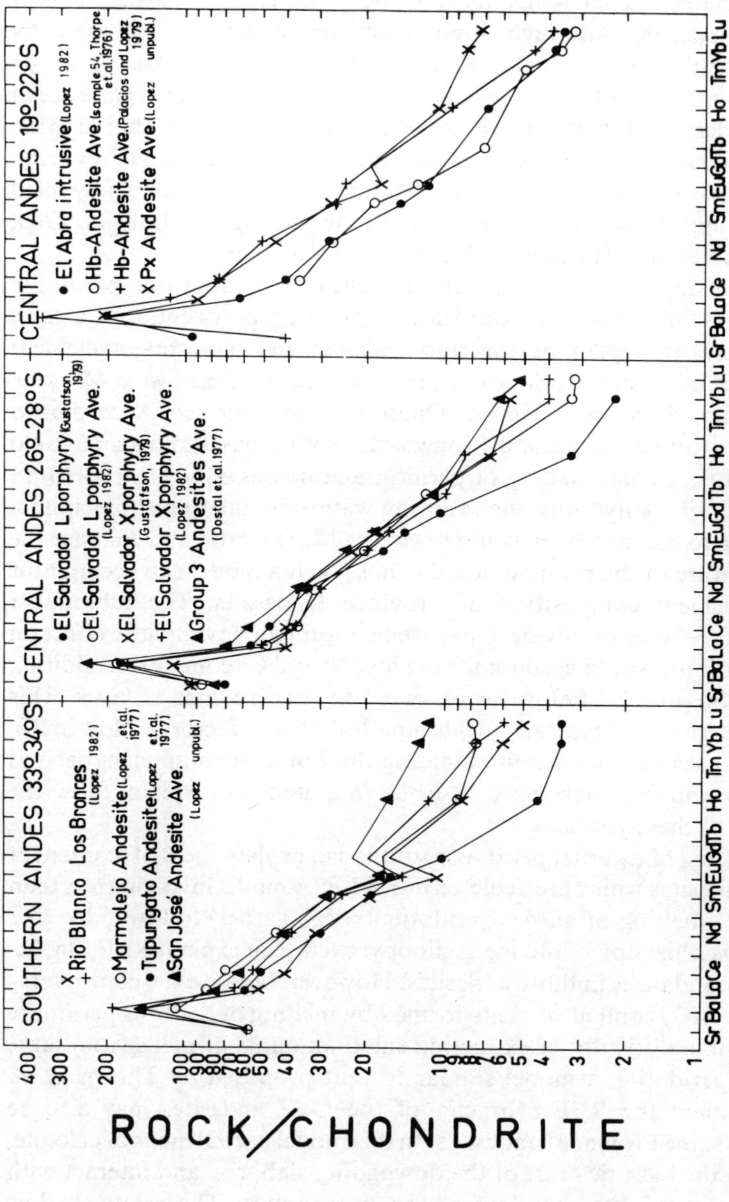

Figure 8 REE abundance distribution patterns, normalized to chondrites, of granodioritic rocks associated with some Chilean porphyry-copper deposits. As a comparison, this figure also includes the REE patterns of Pliocene-Quaternary volcanics geographically associated with those copper ore deposits (data from Thorpe *et al.*, 1976; López-Escobar *et al.*, 1977; Dostal *et al.*, 1977b; Gustafson, 1979; Lopez-Escobar, 1982; and this study)

plate basalts (Hart, 1976) are insufficient to generate the Rb-Sr-Ba abundances of Province I and II basalts and andesites except by less than 5% melting. However, to explain the andesitic major element composition at least 30% melting of such a source is required. This large discrepancy in degree of melting to account for major and some incompatible trace elements precludes a petrogenesis for SVZ calc-alkaline volcanic rocks as direct partial melts of ocean floor basalts. Although these problems could be overcome by assuming a high degree of alteration in the ocean floor basalts or by contamination from continental crustal material, studies of Nazca plate ocean floor basalts demonstrate that less than 1% of the basaltic core is altered to the degree required to explain the Rb and Ba abundances of andesites (Frey *et al.*, 1974) and contamination from continental crustal material is practically ruled out by the O and Pb isotope compositions of the rocks (Deruelle *et al.*, 1983; Barreiro, this volume; Harmon and Hoefs, this volume).

Petrogenetic models involving partial melting of garnet-free peridotite followed by 10–20% crystal fractionation of olivine ± pyroxene appear to be more successful in explaining the trace element and most major element characteristics of basaltic rocks from the Longitudinal Depression Miocene volcanic belt and Province II Pliocene-Quaternary volcanic belt. According to experimental studies (Green and Ringwood, 1968) a basaltic magma could form by 10–15% partial melting of peridotite at depths equivalent to 10–15 kbar. If peridotite is hydrous, the resulting water-rich melt must precipitate olivine during its ascent which would lower the Mg/Fe ratio and enhance the aluminous nature of the residual liquid. Thus, such a model can account for the major element composition of Province II basalts. The subsequent fractionation (~20%) of olivine ± pyroxene, although having little effect on REE distributions, would account for the low Ni and Cr contents in addition to the low $Mg/(Mg + \Sigma Fe)$ ratios observed in the Province II lavas. This peridotite source model typically yields liquids that are a factor of 2 low in Rb, Sr and Ba, suggesting that a contaminating fluid or melt component derived from the downgoing slab may also be required to explain fully the petrogenesis of these rocks.

Partial melting of a garnet peridotite source can explain the REE content of Province I andesites which are depleted in HREE. A model involving less than 5% fractional melting of garnet peridotite (~5% garnet) followed by 20% fractional crystallization of olivine ± clinopyroxene can explain REE, Ba, Sc, Cr and Co abundances in those andesites. However, because the controversy regarding the SiO_2 content of melts formed by melting of hydrous peridotite and the possible difficulty of extracting small amounts of melts from large amounts of peridotite, a model similar to that proposed by Thorpe *et al.* (1976) to explain the REE character of the CVZ andesites may also be applicable to some Province I andesites. In this model partial melts of eclogite, formed from the basaltic crust of the downgoing slab, rise and interact with the overlying peridotite mantle forming pyroxenites. The relatively low density pyroxenites subsequently forms diapirs which undergo partial melting at various depths. These melts, possibly further modified by shallow fractional crystallization, form the compositional range seen in calc-alkaline rocks (Nicholls and Ringwood, 1973). Model calculations show that 15–20% partial melting of such a contaminated garnet peridotite are required to generate primary magmas similar in trace element composition to those generated by 3-5% melting of uncontaminated garnet peridotite (López-Escobar, 1978).

The fact that Marmolejo-San José volcanic group (33°45′S; Province I) contain lavas with a relatively large variability in HREE contents (~15× chondrites to less than 7× chondrites) suggests that the problem of HREE depletion may not be due only to the involvement of garnet as a source refractory phase, but also may be a product of fractional crystallization of hornblende at shallow levels. In liquids of andesitic to dacitic composition, hornblende has solid/liquid partition coefficients for HREE large than unity (Arth and Barker, 1976; Nicholls and Harris, 1980).

The origin of dacites in the SVZ is not well understood. In some volcanic centres, particularly those of Province I, dacites can be explained as a fractionation product of a basaltic or andesitic magma. The Province II dacites, and the scarce rhyolites of the SVZ, are still under study and will be the subject of future reports.

Summary and conclusions

According to present understanding, based upon petrographic, chemical (major and trace elements) and isotopic relationships, the Pliocene-Quaternary calc-alkaline volcanism of the SVZ may be divided into three provinces. Province I (33–37°S) is dominated by hornblende-bearing andesitic and dacitic rocks that follow a typical calc-alkaline trend on an AFM diagram. These rocks exhibit fractionated REE abundances distribution patterns with La/Yb normalized ratios that show a general trend to decrease from about 17 at 33°S to about 5 at 37°S. $^{87}Sr/^{86}Sr$ ratios also decrease from values that are similar to the lowest part of the CVZ Pliocene-Quaternary volcanics range (0.7051–0.7055 at 33°45′S) to values that are close to those characteristic of fresh ocean floor basalts (0.7037–0.7039 at about 37°S). By contrast, Province II (37°46′S) is dominated by basalt and basaltic andesites that, although calc-alkaline in many aspects (high Al_2O_3, high CaO, low TiO_2 contents and low Mg-numbers), follow a rather tholeiitic trend on the AFM diagram. These rocks, as well as the more acid volcanic rocks of this province, particularly those of the western chain between 37–42°S, practically lack hornblende and exhibit flatter REE patterns than rocks of Province I, with La/Yb normalized ratios that rarely exceed 5. Within this province, the La/Yb ratio does not seem to follow any trend with latitude, but the $^{87}Sr/^{86}Sr$ ratios seem to show a general trend to increase from 37°S (~0.7037) to 46°S (~0.7045). Sr isotope ratios higher than 0.7050 have not yet been reported for volcanic rocks of Province II. However, more data are required to evaluate the validity of these trends. The lowest $^{87}Sr/^{86}Sr$ ratios are exhibited by rocks located around 37°S. It is interesting to note that the Laguna del Maule (36°S) and Antuco (37°25′S) volcanics fall within the mantle array on a $^{143}Nd/^{144}Nd$ versus $^{87}Sr/^{86}Sr$ diagram, but basalts from Mirador volcano (40°21′S; western chain) are enriched in radiogenic Sr relative to mantle-derived rocks of areas where continental crust is absent. Nevados de Chillán at 36°50′S may be considered a transitional volcanic group, with lavas of andesitic to dacitic composition like Province I, but lacking hornblende like Province II western chain lavas.

Mt. Burney is the best studied stratovolcano in Province III. It is comprised of hornblende-bearing andesitic to dacitic lavas. Like Province I rocks, these lavas follow a calc-alkaline trend on the AFM diagram and exhibit fractionated REE patterns with La/Yb normalized ratios about 5, similar to the SVZ volcanics at latitude 35°S. They also have similar $^{87}Sr/^{86}Sr$ ratios ($\sim$0.7042).

Superimposed on these general trends each volcano exhibits its own peculiarities. For example, some volcanic suites have very subparallel REE patterns (e.g. Nevados de Chillán and Puyehue); others exhibit REE patterns of varying slope with poor correlation of La/Sm, La/Yb and total REE with major elements, especially SiO_2 content.

Miocene volcanism in the Southern Andes is distributed along the Longitudinal Depression (33–42°) and High Andes Cordillera (33–38°S). Hornblende- and pyroxene-bearing lavas are typical within the northern sector, particularly between 33–34°S. The northern volcanoes are characterized by strongly fractionated REE patterns with La/Yb normalized ratios about 15; pyroxene–bearing rocks have flatter REE patterns with La/Yb ratios about 2. Within the southern sector (37–42°S), Miocene volcanics have REE patterns of varying slope, and some of them are similar to those of Province II Pliocene-Quaternary rocks.

In general, igneous rocks associated with porphyry-copper deposits have modal hornblende and fractionated REE patterns with HREE abundances less than 8× chondrites. The porphyry-copper deposits are actually distributed in areas where modal hornblende is common in Cenozoic calc-alkaline volcanics.

The petrogenesis of calc-alkaline volcanics of the SVZ is best explained by a model involving melting of peridotitic upper mantle material followed by a crystal fractionation involving mainly olivine ± clinopyroxene or plagioclase ± amphibole in addition to the previous phases. The first model generally accounts for the most basic rocks and the second one for andesites and more evolved members. Garnet, as a residual phase, has been used to explain the petrogenesis of some Province I rocks but, because hornblende is common in this sector, the observed HREE depletion could be also due to a shallow level crystal fractionation of this phase. Contamination with subducted lithospheric material seems to be important in some cases whereas contamination with crustal material seems to be much less important than in the Central Andes.

ACKNOWLEDGEMENTS

This work was financed by the Departamento de Geología y Geofísica de la Universidad de Chile and by Grant E1703–8315 of the Departamento de Desarrollo de la Investigación de la Universidad de Chile. Nuclear irradiations were made at the Massachusetts Institute of Technology and Comisión Chilena de Energía Nuclear nuclear reactors. I thank Drs F. A. Frey, D. Gerlach, R.S. Harmon and H. Moreno for constructive discussions and Miss G. Morales for technical assistance.

Geochemical Variations in Volcanic Rocks from Central-south Chile (33–42°S)

R. L. HICKEY, D. C. GERLACH and F. A. FREY

Abstract

Pliocene to Recent volcanic rocks from central-south Chile (33–42°S) have $^{143}Nd/^{144}Nd$ (0.51269–0.51291), $^{87}Sr/^{86}Sr$ (0.70381–0.70464) and $^{18}O/^{16}O$ (5.2–6.8‰) ratios which overlap with those in oceanic-island basalts. Although individual volcanic centres have distinct isotopic ratios, variations in these ratios do not vary systematically along or across the strike of the Andes. At two volcanic centres, Laguna del Maule (36°S) and Puyehue (40°S), rhyolites have Sr, Nd and O isotope ratios similar to associated basalts. As in other volcanic arcs related to subduction, basalts and basaltic andesites from central-south Chile have high K/La, Rb/La and Ba/La and low Rb/Cs abundance ratios relative to ocean floor and island basalts. Although K, Rb, Ba and La abundances in basalts from 12 central-south Chilean volcanic centres vary by factors of 3 to 8, ratios of these elements such as Ba/La (15–28) and K/La (330–720) vary by only a factor of two and do not correlate with variations in elemental abundance.

We conclude that in central-south Chile there is no isotopic evidence that crustal contamination is an important process in creating calc-alkaline sequences ranging from basalt to rhyolite. Although isotopic characteristics of central-south Chilean lavas are consistent with partial melting of mantle sources similar to the sources of oceanic-island basalts, variations in incompatible trace element abundances and ratios are more consistent with a two-stage model involving addition of K, Rb, Ba and Cs-rich fluids derived from the subducted Nazca plate to the peridotite sources of the basalts.

Introduction

It has long been recognized that there is an intrinsic connection between calc-alkaline volcanism and the subduction of oceanic plates. However, the nature of the connection between subduction and the generation of magma is not well understood and is a source of considerable debate among geologists of all specialities. Important questions include the relationship between subduction and the initiation of melting to form magma, and the relative contributions of mantle peridotite, subducted oceanic crust and crust of the overlying plate to the composition of subduction-related magmas. In order to avoid complexities caused by continental crust in the overriding plate, many petrologic and geochemical studies of calc-alkaline rocks have focussed on intra-oceanic island arcs. However, convergence between oceanic and continental plates has been a dominant tectonic setting along the western edges of the South and North American plates, and the geological consequences of such convergence have determined the recent evolution of the western margins of the American continents. Moreover, western South America is the best example of active convergence between oceanic and continental plates; therefore studies of the Andean volcanic chain provide a direct approach for understanding the volcanic consequences of oceanic-continental plate convergence.

Many tectonic and geophysical features vary along the length of the Andes: the dip of the subducted oceanic plate, the thickness of the South American continental crust and the age and structure of the subducted seafloor (Lomnitz, 1962; Stauder, 1973; Barazangi and Isacks, 1976; Sacks, 1977; Hanus and Vanek, 1978; Lowrie and Hey, 1981; Herron *et al.*, 1981). Thus, studies of volcanic rocks erupted along the length of the Andes enable evaluation of correlations between geophysical characteristics of the converging plates and the petrologic-geochemical nature of Andean volcanic rocks erupted on the overriding continental plate. Such comparisons have led to several observations important to the understanding of subduction-related volcanism. Active volcanism in the Andes is restricted to areas where the dip of the subducted plated is greater than about 20°, thus subduction geometry is critical to the initiation of melting. The Andes also have a large scale geochemical zonation, with relatively low $^{87}Sr/^{86}Sr$ and low $^{18}O/^{16}O$ ratios occurring in northern (5°N–2°S; Venezuela and Ecuador) and southern (33–52°S, Chile and western Argentina) active volcanic zones where continental crust is of normal thickness, and high $^{87}Sr/^{86}Sr$ and $^{18}O/^{16}O$ ratios occurring in the central Andes (16–28°S; Peru, Bolivia, northern Chile and Argentina) where crustal thicknesses reach 70 km (Klerkx *et al.*, 1977; Hawkesworth *et al.*, 1979b, 1982; Thorpe and Francis, 1979; Deruelle *et al.*, 1983). Thus the thick crust in the central Andes is clearly manifested in the composition of the volcanic rocks.

The area of the Andes discussed in this paper is central-south Chile in latitudes 33–42°S (Figure 1). Compared to the central Andes, this region is characterized by a greater abundance of basic volcanic rocks, basalts and basaltic andesites, although silicic rocks, including rhyolites, occur at several volcanoes. An immediate aim of our research in this area is to determine the petrogenesis of calc-alkaline basalts and associated silicic rocks within a continental setting, and to compare their evolution with subduction-related magmas from purely oceanic regimes, with the ultimate objective of using such comparisons to understand the links between subduction and volcanism. Within the region 33–42°S, like the Andes as a whole, tectonic and geophysical features vary with latitude: crustal thickness decreases southward from 45 km to 30 km (Lomnitz, 1962; Lowrie and Hey, 1981), maximum depth of earthquake epicentres, possibly indicating the maximum depth of slab penetration, decreases southward from 200 to <100 km (Hanus and Vanek, 1978), and the nature of the subducted oceanic crust changes at fracture zones at about 39°S, with >35 Ma crust to the north and <20 Ma crust to the south (Herron, 1981). Another aim of this study is to identify geochemical variations between individual volcanic centres in central-south Chile and to interpret the source

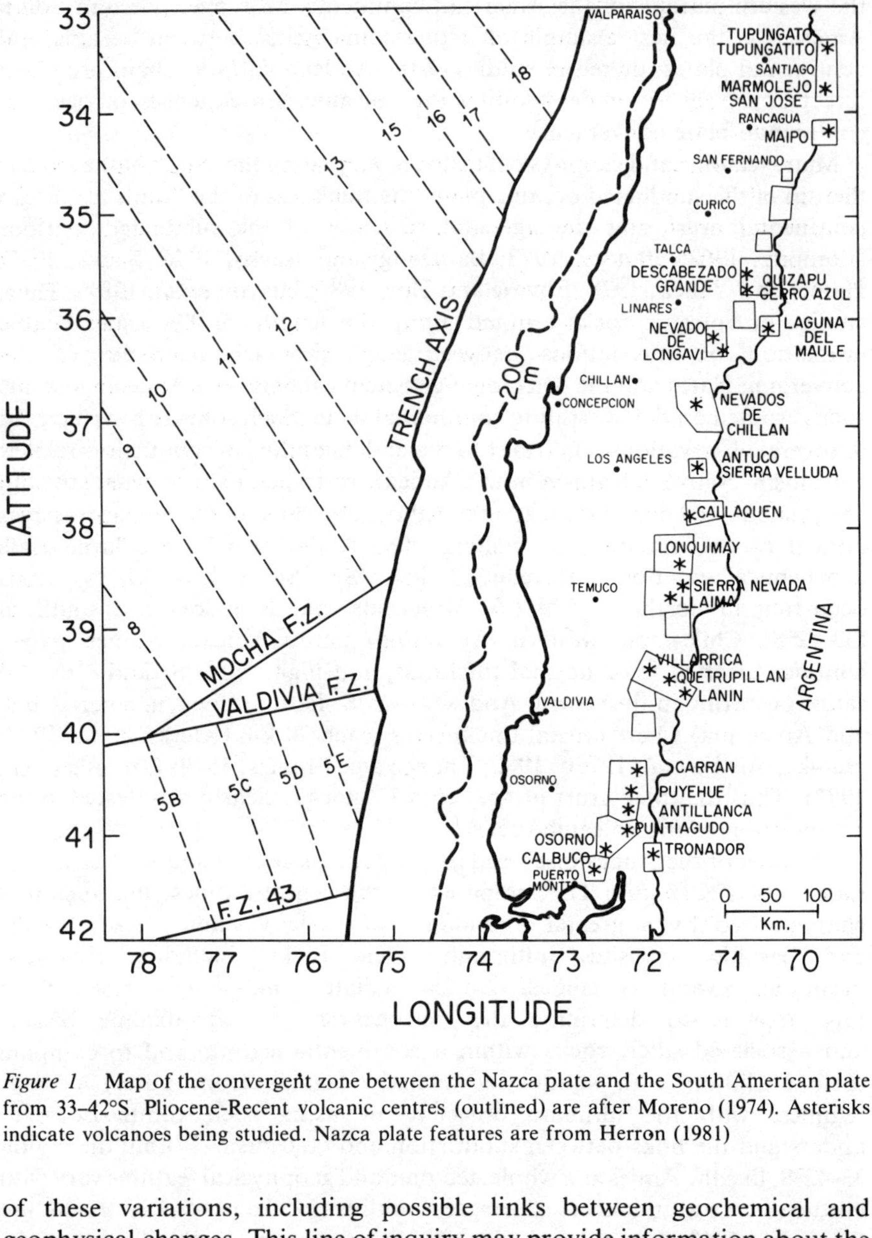

Figure 1 Map of the convergent zone between the Nazca plate and the South American plate from 33–42°S. Pliocene-Recent volcanic centres (outlined) are after Moreno (1974). Asterisks indicate volcanoes being studied. Nazca plate features are from Herron (1981)

of these variations, including possible links between geochemical and geophysical changes. This line of inquiry may provide information about the cause of large-scale geochemical variations along the Andes as a whole. Our approach in this research has been to sample each young volcanic centre in sufficient detail in terms of age and sample location so that we may define petrologic and geochemical variations within individual volcanic centres. Definition of petrologic and geochemical variations within single volcanic centres is required before regional variations can be discerned and interpreted.

Geological background

Figure 1 shows the location of Pliocene-Recent volcanic centres between 33–42°S. These volcanic centres form a chain along the Chile-Argentina border approximately 300 km from the Peru-Chile Trench. An older Eocene-Miocene belt of calc-alkaline volcanics is situated west of the active volcanic chain (Vergara and Munizaga, 1974). The width of the recently active volcanic chain ranges up to ~70 km and in several areas, such as Laguna del Maule (36°S), Lanin (39.5°S) and Tronador (41°S), major volcanic centres have developed at varying distances from the trench. Oldest basement rocks exposed within the volcanic chain are Jurassic-Cretaceous in age, and mid-late Palaeozoic basement is exposed along the coast and eastern Andes (Drake, 1976, 1983). Individual volcanic centres in this region of the Andes are dominated by one or more stratovolcanoes in varying stages of activity and erosion. Descabezado Grande, Quizapu, Nevados de Chillan, Llaima, Villarrica, Carran, Puyehue and Calbuco volcanic centres have erupted within this century (Moreno, 1974). Detailed mapping of volcanoes in this region is an ongoing project of researchers at the Universidad de Chile, Santiago (e.g. Moreno, 1974, 1976; Munizaga, 1978) and the University of California (e.g. Drake, 1976; Hildreth *et al.,* 1983), and our sample selection is based on their mapping.

Geochemical characteristics of volcanic rocks from central-south Chile

The discussion which follows is based on our ongoing study of recent volcanic rocks from central-south Chile. Here, we present aspects of our current data set which we will utilize in subsequent discussion to constrain the petrogenesis of these volcanic rocks.

MAJOR ELEMENTS: DISTRIBUTION OF ROCK TYPES AND REGIONAL TRENDS

In Figure 2, the range of SiO_2 contents found in individual volcanic centres between 33–42°S is plotted versus latitude. The major geographic variation in distribution of rock types is the predominance of andesite and absence of basalt in volcanoes north of 36°S, and the occurrence of basalt or basaltic andesite in volcanoes south of 36°S. Within the region 36–42°S individual volcanoes exhibit different distributions of rock types, but there is no correlation with latitude: Laguna del Maule (36°S) and Puyehue (40.5°S) contain basalt through rhyolite, while the Osorno (41.1°S), Llaima (39°S) and Antuco (37.5°S) volcanic centres contain basalt and andesite. Quetrupillan and Lanin volcanoes (39.5°S) have a bimodal distribution of basalt and dacite.

In Figure 3, K_2O is plotted versus SiO_2 for rocks from volcanic centres grouped by latitude from north to south. In the classification scheme of Peccerillo and Taylor (1975), volcanic rocks from central-south Chile range

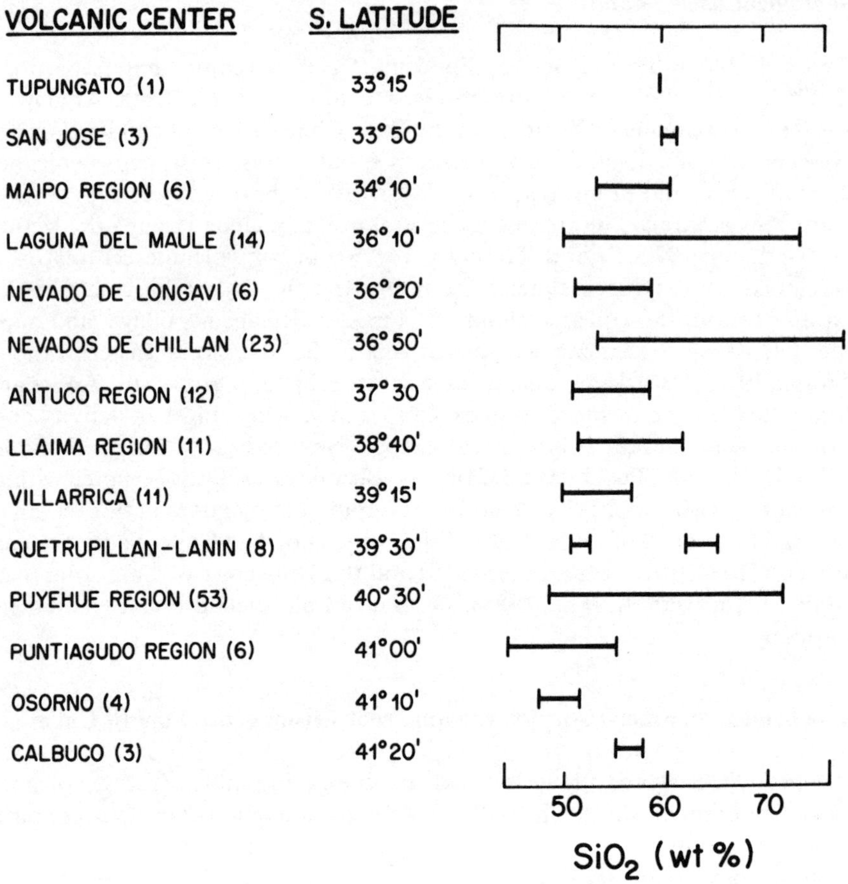

Figure 2 Range of SiO_2 contents in rock from individual volcanic centres plotted versus latitude. Numbers in parentheses indicate number of samples. Northern volcanic centres (Tupungato, San Jose, Maipo) are under-represented, but their silicic character is supported by petrographic studies (e.g. Moreno, 1974; Lopez-Escobar *et al.*, 1977) and study of our collections from San Jose, Maipo and Descabezado Grande which have not yet been analysed for major elements. In addition, Hildreth and Drake (1983) report that recent eruptions from Quizapu are dominantly dacitic (64–70% SiO_2)

from basalt to high-K rhyolites. In this paper we emphasize our data for basic rocks, and in subsequent discussion we use the term basalt to designate all rocks with <54% SiO_2. An important north-south trend in volcanics from this region, noted by Moreno (1976), is the southward decrease in K_2O content in rocks of equal SiO_2. For example, high-K andesites occur at northern volcanoes San Jose and Maipo (Figure 3a), andesites of normal K_2O content occur at Nevados de Chillan and Puyehue, while andesites from Calbuco border on low-K compositions (Figure 3d). Surprisingly, we have no evidence that K_2O increases with increasing distance from the

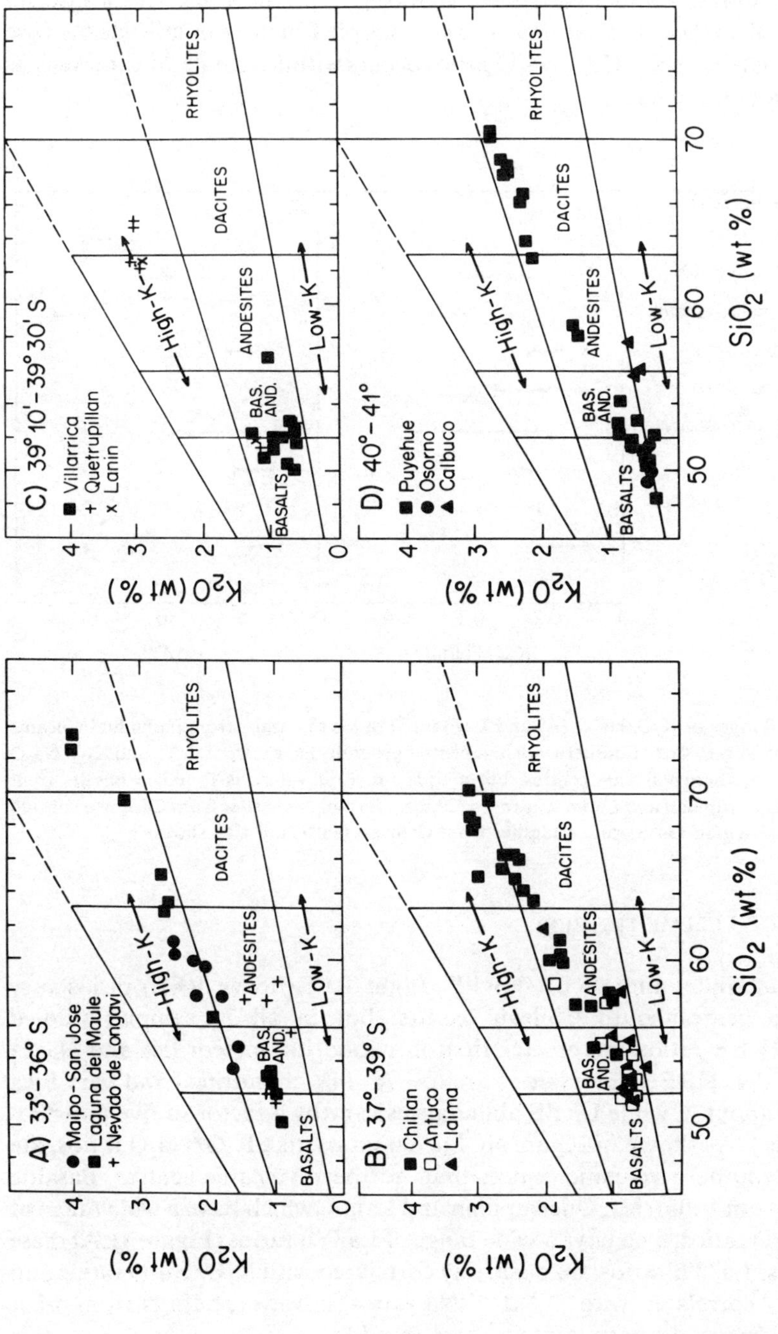

Figure 3 Plots of SiO_2 versus K_2O in rocks from individual volcanic centres grouped by latitude. Note the southward decrease in K_2O contents for rocks of equal SiO_2 content in (A), (B) and (D). Basalts from Villarrica, Quetrupillan and Lanin (C) have a wide range of K_2O contents. Classification boundaries from Peccerillo and Taylor (1975)

trench, as it does in many arc systems, including Peru and north Chile (e.g. Gill, 1981, p. 209). Although K_2O/Na_2O ratios are lower in southern volcanoes than northern volcanoes, ranging from >0.4 in basaltic andesites from Casimiro (34°S, a small cone near Maipo) to <0.2 in basalts from Osorno (41.1°S), K_2O/Na_2O variations are not a simple function of latitude because the complete range of K_2O/Na_2O ratios occurs within a single volcano such as Villarrica (Figure 4a).

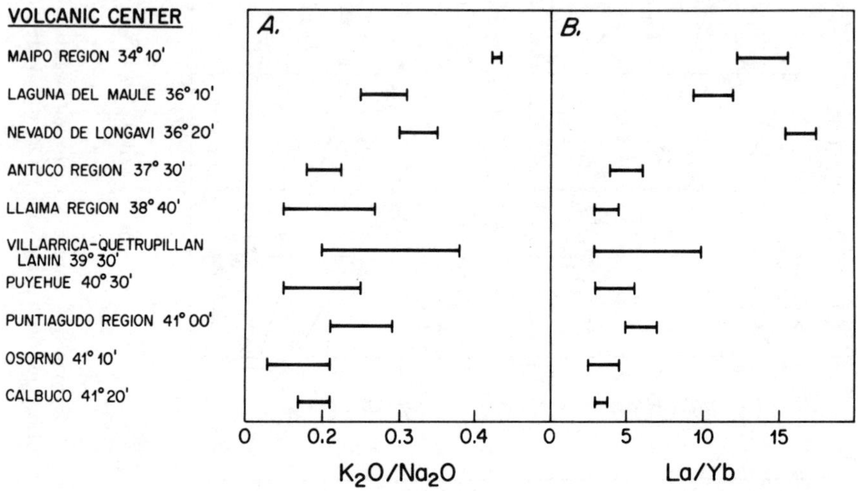

Figure 4 Ranges of K_2O/Na_2O (A) and La/Yb (B) ratios in basalts from individual volcanic centres from 34°S to 41°S. Southern volcanic centres generally have lower La/Yb and K_2O/Na_2O ratios than northern volcanic centres, but a wide range of values is found in basalts from Villarrica, Quetrupillan and Lanin volcanoes (39.5°S). Basaltic andesites from Casimiro volcano in the Maipo region (34°S), and andesites from Calbuco (41°S) are also shown

RARE EARTH ELEMENTS (REE)

REE abundances and LREE/HREE (light REE/heavy REE) ratios vary widely in central-south Chilean basalts, but in all 74 samples studied LREE/HREE ratios are greater than in chondritic meteorities and Nazca plate basalts. HREE abundances are low (6–14× chondrites) and vary by a factor of about 2, while LREE abundances vary by a factor of five (16–80 × chondrites). As shown in Figure 4b, La/Yb ratios, like K_2O/Na_2O ratios, are lower in southern volcanic centres than northern volcanic centres. Basaltic samples from Villarrica, Quetrupillan and Lanin, which have a wide range of K_2O/Na_2O ratios, also have a wide range of La/Yb ratios (Figure 4). At these volcanoes, La/Yb ratios are positively correlated with K_2O/Na_2O ratios and negatively correlated with $^{143}Nd/^{144}Nd$ ratios, however, basalts from other volcanic centres do not maintain these trends.

The variation of HREE abundances in central-south Chilean basalts has no regional trend; low HREE contents are found in both highly LREE-enriched basalts from the north and less LREE-enriched basalts from southern volcanic centres. In contrast, the La abundance of basalt decreases from north to south. At some volcanoes, La abundances in basalts may vary by a factor of two at a relatively constant La/Yb (e.g. Puyehue), however, at other volcanoes, such as Villarrica and Casimiro, La abundances are positively correlated with La/Yb ratio (Figure 5).

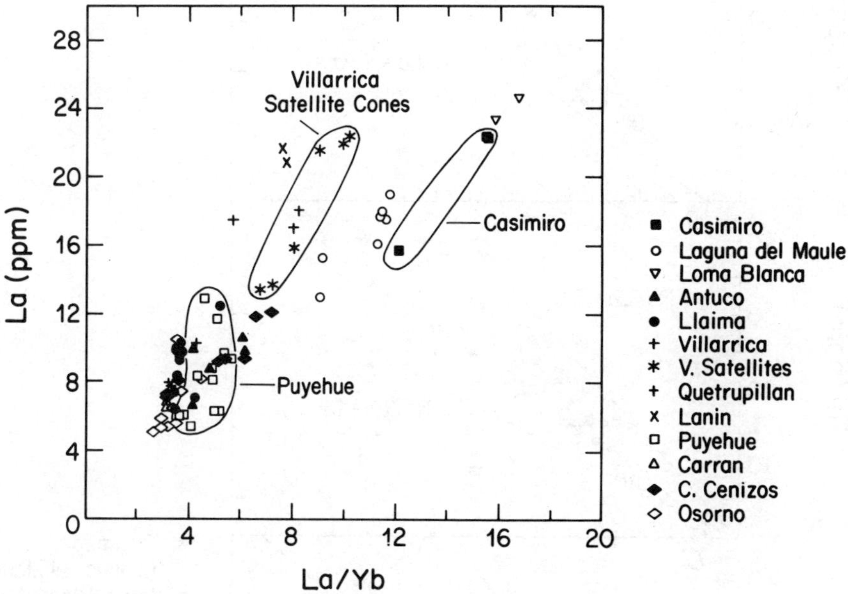

Figure 5 Plot of La abundance versus La/Yb ratios in basalts from central-south Chile, showing variations within and between volcanoes. Casimiro volcano is located in the Maipo region (34°S), Loma Blanca is a cone at Nevado de Longavi (36°S), and Cordon Cenizos is located in the Puntiagudo region (41°S) (see Figure 1). 'V satellites' indicates cones in the vicinity of Villarrica

ALKALI AND ALKALINE EARTH ELEMENTS: Rb, Cs, Sr, Ba

Like K, abundances of Rb, Cs and Ba vary widely in basalts from central-south Chile. An important geochemical feature of the basalts is that, with few exceptions, abundance ratios involving K, Rb, Ba and La have relatively narrow ranges, despite wide variations in their abundances, La/Yb and K_2O/Na_2O ratios and geographic location (Figure 6). The Villarrica satellite cones have lower K/La, Ba/La and Rb/La ratios than other central-south Chilean basalts (Figure 6). Excluding these samples, ranges for interelement ratios in basalts from central-south Chile are: K/La = 480–720, Ba/La =

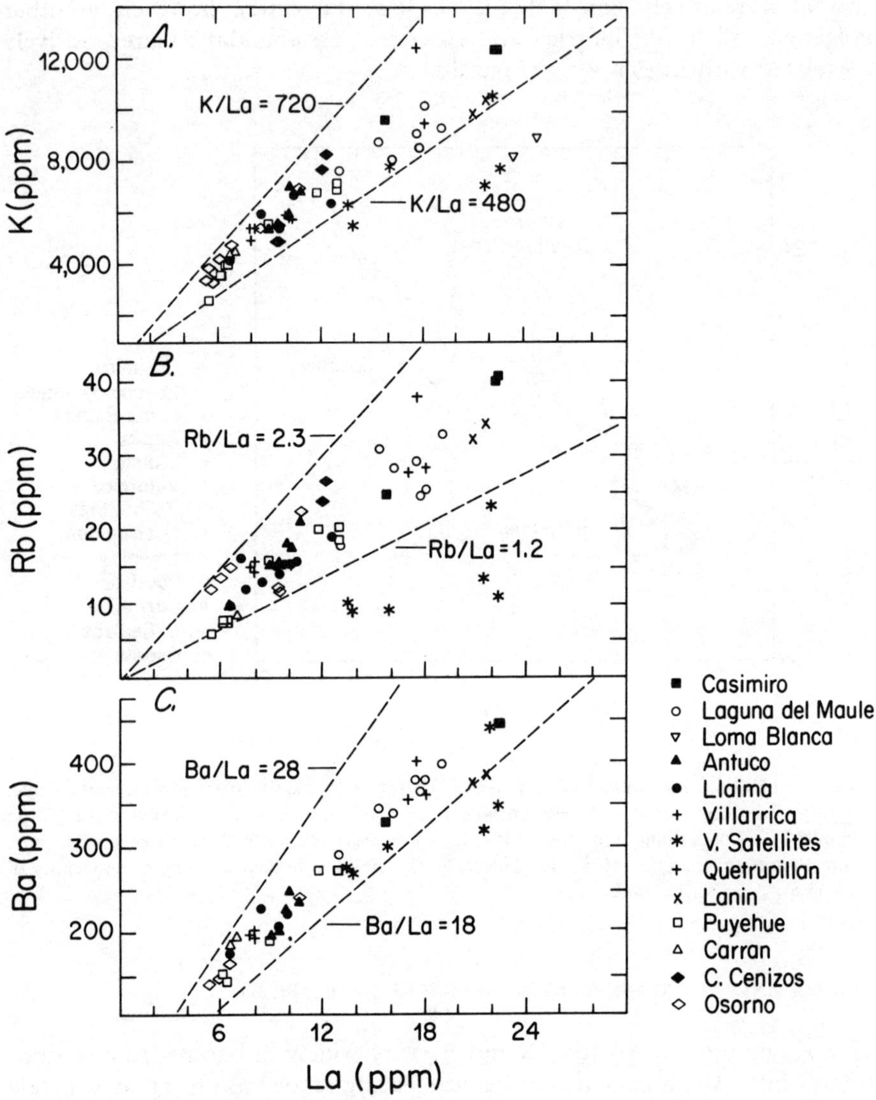

Figure 6 K, Rb and Ba versus La abundances in basalts from central-south Chile. Dashed lines indicate K/La, Rb/La and Ba/La ratios bracketing all volcanic centres except the Villarrica satellite cones. Values for average normal and enriched MORB are: K/La = 294–366, Rb/La = 0.3–0.6, and Ba/La = 3.6–10.2 (Sun *et al.,* 1979). See Figure 5 caption for locations of Casimiro, Loma Blanca and Cordon Cenizos

18–28, Rb/La = 1.2–2.3, K/Rb = 300–530 and K/Ba = 23–31. The K/La, Rb/La and Ba/La ratios are significantly higher than those for oceanic basalts and are typical of subduction-related basalts, however, the range of these ratios is unusually small compared to other volcanic arcs (e.g. Figure 7).

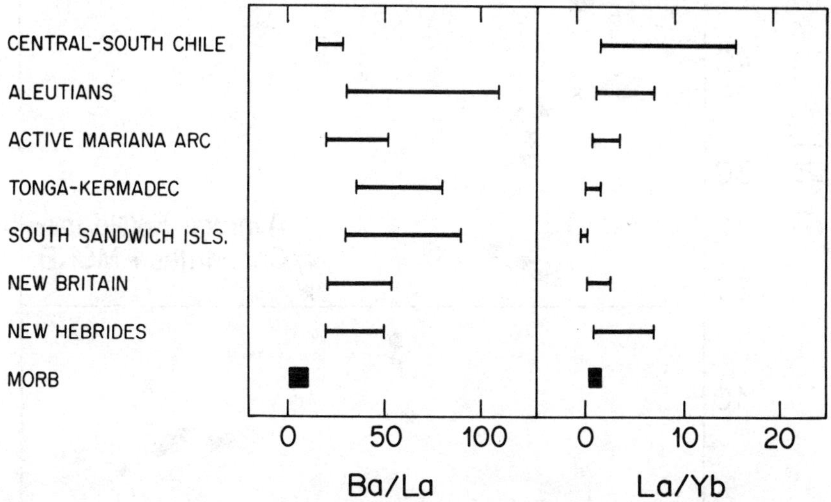

Figure 7 Ranges of Ba/La and La/Yb ratios in central-south Chilean basalts compared with ranges for basalts from other orogenic arcs. Note that relative to other magmatic arcs the central-south Chile basalts have a narrow range in Ba/La and a wide range in La/Yb. All arc basalts shown have higher Ba/La than MORB. Data from Kay (1977) and McCulloch and Perfit (1981) for the Aleutians; Dixon and Batiza (1979) for the Mariana Arc; Ewart *et al.* (1977) for the Tonga-Kermadec Arc; Hawkesworth *et al.* (1977) South Sandwich Islands; Johnson (1981) for New Britain, and Gorton (1977) and Dupuy *et al.* (1982) for the New Hebrides Arc. Values for average normal and enriched MORB (Sun *et al.*, 1979) are indicated by filled boxes

Within individual volcanoes, the variation in K/La, Rb/La and Ba/La ratios is nearly equivalent to that in basalts from all the volcanic centres (e.g. K/La = 550–710, Rb/La = 1.4–2.3 and Ba/La = 22–27 in basalts from Llaima volcano), therefore there are no systematic variations between these ratios and K, Rb, Ba or La abundances. A possible exception to this observation are Ba/La ratios, which are usually higher in basalts with low Ba and La contents (Figure 6c).

Ratios involving the Sr contents of basalts and basaltic andesites from central-south Chile are more variable than ratios such as K/Rb and Ba/La; e.g. Rb/Sr ranges from 0.01 to 0.06 and Sr/Nd ranges from 19 to 54 in the basaltic samples shown in Figure 6. Specifically, Sr/Nd ratios in volcanics from Laguna del Maule and Puyehue volcanoes decrease dramatically with increasing SiO_2, from values higher than those in MORBs in basalts and basaltic andesites to values lower than those in MORBs in andesites through rhyolites (Figure 8). Because plagioclase is an important phenocryst phase in all basaltic volcanics from this region, and it is enriched in Sr relative to

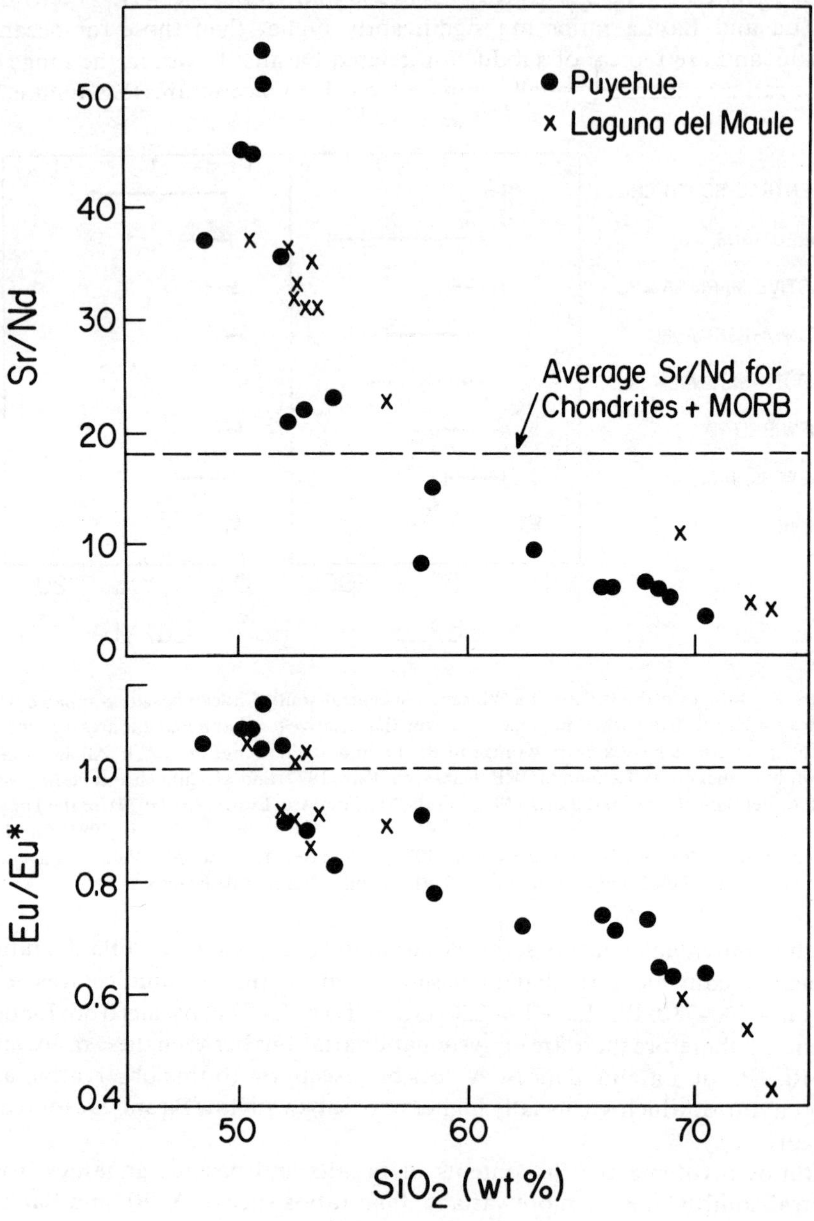

Figure 8 Plots of Sr/Nd and Eu/Eu* ratios versus SiO_2 content in basalts through rhyolites from Laguna del Maule and Puyehue volcanoes. Eu/Eu* is the ratio of the measured Eu content of a rock to the Eu concentration which would produce a smooth chondrite-normalized REE pattern. Plagioclase preferentially incorporates Sr and Eu (as Eu^{2+}) relative to a coexisting melt and the decreasing Sr/Nd and Eu/Eu* with increasing SiO_2 content are consistent with an important role for plagioclase fractionation

coexisting melt, we interpret that the large variations in ratios involving Sr result from plagioclase fractionation. Corroborating evidence for this conclusion are the negative Eu anomalies in some of the basaltic andesites and all of the dacites and rhyolites (Figure 8).

Cs abundances do not follow trends defined by other highly incompatible elements in central-south Chile basalts, e.g. abundances do not decrease from north to south and La/Cs, K/Cs, Rb/Cs and Ba/Cs ratios differ in basalts from different volcanic centres (Figure 9a). The range of K/Cs ratios in central-south Chilean basalts (3000–14 000) is consistent with values found in other subduction-related volcanics (~3000–24 000, Morris and Hart, 1983), but significantly lower than K/Cs ratios in oceanic basalts (>30 000).

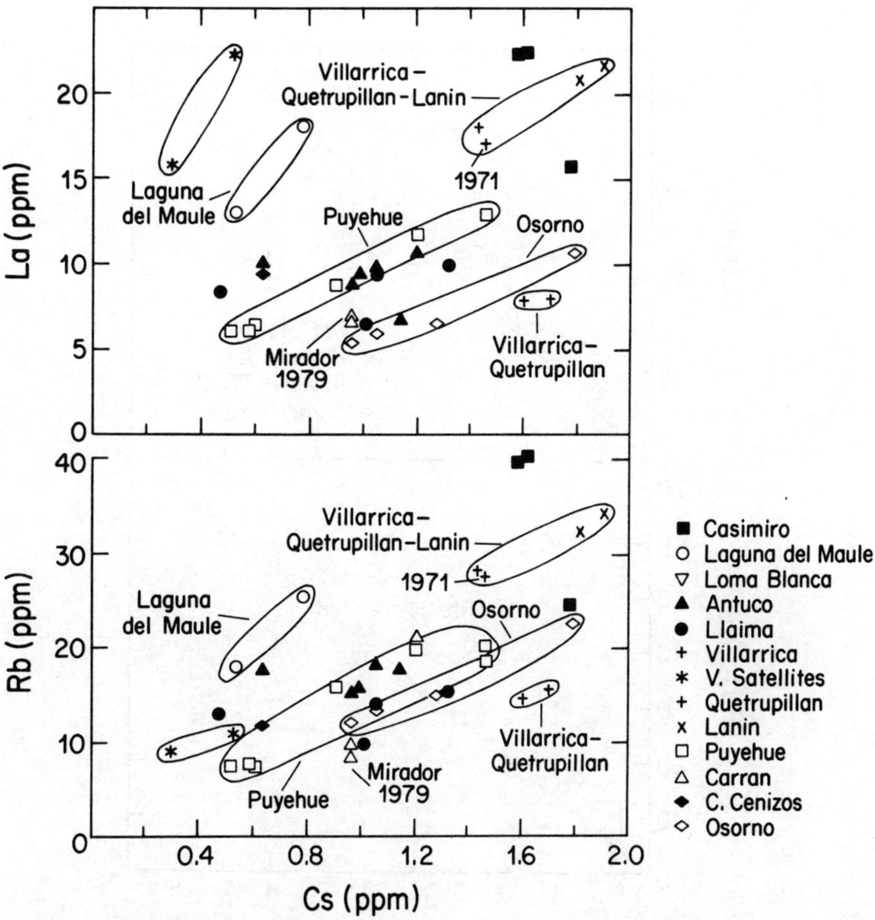

Figure 9a La and Rb versus Cs abundances in basalts from central-south Chile. Samples from recent eruptions of Mirador (Carran volcanic region) and Villarrica are labelled. See Figure 5 caption for locations of Casimiro, Loma Blanca and Cordon Cenizos

A possible explanation for the variability in K/Cs, La/Cs, and Rb/Cs ratios is that Cs abundances in the basalts have been affected by post-eruptive alteration. Cs is known to be a very mobile element during low temperature alteration (Hart *et al.*, 1974; Hofmann and White, 1983) and may be redistributed by post-eruptive ground water-rock interaction; in this case, variations in Cs abundances would not be useful in constraining petrogenetic processes. However, two features suggest that variations in Cs abundances in central-south Chilean basalts from different volcanic centres are primary igneous characteristics. First, within suites from individual volcanoes, Cs, Rb and La abundances form coherent trends; for example, at Puyehue, Cs, Rb and La abundances in pre- and post-glacial volcanic samples are positively correlated (Figure 9b). Because La abundances in relatively fresh volcanic

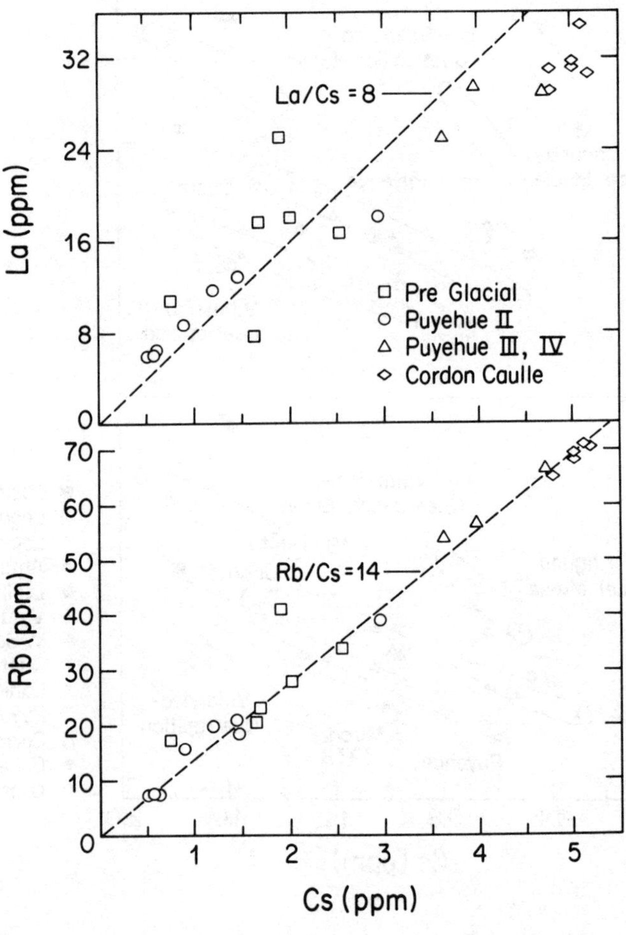

Figure 9b. La and Rb versus Cs abundances in basalts and evolved rocks from the Puyehue-Cordon Caulle region. Dashed lines show average La/Cs and Rb/Cs ratios

rocks are not affected by late-stage alteration, the positive correlation between La and Cs appears to be a magmatic feature. A second line of evidence that inter-volcano variations in La/Cs and Rb/Cs ratios may be primary is that basalts from very recent eruptions from two volcanoes have different ratios (Figure 9a, Mirador 1979 eruption: Rb/Cs = 9–10, La/Cs = 7; Villarrica 1971 eruption: Rb/Cs = 19, La/Cs = 12).

ISOTOPIC RESULTS

In Figure 10, Sr and Nd isotope data for eight central-south Chilean volcanic centres are shown compared to MORBs and central and northern Andean volcanics. Compared to other Andean volcanics, central-south Chilean rocks are distinct in that they plot within the oceanic 'mantle array' (e.g. O'Nions *et al.*, 1977) and overlap with values found in oceanic-island basalts. In addition, $^{18}O/^{16}O$ ratios for samples from Puyehue and Mirador volcanoes ($\delta^{18}O$ = 6.1–6.3‰, Gerlach *et al.*, 1983), and Nevados de Chillan, Antuco, Lonquimay, Llaima, Villarrica, Choshuenco, Osorno and Calbuco volcanoes ($\delta^{18}O$ = 5.2–6.8‰, Deruelle *et al.*, 1983) are low and also coincide with values for unaltered oceanic basalt. In contrast, Pb isotopic ratios for Tupungato, Llaima, Callaquen, Lanin, Antuco and Villarrica volcanoes (Barreiro, 1982b)

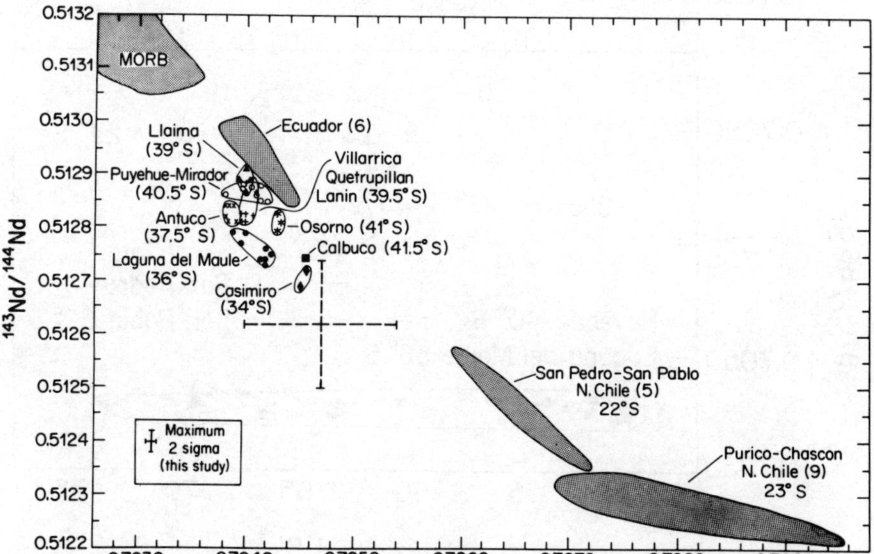

Figure 10 $^{143}Nd/^{144}Nd$ versus $^{87}Sr/^{86}Sr$ for volcanics from central-south Chile compared to MORBs and volcanics from other areas of the Andes. Data for Ecuador and north Chile from Hawkesworth *et al.* (1979b, 1982), number of samples indicated in parentheses. Dashed cross indicates range for bulk earth from Zindler (1982)

have higher $^{207}Pb/^{204}Pb$ than the majority of oceanic basalts with similar $^{206}Pb/^{204}Pb$.

Ratios of $^{143}Nd/^{144}Nd$ and $^{87}Sr/^{86}Sr$ in central-south Chilean volcanics do not vary systematically with latitude. In fact, the lowest $^{143}Nd/^{144}Nd$ and highest $^{87}Sr/^{86}Sr$ ratios are found in the northernmost (33–36°S) and southernmost (41–42°S) volcanoes, while the highest $^{143}Nd/^{144}Nd$ and lowest $^{87}Sr/^{86}Sr$ ratios are found in volcanoes between 37–40°S. Consequently, there is no systematic variation between $^{143}Nd/^{144}Nd$ and relative LREE enrichment in volcanics from this region.

As shown in Figure 10, several volcanoes have small but significant

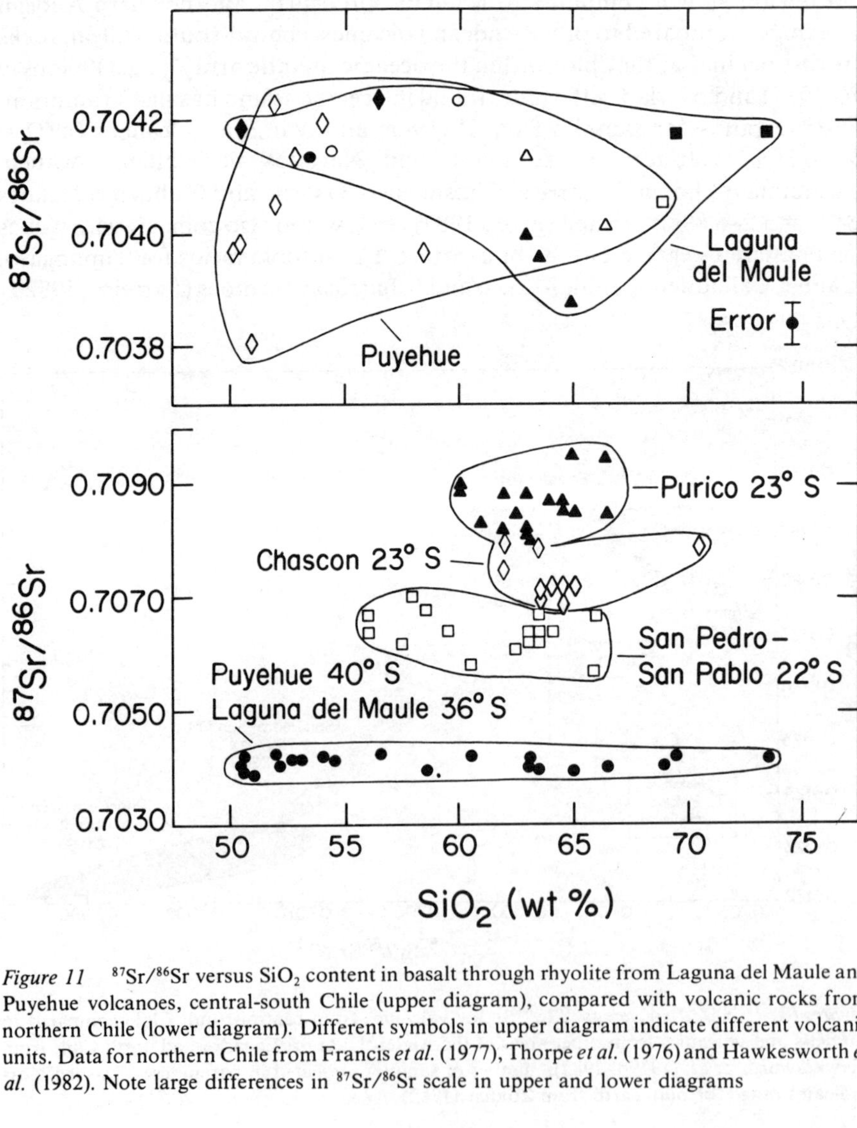

Figure 11 $^{87}Sr/^{86}Sr$ versus SiO_2 content in basalt through rhyolite from Laguna del Maule and Puyehue volcanoes, central-south Chile (upper diagram), compared with volcanic rocks from northern Chile (lower diagram). Different symbols in upper diagram indicate different volcanic units. Data for northern Chile from Francis *et al.* (1977), Thorpe *et al.* (1976) and Hawkesworth *et al.* (1982). Note large differences in $^{87}Sr/^{86}Sr$ scale in upper and lower diagrams

variations in isotopic composition. For example, volcanics from Laguna del Maule and Puyehue vary in $^{87}Sr/^{86}Sr$ by up to 0.00040, compared to typical uncertainties of ±0.00005 at the 95% confidence limit, whereas volcanics from Laguna del Maule, Villarrica and Quetrupillan vary in $^{143}Nd/^{144}Nd$ by up to 0.00008, compared to typical uncertainties of ±0.00002 at the 95% confidence limit. At Laguna del Maule, relatively high $^{143}Nd/^{144}Nd$ and low $^{87}Sr/^{86}Sr$ ratios occur in a group of dacitic volcanics which are older than other Laguna de Maule volcanics (Figure 11). At Villarrica and Quetrupillan volcanoes, low $^{143}Nd/^{144}Nd$ ratios are correlated with high La/Yb ratios. For Laguna del Maule and Puyehue, samples analysed include basalts through rhyolite, however, in these volcanoes, changes in $^{87}Sr/^{86}Sr$ ratio do not correlate systematically with SiO_2, content (Figure 11). Basalts through dacites from Puyehue also have constant $^{18}O/^{16}O$ ratios (Gerlach *et al.,* 1983).

SUMMARY OF IMPORTANT GEOCHEMICAL FEATURES

Basalts from central-south Chile have $^{87}Sr/^{86}Sr$ and $^{143}Nd/^{144}Nd$ ratios which lie within the mantle array established by oceanic basalts and these ratios overlap with those in oceanic-island basalts, such as at Hawaii. Although most volcanoes from 33–42°S have distinctive isotopic characteristics (Figure 11), there are no correlations of isotopic ratios with volcano location. Except for basalts from the closely associated Villarrica, Quetrupillan and Lanin volcanic centres, elemental abundances and ratios of incompatible elements in basalts from central-south Chile follow well-defined large scale trends: La/Yb and K_2O/Na_2O ratios and K, Rb, Ba and La abundances vary widely and are higher in northern volcanic centres than southern volcanic centres (see Figures 3, 4 and 5), but ratios of K, Rb, Ba and La are similar (within a factor of 2) at all volcanic centres (see Figure 6). In contrast, on a regional basis Cs abundances do not correlate with La, K, Rb or Ba contents (see Figure 9a) although at individual volcanic centres which range from basalt to rhyolite, e.g. Puyehue, Cs abundances are strongly correlated with abundances of other highly incompatible elements (see Figure 9b). The variation in ratios of Cs to other highly incompatible elements at different volcanoes is one of the more puzzling aspects of the geochemistry of central-south Chilean basalts.

Discussion

ROLE OF CONTINENTAL CRUST IN ANDEAN MAGMATISM: EVIDENCE FROM CENTRAL-SOUTH CHILE

An important problem in understanding the generation of continental arc volcanics is identifying the role of continental crust in determining the composition of the volcanic rocks. Silicic rocks derived in part by fractional crystallization of more basic magmas are especially sensitive to assimilation of continental crust because fractionation of plagioclase, often a major phenocryst in calc-alkaline lavas, combined with assimilation leads to rapid increase in $^{87}Sr/^{86}Sr$ of the evolving magma (e.g. De Paolo, 1981; James, 1982a). In fact, in the central Andes where silicic volcanic rocks are abundant and have relatively high $^{87}Sr/^{86}Sr$ ratios and $\delta^{18}O$ values, there is a consensus that assimilation of continental crust has been a significant process (e.g. Francis *et al.*, 1977; Francis *et al.*, 1980; James, 1982a; Deruelle *et al.*, 1983; Harmon *et al.*, 1983). However, few Andean volcanic centres exhibit well-defined correlations between $^{87}Sr/^{86}Sr$ or $\delta^{18}O$ and SiO_2 (see Figure 11). Correlations between $^{87}Sr/^{86}Sr$ and Sr abundances or Rb/Sr ratios, possibly reflecting the combined effects of crustal assimilation and plagioclase fractionation, are found in volcanic rocks from Cerro Galan, northwest Argentina (Francis *et al.*, 1980) and the Arequipa and Barroso volcanics, southern Peru (Briqueu and Lancelot, 1979; James, 1982a). In contrast, basalts through rhyolites from Laguna del Maule and Puyehue volcanic centres in central-south Chile have similar $^{87}Sr/^{86}Sr$ ratios and $\delta^{18}O$ values. In addition, trace element characteristics of evolved rocks from these volcanic centres can be modelled in terms of fractional crystallization from basaltic precursors without input from crustal wall rocks (Gerlach *et al.*, 1982; Frey *et al.*, 1984).

The lack of evidence for crustal assimilation in volcanic rocks from central-south Chile is surprising because the wide range of rock types require a large degree of shallow-level fractional crystallization. Moreover, $^{87}Sr/^{86}Sr$ ratios and $\delta^{18}O$ values in basaltic rocks from central-south Chile have low, mantle-like values. Thus the isotopic contrast between fractionating magma and crustal wall rocks was greater than in central Andean volcanic centres, where 'mafic' precursors typically are enriched in ^{87}Sr and ^{18}O; consequently, in central-south Chile, even small amounts of assimilation of crust with relatively high $^{87}Sr/^{86}Sr$ and $^{18}O/^{16}O$ ratios would significantly increase $^{87}Sr/^{86}Sr$ ratios and $\delta^{18}O$ values in the evolving magmas. Because there are no detectable correlations between these isotopic parameters and lava composition at Laguna del Maule and Puyehue we conclude that at these volcanic centres wall rock assimilation did not occur during shallow level fractional crystallization, or that crustal wall rocks enclosing fractionating magmas at each site were not significantly different from the magmas in Sr, Nd and O isotope ratios.

Even the most mafic volcanic rocks from the central Andes typically have isotopic features suggestive of crustal assimilation (Klerkx *et al.*, 1977; Thorpe *et al.*, 1979a; Francis *et al.*, 1980; James, 1982a, Deruelle *et al.*, 1983). In contrast, there is no Sr, Nd or O isotope evidence for crustal contamination of basaltic lavas from central-south Chile because these lavas have Sr, Nd and O isotope ratios similar to basaltic lavas from intra-oceanic volcanic arcs where continental crust is absent. This contrast in the role of the continental crust has been attributed to the differences in crustal thickness between the central and southern Andes (e.g. Klerkx *et al.*, 1977; Deruelle, 1982; Deruelle *et al.*, 1983; Harmon *et al.*, 1983). However, basaltic magmas from central-south Chile ascended through 30 km of crust which includes rocks as old as

Palaeozoic. It is not readily explicable why passage through 30 km of crust should result in *no* apparent contamination, while passage through 70 km of crust in the central Andes should result in extensive contamination. Possibly, the evidence for crustal contamination of magmas in the central Andes results from different crustal compositions in the central and southern Andes; for example, Precambrian basement probably underlies many central Andean volcanoes whereas such ancient basement has not been found in central-south Chile. In addition, temperatures in the lower part of the thick crust in the central Andes may be sufficiently elevated so that intrusion of mantle-derived magma enables partial melting of the lower crust; consequently, mixing of the crustal and mantle-derived melts is a likely mechanism for crustal contamination. If the thinner crust in the central-south Andes is not as readily melted by intrusion of mantle-derived magmas, this type of crustal contamination may not occur.

TRACE ELEMENT CHARACTERISTICS OF BASALTS FROM CENTRAL-SOUTH CHILE: THEIR BEARING ON MODELS FOR ARC BASALT GENESIS

Volcanic rocks from central-south Chile have Nd, Sr and O isotope compositions which overlap with the ranges found in oceanic-island volcanics; thus, they are similar to volcanics from several intraoceanic island arcs, e.g. the active Mariana Arc (Stern, R. J., 1981; Stern and Ito, 1983), the Aleutians (McCulloch and Perfit, 1981; Morris and Hart, 1983) and the Bismarck Arc (De Paolo and Johnson, 1979; Johnson *et al.*, 1984). The similar isotopic characteristics of volcanics from these arcs and oceanic-island volcanics have led to conclusions that the sources of arc and oceanic-island volcanics are closely related or identical (De Paolo and Wasserburg, 1977; Stern, R. J., 1981; Stern and Ito, 1983; Morris and Hart, 1983).

Despite these isotopic similarities, arc basalts differ from oceanic-island basalts in several trace element characteristics; for example, arc basalts have: (a) lower contents of TiO_2, HREE, Y, Zr, Hf, Nb and Ta (e.g. Gill, 1981), (b) higher, often much higher, ratios of the elements K, Rb, Cs, Sr and Ba to REE elements (Gill, 1981), and (c) higher ratios of Cs to Rb and K (Morris and Hart, 1983). These differences have led to two basic types of models regarding the origin of arc basalts: (a) models which explain the high abundances of alkali and alkaline earth elements compared to REE in arc basalts by partial melting of an oceanic mantle source containing a refractory residual phase which preferentially retains REE (Morris and Hart, 1983; Stern and Ito, 1983), and (b) models which explain the high abundances of alkali and alkaline earth elements compared to REE in arc basalts by preferential addition of these elements to mantle sources of arc magmas or to the arc magmas (De Paolo and Johnson 1979; Kay, 1980; Arculus and Johnson, 1981).

Relative to oceanic basalts, basalts from central-south Chile have trace element characteristics which are consistent with other arc basalts: low HREE contents, high Ba/La, K/La and Rb/La ratios (see Figures 6 and 7) and low

Rb/Cs and K/Cs ratios (see Figure 9). The unique feature of central-south Chilean basalts compared to basalts from other arcs is that despite wide variations in abundances of K, Rb, Ba and La and ratios of La/Yb and K_2O/Na_2O, the Ba/La, K/La and Rb/La ratios vary over only a narrow range (see Figures 6 and 7) along this 900 km segment of the Andes. This feature puts important limits on models explaining the high Ba/La, K/La and Rb/La ratios in arc basalts.

For example, in models which explain high Ba/La, K/La and Rb/La ratios as the result of a residual mineral which preferentially retains LREE during partial melting of mantle peridotite to form basaltic magma, K, Rb and Ba are highly incompatible elements, while La is moderately or slightly incompatible. In central-south Chilean basalts, La, Ba, Rb and K abundances vary by factors of 3 to 8, and La/Yb ratios vary by a factor of 8. As discussed below and by Lopez-Escobar *et al.* (1977) these variations may be produced by differing degrees of partial melting of sources with similar trace element abundances; alternatively, the sources may have had variable trace element contents. If the wide range of K, Rb, Ba and La abundances and La/Yb ratios in central-south Chilean basalts results from varying degrees of partial melting, the similar K/La, Rb/La and Ba/La ratios require that these elements were all incompatible during partial melting or that they all had very similar solid/melt partition coefficients. This result is not consistent with a model which requires relatively more compatible behaviour of La: moreover, the trend in Figure 6c opposes the behaviour expected if La is more compatible than Ba: Ba/La ratios are lowest in basalts with high Ba and La concentrations (low degree of partial melting) and highest in basalts with low Ba and La concentrations (high degree of partial melting). Therefore, models requiring variable degrees of melting of a compositionally homogeneous source containing a phase which preferentially retains LREE are not consistent with the data.

If the wide range in K, Rb, Ba and La abundances and La/Yb ratios in central-south Chilean basalts are inherited from source heterogeneities, and the high ratios of K, Rb and Ba to La result from a residual LREE-retaining mineral, the similar K/La, Rb/La and Ba/La ratios in basalts with different abundances of these elements can be produced only if the basalts were generated by similar degrees of partial melting of sources having similar K/La, Rb/La and Ba/La ratios and similar amounts of the LREE-rich accessory phase. Therefore, a one-stage partial melting model for the generation of central-south Chilean basalts requires mantle peridotite sources with very specific geochemical characteristics. A more geologically realistic model is that the degree of partial melting and the amount of a LREE-rich accessory phase vary along the length of the volcanic arc. In this case it would be extremely fortuitous for basalts with similar K/La, Rb/La and Ba/La ratios to be generated.

Models for arc basalts involving selective enrichment in K, Rb and Ba to account for high abundances of these elements compared to REE also require special conditions to account for the trace element characteristics of central-south Chilean basalts; that is, the relatively constant K/La, Rb/La and Ba/La ratios in basalts from different volcanic centres require that a selective enrichment model explain the creation of mantle sources with similar K/La, Rb/La and Ba/La ratios. In order to create such sources, a material with high abundances of K, Rb and Ba relative to La must be generated and this material must be added to the mantle sources of the basalts such that K/La, Rb/La and Ba/La ratios are similar (within a factor of 2) in all areas. These conditions could be satisfied if the added material had constant K/La, Rb/La and Ba/La ratios, and if this component contributes nearly all of these elements to the mixed source. Commonly, this component has been equated with fluids derived by dehydration or melting of a subducted slab containing altered basalt and sediments (e.g. Kay, 1980; Hawkesworth, 1982), but important characteristics of such a process are not established. Quantitative evaluation of this model requires identification of the residual minerals and derived fluid (e.g. H_2O-rich vapour or silicate melt) and knowledge of the partitioning behaviour of K, Rb, Ba and REE between these phases at the relevant pressures and temperatures.

Experimental studies of REE partitioning between water vapour and silicate minerals indicate that the vapour is preferentially enriched in LREE, and Mysen (1979) has proposed that the LREE/HREE ratios of calc-alkaline magmas reflect metasomatism of their source by a H_2O-rich vapour derived from subducted oceanic crust. Intuitively, we expect that such H_2O-rich vapours would also have high alkali metal/LREE ratios; therefore, they might be the cause of high K/La, Rb/La and Ba/La in arc volcanics. However, the limited range of such ratios in central-south Chile basalts (see Figure 6) coupled with the variability of REE vapour/solid partition coefficients (e.g. Mysen, 1983) requires uniform vapour compositions (e.g. CO_2/H_2O ratio) and derivation at the same pressure from sources with similar alkali metal/LREE ratios.

In contrast, derivation of a K, Rb and Ba enriched component by partial melting of the subducted slab requires a LREE-retaining residual mineral; therefore, the same restrictions which applied to the one-stage partial melting model are necessary to explain relatively constant Ba/La, K/La and Rb/La ratios in an enriched component formed by partial melting. For example, the slab must contain uniform amounts of the LREE-retaining phase and undergo the same degree of melting in all areas. However, unlike the one-stage model, the basalts could be derived by varying degrees of partial melting of the peridotite source if K, Rb, Ba and La were all highly incompatible during partial melting of the peridotite.

In order to explain the geochemical characteristics of central-south Chilean basalts, both a one-stage partial melting model and a two-stage model involving source enrichment of K, Rb and Ba require source materials with special characteristics and unusually uniform processes. In general, a one-stage model has the advantage of simplicity: Nd and Sr isotope characteristics and K/Rb, Ba/Rb and K/Ba ratios in central-south Chilean basalts overlap with those of oceanic-island basalts; therefore, partial melting of a similar mantle source is a direct explanation for these geochemical similarities. Explanation of these characteristics by an enriched component derived from the subducted slab requires that the isotopic characteristics of the subducted oceanic crust overlap with oceanic-island basalts and the features such as the degree of alteration and amount and type of sediment subducted do not vary significantly along the length of this 900 km segment of the Andes.

However, derivation of materials enriched in K, Rb and Ba relative to La from the subducted oceanic crust may be more feasible than derivation of basalts with these characteristics from mantle peridotite. The single-stage model requires a LREE-rich accessory mineral in the peridotite source. However, REE-rich accessory minerals are rare in peridotites and their presence reflects unusually high REE abundances which were probably created by metasomatism (e.g. Menzies, 1983; Roden *et al.*, 1984). As discussed earlier, the single-stage model for central-south Chilean basalts requires sources with variable REE contents but similar amounts of the REE-rich accessory phase. Such a correlation is unlikely. On the other hand, in some depth range, subducted oceanic crust must dehydrate and partially melt. Therefore, derivation of a component with high K/La, Rb/La and Ba/La from the subducted slab may be the link between calc-alkaline volcanism and subduction, and specifically provides an explanation for eruption of basalt with high K/La, Rb/La and Ba/La in subduction zone environments.

Finally, in order to explain the similar K/La, Rb/La and Ba/La ratios in these basalts, the one-stage model requires uniform degrees of partial melting of a peridotite source to generate the basalts, whereas the two-stage model requires uniform fractionation of these elements during generation of the enriched component but permits variable degrees of partial melting of a peridotite source to generate the basalts. Although we find the two-stage model more plausible, we can not unambiguously discriminate between these models with the data presented here. However, we expect that additional petrogenetic constraints will arise from other types of geochemical data (acquisition in progress), in particular, Pb isotope ratios and abundances of Zr, Hf, Nb and Ta.

DISTRIBUTION OF Cs IN CENTRAL-SOUTH CHILEAN BASALTS

The low K/Cs, Rb/Cs and La/Cs ratios in central-south Chilean basalts are characteristic of arc volcanics and distinct from those of ocean floor and oceanic-island basalts. In addition, the distribution of Cs in central-south Chilean basalts differs markedly from that of other incompatible elements. Cs abundances follow no regional trends and K/Cs, Rb/Cs and La/Cs ratios differ in basalts from different volcanoes and in basalts with similar ratios of K/La and K/Rb. This variation is surprising because Rb and Cs are the most incompatible elements studied, and they are difficult to fractionate by solid-melt segregation processes. For example, Puyehue samples ranging from basalt to rhyolite are nearly isotopically homogeneous (see Figure 11), and it is likely that fractional crystallization was a significant process creating the compositional variability; however, Rb/Cs ratios are uniform: 15±3 in 23 basaltic to rhyolitic samples from Puyehue (see Figure 9b). We also expect that Rb/Cs ratios in primary melts of peridotites are independent of the degree of melting; therefore, a single-stage model is not capable of explaining the different Rb/Cs ratios in arc basalts and oceanic basalts.

Minerals which fractionate Rb from Cs are mainly clays and micas; e.g. pelagic sediments from the southwest Pacific ocean have low K/Cs and Rb/Cs ratios (Morris and Hart, 1983) possibly reflecting a preference for Cs. If the subducted Nazca plate is the source for enrichment in K, Rb and Ba in central-south Chilean basalts, then it is conceivable that subducted sediment is a source for enrichment in Cs compared to Rb and K, although the amount of subducted sediment present in the source material is limited by the mantle-like $^{87}Sr/^{86}Sr$ ratios and $\delta^{18}O$ values in these basalts. Cs can also be fractionated from K and Rb by H_2O vapour-solid partitioning; e.g. H_2O/muscovite and H_2O/sanidine partition coeffients for Cs are an order of magnitude greater than for Rb (Volfinger, 1976). Therefore, enrichment in alkali and alkaline earth elements in central-south Chilean basalts by addition of a slab-derived, volatile-rich fluid may also account for enrichment in Cs relative to Rb and K. In these cases the inter-volcano differences in Rb/Cs ratios may reflect: (a) different amounts of sediment, although variations in amount and type of sediment are restricted by the narrow range in K/La, Rb/La and Ba/La, or (b) variations in intensive parameters which affect Cs partitioning more than K, Rb or Ba.

Alternative explanations for the independent behavior of Cs are post-partial melting processes. On the basis of coherent La, Rb and Cs abundance variations in basalts to rhyolites from single volcanoes (e.g. Puyehue) we eliminated late-stage processes as a general explanation for Rb/Cs variability. However, interaction of basaltic magma with crustal wall rocks is another possible mechanism for affecting Cs abundances without significantly changing Sr and O isotope ratios. For example, recent experimental results by Watson (1982) and Watson and Jurewicz (1983) on basalt magma-crust interaction suggests that selective contamination may occur. These results indicate that alkali elements are particularly susceptible to selective contamination effects because of their high diffusivity compared to SiO_2. Depletion or enrichment in the basaltic magma depends on the relative preference of the element for basaltic and silicic melts and the magnitude of the concentration contrast between the basaltic and silicic melts. In the experiments of Watson (1982) and Watson and Jurewicz (1983) K, which has a high preference for silicic melts, diffused selectively into basalt magma because of the high contrast in K concentrations between basaltic and granitic melts. We have no evidence for selective contamination of K and Rb in central-south Chilean basalts: K/La, K/Rb and Rb/La ratios are relatively constant. However, alkali metal contamination may have occurred to a small extent, and be reflected only in Cs abundances because of the higher contrast between basalt and crustal rock Cs abundances. The estimated lower crustal average of Taylor and McLennan (1981) has low K/Cs (7000) and Rb/Cs (11) ratios; therefore, selective assimilation of lower crustal rocks might be reflected in Cs abundances.

Although crustal contamination may explain the variable Rb/Cs ratios in central-south Chilean basalts, this is not a general explanation for the low Rb/Cs ratios in arc volcanics compared to oceanic volcanics, i.e. low Rb/Cs ratios are also found in arcs which lack continental crust (Morris and Hart, 1983; Stern and Ito, 1983). The most suitable source for Cs-enrichment in arc volcanics appears to be the subducted oceanic crust. Such a source for Cs-enrichment favours the two-stage model, i.e. enrichment of the source in K, Rb, Cs and Ba, over the one-stage partial melting model for central-south Chilean basalts.

Conclusions

Basaltic rocks from central-south Chile have Sr, Nd and O isotope ratios which overlap with oceanic-island basalts. In addition, dacites and rhyolites from the two volcanic centres studied in detail have Sr, Nd and O isotope ratios which overlap with their associated mafic rocks; therefore, despite passage through ~30 km of crust it seems unlikely that upper crustal contamination significantly affected the isotopic composition of volcanic rocks in this region of the Andes. This conclusion contrasts with the abundant evidence for crustal contamination in the central Andes where the crust is considerably thicker.

Basalts from volcanic centres between 33–42°S in Chile have widely varying La/Yb and K_2O/Na_2O ratios and La, Ba, K and Rb abundances. Despite these variations, K/La, Ba/La, Rb/La, K/Rb and K/Ba ratios vary by less than a factor of two in basaltic rocks from this 900 km segment of the Andes. Although Sr, Nd and O isotope ratios for these basalts are similar to those in oceanic-island volcanics, K/La, Rb/La and Ba/La ratios are higher, as is typical of arc rocks. Because of the relatively narrow range of these ratios in central-south Chile basalts, models explaining the high K/La, Rb/La and Ba/La either by a residual LREE-retaining phase during partial melting of mantle peridotite, or by selective enrichment of K, Rb and Ba, require unique sources. Single-stage partial melting models require sources with heterogeneous trace element abundances but uniform contents of a LREE-rich phase, and similar degrees of partial melting at all volcanic centres. Two-stage models require mixing of a peridotite with a component having relatively high abundances of K, Rb, Ba and La and high and uniform ratios of K/La, Rb/La and Ba/La. Like other arc volcanics, central-south Chilean basalts have lower Rb/Cs ratios than oceanic basalts. However, unlike K/La, Rb/La and Ba/La ratios, Rb/Cs ratios are different in volcanic centres from central-south Chile. It seems unlikely that single-stage melting of peridotite sources suitable for generating oceanic-island basalts can satisfy the geochemical constraints, especially the variable Rb/Cs and high Rb/La ratios, established by central-south Chilean basalts. However, fluids such as H_2O and CO_2-rich vapours or silicate melts, derived from the subducting Nazca plate may have relative abundances of K, Rb, Cs, Ba and La consistent with those in central-south Chilean basalts. Therefore, we favour a two-stage model involving melting of a source formed by mixing peridotite with a fluid component derived from the subducting Nazca plate.

ACKNOWLEDGEMENTS

This research is being financially supported by US National Science Foundation Grant EAR-8115850. We thank S. R. Hart for access to his mass spectrometers and laboratory facilities and the MIT nuclear reactor for neutron irradiations. None of these geochemical and petrologic studies could have been done without the support of our Chilean colleagues L. Lopez-Escobar and H. Moreno-Roa and the Departmento de Geologia, Universidad de Chile, Santiago.

Sr, Nd, Pb and O Isotope Composition of Late Cenozoic Volcanics, northernmost SVZ (33–34°S)

C. R. STERN, K. FUTA, K. MUEHLENBACHS, F. M. DOBBS, J. MUÑOZ, E. GODOY and R. CHARRIER

Abstract

The three northernmost (33–34°S) volcanic centres in the SVZ share more petrochemical similarities with the volcanoes of the CVZ than the volcanoes further south in the southern Andes, including the volumetric predominance of intermediate rather than basic compositions, the presence of rhyolite and the common occurrence of hydrous minerals. Rhyolites from these volcanoes have radiogenic Sr and Nd isotope compositions and high $\delta^{18}O$ values compared to spatially associated basalts and andesites, indicating incorporation of significant amounts of crustal components during their petrogenesis. Basalts and andesites from these volcanoes have similar Sr and Nd isotope compositions but contain more radiogenic Sr and less radiogenic Nd than rocks of equivalent bulk composition from the southern SVZ. However, $\delta^{18}O$ values of rocks from both regions are similar. These data are consistent with the incorporation of crustal components either during passage of mantle-derived mafic magmas through the lower continental crust, or by their introduction by subduction processes into the mantle source region where these magmas were generated. Similar suggestions have been made concerning the petrogenesis of the volcanoes of the CVZ. The available geochemical data can not distinguish unambiguously between these two models; both processes may be important.

Introduction

The SVZ between 33–46°S in central-south Chile can be subdivided into two sections (Figure 1): a southern segment between 36–46°S in which the volcanic centres are composed dominantly of olivine-bearing high-Al basalts and associated dacite differentiates (Moreno, 1976; Stern *et al.*, 1976), and a more northerly segment from 33–36°S in which andesites, commonly with hornblende and occasionally with biotite, are volumetrically more important, basalts less important and rhyolites of significant volume are present (Hildreth *et al.*, 1981; Hildreth *et al.*, in press; Stern *et al.*, in press). The volcanic centres of the northern segment of the SVZ are thus more similar to the volcanic centres of the CVZ between 16–28°S than to the volcanic centres further south in the SVZ with respect to both the common presence of hydrous minerals and the volumetric predominance of intermediate and silicic rocks (Deruelle, 1982). Also, large-scale caldera structures and associated

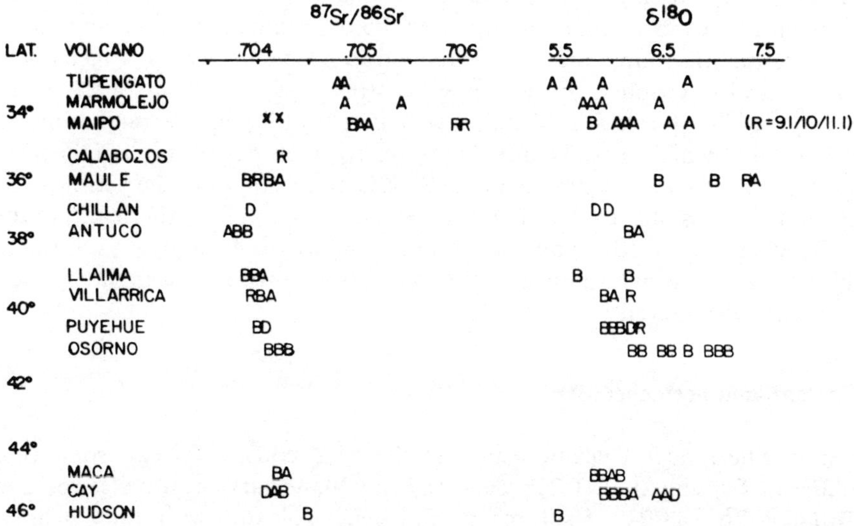

Figure 1 Plot of $^{87}Sr/^{86}Sr$ ratios and $\delta^{18}O$ values for basalts (B), andesites (A), dacites (D), and rhyolites (R) from the volcanic centres of the SVZ. The figure also shows the Sr isotope composition of alkali basalts outcropping just east of the Maipo volcano at 34°S (X—data of Stern *et al.*, 1983)

pyroclastic outflow units, which are common in the CVZ, have been identified in the northern section, but not in the southern section, of the SVZ (Hildreth *et al.,* 1981; Stern *et al.,* in press). The limited amount of geochemical data available suggests other fundamental similarities between the intermediate rocks in the northern section of the SVZ and in the CVZ. For example, samples from both these regions have more fractionated rare earth element concentrations and lower heavy REE than samples in the southern part of the SVZ (Lopez-Escobar *et al.*, 1977; Lopez-Escobar, this volume).

Physiographically, the two subsections of the SVZ are distinguished by the elevation of both summits and bases of the volcanic centres. In the southern segment of the SVZ the bases of the volcanic centres occur at between sea level and 1000 m and their summits rarely exceed 3000 m, whereas in the northern segment of the SVZ the bases of the volcanic centres generally occur between 2–3000 m and their summits exceed 4000 m and, in some cases, 5–6000 m. These differences in elevation reflect the increased crustal thickness below the volcanoes in the northern section of the SVZ, which occur above the southern end of the keel of greater-than-normal crustal thickness underlying the central Andes (Lowrie and Hey, 1981).

This paper discusses isotope data for samples from the three most northerly volcanic centres in the SVZ; Tupungato/Tupungatito, Marmolejo/San José, and Maipo. The samples range from basalt through rhyolite in composition. The data indicate that the petrogenesis of the mafic and intermediate rocks erupted from these three volcanoes involved either incorporation of crustal material into mantle-derived magmas by assimilation during the passage through the lower continental crust, or the recycling of crustal components into the mantle source region by subduction processes. The petrogenesis of silicic rocks erupted from these centres has involved near-surface incorporation of crustal material as well as crystal-liquid differentiation. To the extent to which crustal components are involved, petrogenesis of the three northernmost volcanic centres in the SVZ is more similar to the petrogenesis of the volcanics of the CVZ than to the volcanoes in the southern part of the SVZ, where intermediate and silicic rocks apparently form by crystal-liquid differentiation from basaltic parental magmas without significant interaction with crustal material.

Geology and petrochemistry

The northernmost volcanic centre in the SVZ consists of two coalescent edifices; Tupungato (33°22′, 6650 m), a Quaternary stratovolcano, and Tupungatito (33°24′, 5640 m) a Holocene volcano with nine historic eruptions, the last occurring in 1968 (González-Ferrán, 1972), and continuing fumarolic activity associated with a well-preserved summit crater. The next centre to the south consists of several coalescent volcanoes; Marmolejo (33°44′, 6110 m), a Quaternary stratovolcano, and San José (33°47′, 5890 m), a Holocene stratovolcano complex capped by pyroclastics of several small craters, from which six historic eruptions and sporadic fumarolic activity have been reported. Samples collected from the Chilean side of both Tupungato/Tupungatito and Marmolejo/San José are pyroxene andesites with either olivine or hornblende±biotite (Thiele and Katsui, 1969), but olivine basalts and basaltic andesites have been reported from the Argentine side of both these volcanic centres (Gonzalez-Diaz, 1961; Polanski, 1964).

The next volcanic centre to the south, Maipo volcano (34° 10′, 5290 m) is more complex. The Maipo volcano seems to have been the source, during the Pleistocene, of greater than 500 km^3 of crystal-poor rhyolite which forms welded and unwelded pyroclastic flows to the east of the pampas of Argentina and to the west in the central valley of Chile between Santiago and Rancagua (Santana, 1971; Stern *et al.*, in press). The central edifice of the Maipo volcano is located within a distinct circular depression (Charrier, 1981, p. 177) which is probably the caldera related to the extrusion of the rhyolite. The central edifice of the volcanic centre, which consists dominantly of pyroxene andesites, some with oxyhornblende, and associated olivine basalts and basaltic andesites, may be in part a stratovolcano and in part a resurgent dome. Many smaller Quaternary to Holocene volcanoes consisting of basalts, andesites and rhyolites surround the Maipo volcano.

Table 1 presents average major and selected trace element data and isotope ratios for representative samples of basalt, andesite and rhyolite from the three northernmost volcanoes in the SVZ. These and other geochemical data are further illustrated in Figures 1, 2 and 3. The basalt is similar in

Table 1 **Average major and trace element and isotope composition of basalts, andesites and rhyolites from the three northernmost volcanoes in the SVZ**

	Basalt	Andesite	Rhyolite
Samples	1–2	4–6	1–3
SiO_2	53.9	60.4	74.7
TiO_2	1.3	0.9	0.1
Al_2O_3	18.2	17.4	12.6
FeO*	7.7	5.9	0.7
MgO	4.5	2.9	0.2
CaO	8.4	6.0	0.4
Na_2O	3.7	3.5	4.1
K_2O	1.3	2.7	3.9
Rb	22	62	141
Sr	600	660	65
La	24	28	18
Yb	1.2	1.1	1.4
$^{87}Sr/^{86}Sr$	0.7049	0.7050	0.7060
$\delta\ ^{18}O$ (‰ SMOW)	5.8	6.0	10.1
$^{143}Nd/^{144}Nd$	0.51262	0.51259	0.51255

composition to other olivine-bearing, high-Al basalts from further south in the SVZ although it has a slightly higher K_2O content and higher K/Rb and $(La/Yb)_N$ ratios. The andesites have similar high $(La/Yb)_N$ ratios, reflecting both higher LREE (80–100 × chondrite) and lower HREE (3–8 × chondrite) than intermediate volcanic rocks from further south in the SVZ (LREE=18–30 × chondrite, HREE=8–14 × chondrite; Lopez-Escobar *et al.*, 1977). In this respect the mafic and intermediate rocks from the northern segment of the SVZ are more similar to mafic and intermediate rocks from the CVZ (Thorpe *et al.*, 1976). Neither basalts nor andesites have detectable Eu anomalies. The high-silica rhyolite has a significant negative Eu anomaly and low LREE and Zr contents.

Isotopically, the samples from the northern section of the SVZ exhibit some significant differences compared to samples from the volcanic centres further south in the SVZ. The volcanic centres from the southern segment of the SVZ are characterized by nearly constant Sr and Nd isotope compositions which are independent of SiO_2 variations through the entire sequence basalt-andesite-dacite- rhyolite (Figures 1 and 2). In the northern section of the SVZ, although the basalts and andesites have $^{87}Sr/^{86}Sr$ and $^{143}Nd/^{144}Nd$ ratios, which are similar to those in the southern segment, the rhyolites are significantly more radiogenic. Also, basalts and andesites from the northern section are more radiogenic with respect to their Sr isotope composition

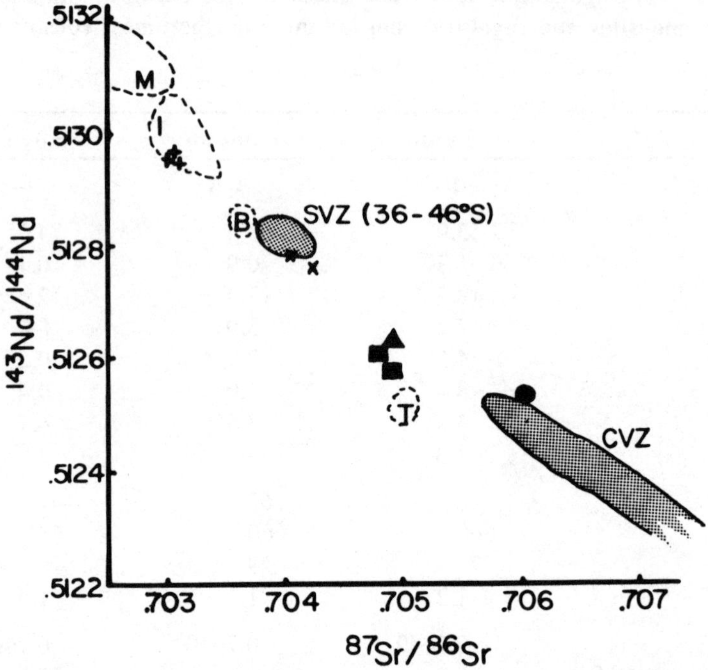

Figure 2 Plot of $^{87}Sr/^{86}Sr$ versus $^{143}Nd/^{144}Nd$ ratios for a basalt (solid triangle), andesites (solid squares) and a rhyolite (solid circle) from the three most nothern volcanoes in the SVZ, compared with fields (shaded areas) for a similar range of rock types from the southern part of the SVZ and the CVZ. The figure also shows the mantle array given by samples from the Atlantic ocean region (M—MORB; I—Iceland; B—Bouvet; T—Tristan de Cuhna) and alkali basalts from just east of the Maipo volcano at 34°S (Xs) and from outcrops of the Patagonian plateau lavas well east of the Andean cordillera close to the Atlantic coast line (+ — data from Stern *et al.*, 1983)

than lavas of equivalent bulk composition samples from further south in the SVZ (see Figure 1). They are not, however, as radiogenic as samples from the CVZ which typically have $^{87}Sr/^{86}Sr$ greater than 0.705 (Francis *et al.*, 1977, 1980; Deruelle *et al.*, 1983).

The rhyolites have significantly higher $\delta^{18}O$ values than associated mafic and intermediate rocks in the northern section of the SVZ. The basalts and andesites from this zone are similar, with respect to O isotope composition, both to each other and to the intermediate and mafic rocks from the southern part of the SVZ (Deruelle *et al.*, 1983), despite their differences with respect to Sr and Nd isotope composition (see Figure 1). The Pb isotope composition of basalts, andesites, dacites and rhyolites are similar throughout both the SVZ and the Austral Volcanic Zone as far south as 52°S (Figure 3; Barreiro and Stern, 1982).

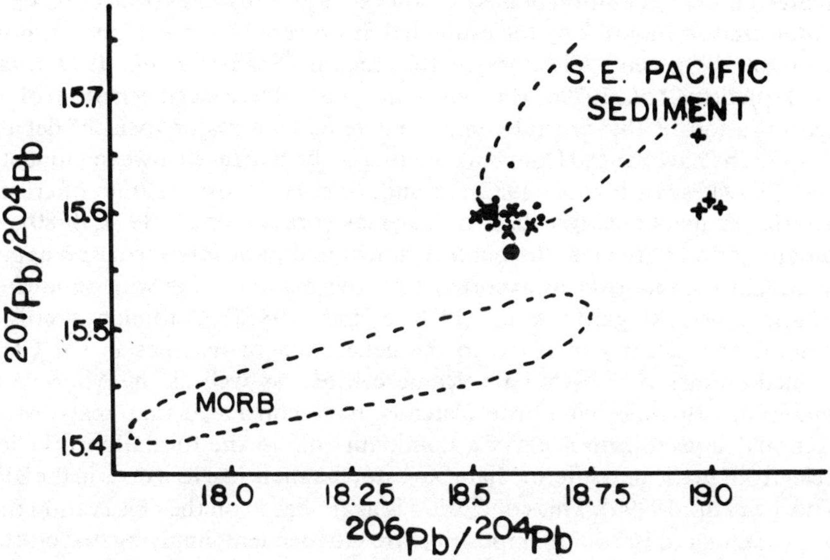

Figure 3 Plot of the Pb isotope composition of an andesite (solid square) and a rhyolite (solid circle) from the three most northerly volcanoes in the SVZ, compared with basalts, andesites and dacites from the southern segment of the SVZ (dots) and alkali basalts from Patagonia just east of the Andean cordillera at 34°S (Xs) and from well east of the Andes near the Atlantic coast (+s). Fields for MORB and sediments of the southeast Pacific are also shown

Discussion

Detailed studies of the volcanic centres of the southern part of the SVZ have concluded that in this region (36–46°S) of the Andes near-surface crystal-liquid fractionation plays a major role in the generation of intermediate and silicic rocks from mantle-derived mafic parental magmas (Lopez-Escobar *et al.*, 1977; Stern, 1979; Muehlenbachs and Stern, 1980; Stern and Futa, 1982a; Deruelle *et al.*, 1983; Frey *et al.*, in press). Observations that support this conclusion include: (a) the similar isotopic composition of rocks with a wide range of SiO_2 contents; (b) the coherent geochemical variations associated with increasing SiO_2 content such as decreasing MgO, CaO, Ni and Sr, and increasing K_2O and LIL element abundances, including REE concentrations which are consistent with fractionation models involving low-pressure mafic phases and plagioclase; and (c) the volumetric predominance of olivine-bearing, high-Al basalts in this region. The mantle source of these basalts is isotopically similar to the mantle source of some oceanic intraplate basalts, consistent with the suggestion that the mantle beneath the southern part of the SVZ has not been significantly modified by the addition of crustal components, including Sr derived from sea water, carried into the mantle by the subduction of oceanic lithosphere.

By contrast, a significant amount of a crustal component is apparently involved in the generation of basalts, andesites and rhyolites from the CVZ as is indicated in Figure 2 by the evaluated Sr, lowered Nd and high O isotope ratios of the recent volcanoes in this region (Francis *et al.*, 1977, 1980; Hawkesworth *et al.*, 1979b; Harmon *et al.*, 1981). The exact mechanism of the incorporation of this crustal component remains a major focus of debate. Suggestions have ranged from bulk melting of the thickened lower continental crust (Hawkesworth *et al.*, 1982), to bulk or selective assimilation of crustal material in mantle-derived basaltic magmas (Francis *et al.*, 1977, 1980), or contamination of the mantle source region of mafic and intermediate magmas by crustal components transported into the mantle along with subducted oceanic crust (Magaritz *et al.*, 1978; James, 1981a). Although a crustal component is clearly involved in the generation of magmas in the CVZ, detailed studies of individual volcanic centres, as well as the Miocene to Quaternary rhyolite ignimbrite plateaus, have concluded that near-surface fractional crystallization plays a dominant role in the formation of silicic rocks from mafic and intermediate parental magmas just as it does in the SVZ (Thorpe *et al.*, 1979a). This conclusion is again based on the observation that SiO_2 content and isotopic composition are independent, implying that crustal components are not incorporated during crystal-liquid fractionation that produces the observed range of silica contents.

Basalts and andesites from the volcanic centres in the northern part of the SVZ (33–34°S) have similar isotopic compositions, a feature which is consistent with the derivation of the andesites from the basalts by crystal-liquid fractionation. The rhyolites, by contrast, have a different and very distinct isotopic composition. They have more radiogenic Sr, less radiogenic Nd and much higher $\delta^{18}O$ values than the basalts and andesites which suggests that incorporation of crustal components was important in the petrogenesis of the rhyolites. Fractional crystallization has also played a major role in the formation of these rhyolites as indicated by their significant negative Eu anomaly, but the data suggest that assimilation of crustal material was associated with the near-surface evolution of these highly evolved rocks.

The isotopic compositions of the basalts and andesites from the northern segment of the SVZ have higher $^{87}Sr/^{86}Sr$ and lower $^{143}Nd/^{144}Nd$ ratios than similar rocks from the southern section, a feature which again suggests the involvement of a component of crustal material in their petrogenesis. On the other hand, the fact that the mafic and intermediate rocks from the northern segment of the SVZ contain less ^{87}Sr and more ^{143}Nd than similar rocks from the CVZ probably reflects instead the younger age and less radiogenic character of the crust in the southern Andes as compared to the central Andes rather than differences in the relative proportions of crustal components involved in the petrogenesis of Recent volcanics in each region.

Despite the differences in Sr and Nd isotope composition between the volcanic centres from the northern and southern parts of the SVZ, basalts and andesites from both these regions have similar O isotope compositions (see Figure 1). This suggests that, except in the case of the rhyolites, near-surface bulk assimilation was not the process by which crustal components were incorporated into the volcanic rocks of the northern part of the SVZ, because near-surface rocks which have reacted with meteoric water generally have high $\delta^{18}O$ values as well as $^{87}Sr/^{86}Sr$ ratios so that near-surface bulk assimilation produces correlated changes in these two isotopes. One possibility for increasing $^{87}Sr/^{86}Sr$ ratios in mantle-derived magmas without increasing $\delta^{18}O$ values would be either bulk or selective assimilation or mixing with anatectic melts derived from lower crustal lithologies such as older calc-alkaline plutons which had never interacted with meteoric water so that they continue to preserve their low $\delta^{18}O$ values. Another possibility is mixing of subducted crustal components with a mantle reservoir which has a low Sr content but a large O content so that the resultant mixture would have a more radiogenic Sr character but be buffered at mantle O isotope ratio (James, 1981a).

The available data do not allow unambiguous discrimination between these two possibilities. The increase in crustal thickness and elevation below the northern segment of the SVZ and the presence of thick continental crust and high elevations in the CVZ has been cited by many authors as the critical factors in causing increased residence time of mantle-derived magmas within the continental crust, thereby leading to assimilation of more significant amount of crustal components compared to the southern part of the SVZ (Francis *et al.*, 1979; Harmon *et al.*, 1981). However, many lines of evidence also indicate that subduction of crustal components, followed by their release into the overlying mantle wedge, is an important process in convergent plate boundary and specifically Andean magmatism. Foremost is the linearity of the volcanic front and the uniformity of the depth to the underlying Benioff zone below the volcanic front in both the Andes and other convergent plate boundary arcs. This uniformity can be attributed to the initiation of melting due to dehydration of subducted oceanic crust. Transport of dissolved solids, particularly crustal Sr, into the overlying mantle wedge by this dehydration process has been documented by combined Sr–Nd isotope studies in the South Sandwich Islands (Hawkesworth *et al.*, 1977) and in the northern Andes (Hawkesworth *et al.*, 1979b). Pb and Be isotope studies have also indicated the apparent cycling of crustal components through subduction zones (Kay, 1980; Barreiro, 1983; Brown *et al.*, 1982).

Although the isotope composition of the high-Al basalts erupted from the volcanic centres of the southern part of the SVZ are similar to intraplate basalts, some suggestion for the influence of crustal components in this region is indicated by the enrichment of Cs in some of these lavas (Hickey *et al.*, 1983). Outcrops of Patagonian plateau lavas which occur just east of the Andean cordillera at 34°S are similar isotopically to the orogenic volcanic centres in the southern segment of the SVZ, but the isotope composition of outcrops of the Patagonian plateau basalts which occur well to the east near the Atlantic coast are significantly different (Figures 2 and 3; Stern *et al.*, 1983) suggesting that the mantle below the Andean cordillera underneath which the Nazca plate is being subducted may have been modified by the addition of small amounts of crustal components. Finally, the similarity of the Pb isotope composition of the rhyolites from the Maipo volcano, which have clearly incorporated continental material, and the mantle-derived alkali basalts which outcrop just east of the Maipo volcano, suggest an overall homogenization of the crust-mantle system in this region which is best accomplished by recycling of crustal components through subduction processes.

If crustal components are incorporated into Andean magmas by being transported into the mantle by subduction, the question arises as to why the significance of crustal components increases in just those zones where the continental crust increases in thickness and elevation. We suggest three possibilities. First, as crustal thickness increases, the thickness of the mantle wedge above the Benioff zone decreases (Figure 4) so that the relative proportion of crustal components being released from the subducted oceanic lithosphere becomes more important. Second, the relative proportion of

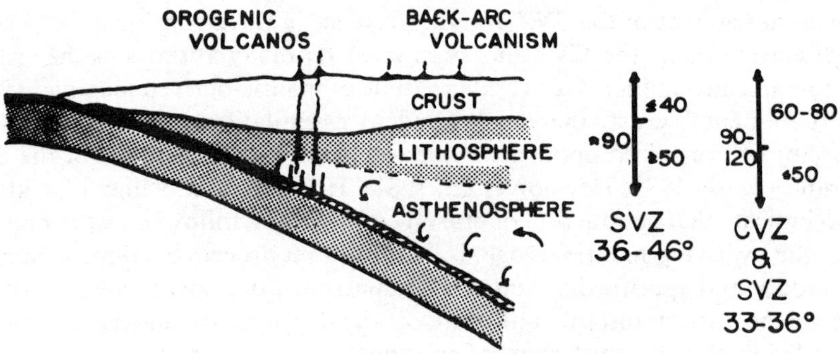

Figure 4 Schematic cross-section through regions of the Andes with active volcanism in which subducted oceanic lithosphere descends at an angle of approximately 25° and asthenospheric mantle is present. Back-arc volcanism has been described from these same regions. The right side of the figure summarizes aspects of the geometry, including the depth below the volcanic front of the Benioff zone, the thickness of the continental crust and the mantle wedge, in the CVZ and the two sections of the SVZ

crustal versus mantle components in the mantle wedge above the Benioff zone may reflect, in part, the vigour of convection in the mantle wedge as a mechanical response to subduction. In South America, Late Tertiary to Quaternary back-arc volcanism, an expression of diapirism in the mantle wedge, is most intense in the region east of the southern part of the SVZ. Such volcanism is much less intense or absent east of the northern segment of the SVZ and east of the CVZ. Lack of convective stirring in these regions may act to increase the significance of crustal components which are being added continually to the mantle wedge by subduction. Thirdly, the subduction process itself might be different in the regions of thicker crust. Specifically, evidence for tectonic erosion of the continental coast in some portions of the Central Andes (Hussong and Wipperman, 1981), has been advanced by both Rutland (1971) and Scholl *et al.* (1977) and this erosion may have lead to the thickening of the continental crust in this region by extensive recycling of continental material through subduction related processes.

In summary, crustal components are clearly involved in the petrogenesis of Andean volcanoes in both the northern segment of the SVZ and in the CVZ, and possibly, but much less significantly, in the southern part of the SVZ. These crustal components may be incorporated into Andean magmas by assimilation or mixing as mantle-derived liquids pass through the continental crust, or they may be tectonically transported into the mantle source regions of orogenic magmas by subduction. Geochemical data distinguishing these two possibilities remain ambiguous. Both processes may occur. The close association of those regions of Andean volcanism which contain significant amounts of crustal components with regions of increased crustal thickness does not constitute proof that assimilation of crustal components occurs by contamination of mantle-derived magmas as they migrate through the crust.

ACKNOWLEDGEMENTS

We want to thank C. Hedge, B. Doe, R. Zartman and Z. Peterman for permission and assistance in obtaining Sr and Nd isotope data at the USGS in Denver, Colorado. B. A. Barreiro determined the Pb isotope data referred to in the text. This study was supported by NSF grant EAR80–20791, and is a contribution to the IGCP Project 120 'Magmatic Evolution of the Andes'.

The Cerro Purico Shield Complex, North Chile

P. W. FRANCIS, W. F. McDONOUGH, M. HAMMILL, L. J. O'CALLAGHAN and R. S. THORPE

Abstract

Cerro Purico is a 1.3 Ma old ignimbrite shield volcano. The first erupted unit was the rhyolitic Toconao ignimbrite ($^{87}Sr/^{86}Sr = 0.7100$), followed after an erosional interval by the dacitic Purico ignimbrite with the same radiometric age ($^{87}Sr/^{86}Sr = 0.7085$). Andesitic to dacitic lavas and extrusive domes were later emplaced in the summit region. The upper part of the Purico ignimbrite contains banded and mafic (andesitic) pumices isotopically indistinguishable from the host. The youngest dome (Chascon) contains basaltic xenoliths characterized by Sr isotope ratios (0.7059) lower than the host dacite (0.7073). While the major and trace element chemistry of the suite is unexceptional, a complex magmatic history is required to explain the origin of mafic pumice clasts and isotopically contrasted mafic xenoliths. Simple crustal anatexis cannot account for the range of compositions observed.

Introduction

Much of the geochemical work carried out on volcanic rocks in the Andes over the last decade has necessarily been of a reconnaissance character. Unfortunately, this has often led to rather sweeping conclusions being drawn from small numbers of analyses from widely scattered volcanoes whose individual evolutionary histories and inter-relationships are poorly known, if at all. Such regional studies have some value in the early stages of investigation of regions such as the Andes, and we have contributed to them ourselves. Far reaching conclusions drawn from such studies may often be misleading, however, and the point is now being reached in Andean studies where detailed studies of individual volcanic centres are required to make further real progress in elucidating the petrological evolution of the region. As an example of the issues involved, we present here the outlines of a study of a single volcanic complex whose rocks express a range in isotopic and trace element chemistry almost as broad as that of the entire Central Andes, and comment on some of the physical processes that may have been responsible for these relationships.

Cerro Purico (Figure 1) is located on the Chile/Bolivia frontier at 67° 45′W, 22° 57′S, northeast of the Salar de Atacama and 30 km east of the village of San Pedro de Atacama. It consists of a roughly circular, gently inclined apron of ignimbrites 25 km in diameter, capped by a summit complex of andesite lavas and dacite domes, which reach a maximum altitude in excess of 5800 m.

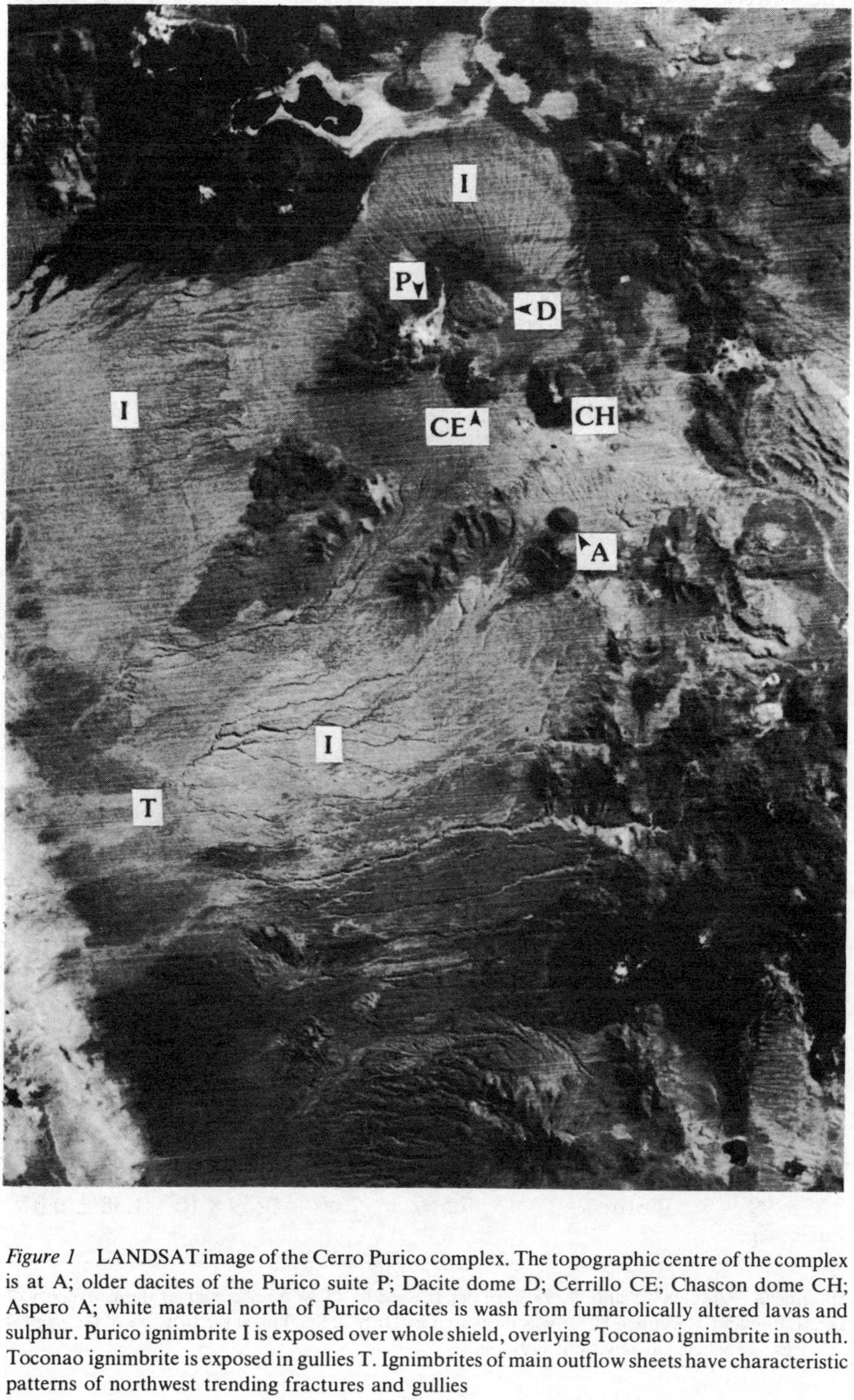

Figure 1 LANDSAT image of the Cerro Purico complex. The topographic centre of the complex is at A; older dacites of the Purico suite P; Dacite dome D; Cerrillo CE; Chascon dome CH; Aspero A; white material north of Purico dacites is wash from fumarolically altered lavas and sulphur. Purico ignimbrite I is exposed over whole shield, overlying Toconao ignimbrite in south. Toconao ignimbrite is exposed in gullies T. Ignimbrites of main outflow sheets have characteristic patterns of northwest trending fractures and gullies

The structure was first identified as an individual entity from LANDSAT imagery (Baker, 1981), and has subsequently been the subject of a geochemical investigation by Hawkesworth *et al.* (1982) who carried out an isotopic and REE study, and by the Chilean Servicio Nacional de Geologia y Mineria as part of a broader study of volcanism and volcanic stratigraphy in north Chile (SNGM, 1982).

Stratigraphy and ages

Two major ignimbrite units are exposed on the flanks of Cerro Purico, the Toconao and Purico ignimbrites. The Purico ignimbrite is a massive unit covering some 800 km^2 and was first described by Guest (1968, 1969) who termed it the Cajon ignimbrite, thinking that its source was located northwest of Purico in the Chaxas area. Since we consider its source to be on the Purico shield, we have named the unit the Purico ignimbrite.

The Toconao ignimbrite is a highly distinctive unit which can be traced in quebradas leading up to Cerro Purico (Guest, 1969) and further south in the Quebrada Talabre. It is quarried extensively as a building and ornamental stone. Radiometric age determinations give an age of 1.3 Ma (Table 1). It is overlain directly by the Purico ignimbrite, which has the same radiometric age, but is separated from it by a small but distinct erosional interval. We conclude that the Toconao ignimbrite was the first major unit to be erupted in the evolution of the Purico complex. The Purico ignimbrite is exposed over the whole of the flanks of the shield and in its summit area. While it overlays the

Table 1 **K-Ar age determinations**

Specimen	Mineral	K% Triplicate	% Rad Ar	Vol $^{40}Ar_{rad}$ in scc/gm	Age and error Ma
P1 (SA78/80) Dacite dome	Biotite	7.34 ±0.114	9.25	2.809×10^{-7}	0.98 ± 0.05
P2 (SA78/81) Dacite dome	Whole rock	2.353 ±0.045	1.86	7.375×10^{-8}	0.81 ± 0.04
82/08 Toconao ig.	Whole rock	3.98	2.00	1.999×10^{-7}	1.3 ± 0.7
SA78/99 Purico ig.	Biotite	7.560 ±0.638	7.00	4.0159×10^{-7}	1.38 ± 0.07

Specimens 78/80, 78/81 and 78/99 were determined by G.A. Kretschmar at the University of Leeds, England. Specimen 82/08 was measured by Dr N.J. Snelling at the Institute of Geological Sciences, Geochemical Division, London.

Figure 2 Western flanks of Purico complex, showing Purico ignimbrite overlying Toconao ignimbrite in quebrada at Toconao. Erosional boundary between the two ignimbrites marked by black line

Toconao ignimbrite in the south, there is no evidence that the Toconao ignimbrite is interstratified between the Purico and Cajon ignimbrites on the north flanks of the shield, suggesting that the Toconao ignimbrite was erupted from a vent on the south side of the present structure (Figure 2).

Amongst the rocks of the summit complex, the lavas of Cerro Purico and Toco are deeply weathered, exposing extensive fumarolic alteration deposits. They are mantled by moraine deposits, and are the oldest post-ignimbrite lavas. Some of the older lavas may predate the ignimbrite, but the stratigraphic relations are equivocal. Cerros El Negro, Aspero and Putas are morphologically younger (i.e. not glaciated and without moraines), but Cerro El Chascon is conspicuously the youngest dome, preserving details of flow morphology and a summit crater. Its age is not determined, but may be post-moraine. On the basis of field relationships, dacite dome D appears to be intermediate in age between Purico/Toco and Chascon. Two age determinations on the dome D yielded K–Ar ages of 0.98 to 0.81 Ma. This suggests that the morphologically much more youthful Cerro El Chascon is likely to be less than 0.5 Ma old, and maybe only a few tens of thousands of years old.

Petrography

IGNIMBRITES

The rhyolite pumice of the Toconao ignimbrite consists of transparent, fresh, straw-textured glass, and possesses a very small proportion of phenocrysts.

Quartz, plagioclase (poorly zoned oligoclase), biotite and sanidine have been identified in pumice clasts, but these phases are present in such small amounts as to prohibit modal analyses. Although carefully looked for, no pumice clasts showing evidence of magma mixing were found.

The dacitic pumice of the Purico ignimbrite consists of a fresh, glassy, vesicular groundmass containing abundant phenocrysts of oligoclase, (zoned, both oscillatory and normal), hornblende, biotite, minor hypersthene and rare corroded quartz and clinopyroxene. Hornblende crystals commonly exceed 1 cm in diameter, and both hornblende and biotite are fresh. The most striking feature of the Purico ignimbrite is the presence, in the uppermost one-third of its thickness, of spectacular banded pumice clasts, the largest of which exceeds 40 cm. The mafic component of the banded pumice has a darker, brown glass, which is intermingled with the lighter dacitic glass on a scale of millimetres, and the vesicles in the darker pumice are somewhat smaller than those in the lighter glass. Phenocrysts in the mafic component are plagioclase (up to 1.5 mm), hornblende (usually rimmed with fine grained opaques), hypersthene, augite and oxides; there is no quartz. The most conspicuous mineralogical differences between the two pumices is the presence of augite, the absence of quartz and the much smaller size of the hornblende crystals in the darker component.

SUMMIT COMPLEX

The extrusive rocks of the summit complex are, in general, ordinary coarsely phenocrystic andesites and dacites. Plagioclase, orthopyroxene and clinopyroxene are the dominant phenocryst phases in the andesites, with accessory Fe–Ti oxides. In the dacites, plagioclase is the dominant phase, with lesser amounts of hornblende and biotite; Fe–Ti oxides are again the most abundant accessory phases.

The dacitic rocks of the Chascon/Aspero centres contain oligoclase, hornblende and biotite (sometimes up to 4 mm), quartz and rare hypersthene phenocrysts in a finely crystalline groundmass. Quartz with clinopyroxene mantles are found in some rocks. Euhedral sphene crystals are common as an accessory phase in the Chascon-Aspero rocks. Abundant small xenoliths and xenocrysts (especially clinopyroxene and olivine, surrounded by reaction rims of hornblende) are common in these flows.

MAFIC XENOLITHS

Numerous xenoliths of basaltic composition are found in the Chascon dacites. There are two textural types of xenolith. The first has a fine grained, felty groundmass, with minor phenocrysts and, overall, a texture that is very similar to extruded basalts. The second has a coarser grained, plutonic texture, with intercrystalline glass. Similar xenoliths of both textural types are described from other arc environments (Wager, 1962; Nicholls, 1971; Lewis, 1973), and the chemistry and origin of such xenoliths have been discussed (Nicholls, 1971; Stern, R.J., 1979; Arculus and Wills, 1980). The basaltic textured xenoliths are glassy, up to 0.25 m across, and have a globular outline. The basaltic textured xenoliths contain plagioclase, hornblende, clinopyroxene, olivine and biotite, in decreasing order of abundance, along with accessory Fe–Ti oxides and acicular apatite, all in a brown glass. Certain large plagioclase phenocrysts are riddled with glass inclusions throughout, whereas others contain melt inclusions only along growth planes. Microcrystalline plagioclase with skeletal growth are found in the brown glass. The plutonic textured xenoliths contain millimetre-sized plagioclase, hornblende and biotite, with minor amounts of clinopyroxene and olivine, accessory Fe–Ti oxides, and intercrystalline brown glass. These xenoliths often have vesicles, some of which are filled with minute, acicular crystals (cristobalite?). Textural features and mineral assemblages similar to those found in the plutonic textured xenoliths have been described by Jakes and Smith (1970), Nicholls (1971), and Arculus and Wills (1980), and are interpreted to be representative of liquidus or near liquidus phases.

Geochemistry

MAJOR ELEMENT CHEMISTRY

Major and trace element data for 38 Purico samples were determined by XRF and INAA techniques. The major element variations are summarized in the variation diagrams in Figure 3. The rocks of the Cerro Purico complex, as a whole, are characterized as low TiO_2, high K, calc-alkalic type (Peccerillo and Taylor, 1976), with major and trace element contents that are closely similar to other Central Andean volcanoes (Francis *et al.*, 1974; Roobol and Francis, 1976; Deruelle, 1982). Uniform decreases in CaO, $FeO + Fe_2O_3$, Al_2O_3, MnO and TiO_2 contents, and an increase in K_2O abundance are seen with increasing SiO_2 content. The mafic xenoliths, found in the Chascon dacites, have basaltic andesite to high K basaltic andesite compositions, with low TiO_2 and Na_2O and high CaO contents compared to similar SiO_2 compositions reported by Ewart (1982) and Deruelle (1982). The covariation of CaO, MnO and K_2O with SiO_2, in the mafic xenoliths, follows the trends observed for the Purico volcanic rocks.

TRACE ELEMENT CHEMISTRY

The trace element data are summarized in Figure 4. The range of trace element concentrations (Table 2) is, in general, similar to those found in other medium to high K andesites and dacites from CVZ (Thorpe *et al.*, 1982; Deruelle, 1982; Ewart, 1982). The positive covariations of Th with SiO_2, K, LREE, Rb, Rb/Sr, Ta, Nb and Pb, and the negative covariations of Th with Sr, Eu/Eu*, Sc, V,

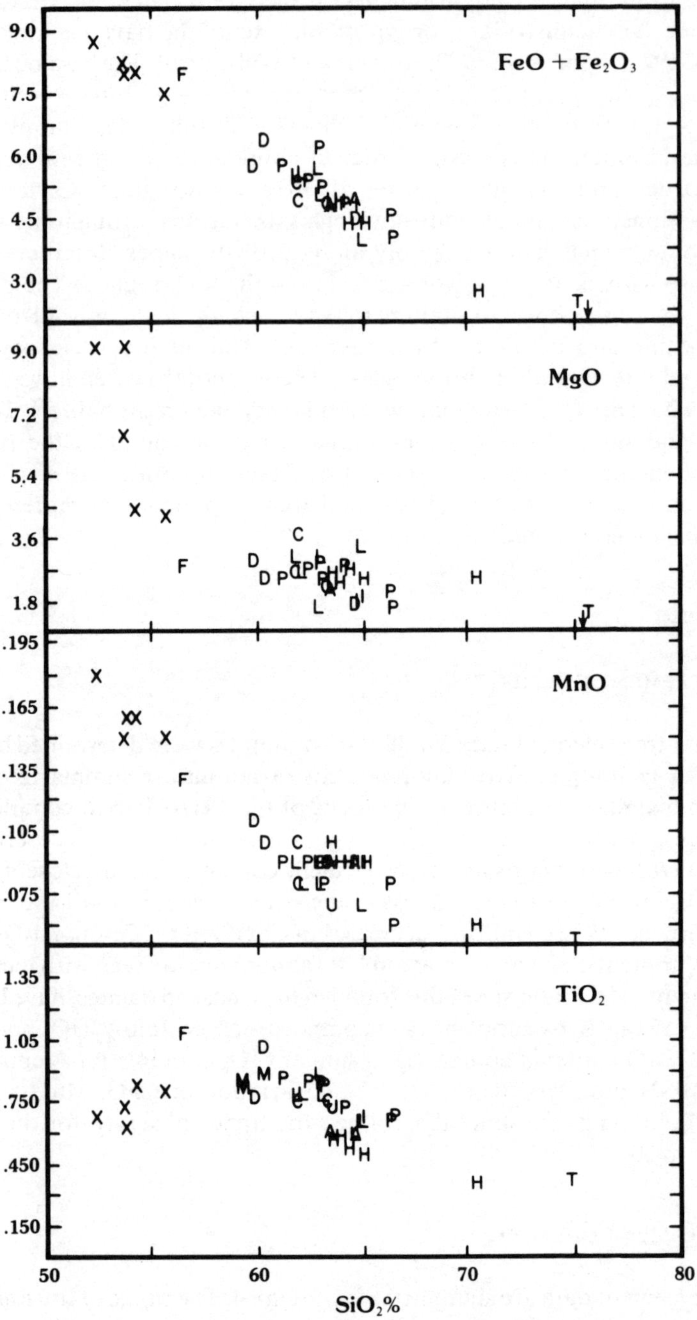

Figure 3 Major element variation diagrams for rocks of the Cerro Purico complex. P—Purico dacites; I—Purico ignimbrite; T—Toconao ignimbrite; D—dacite dome; H—Chascon; C—Cerrillo; A—Aspero; M—mafic clasts in Purico ignimbrite; X—mafic xenoliths in Chascon dacite

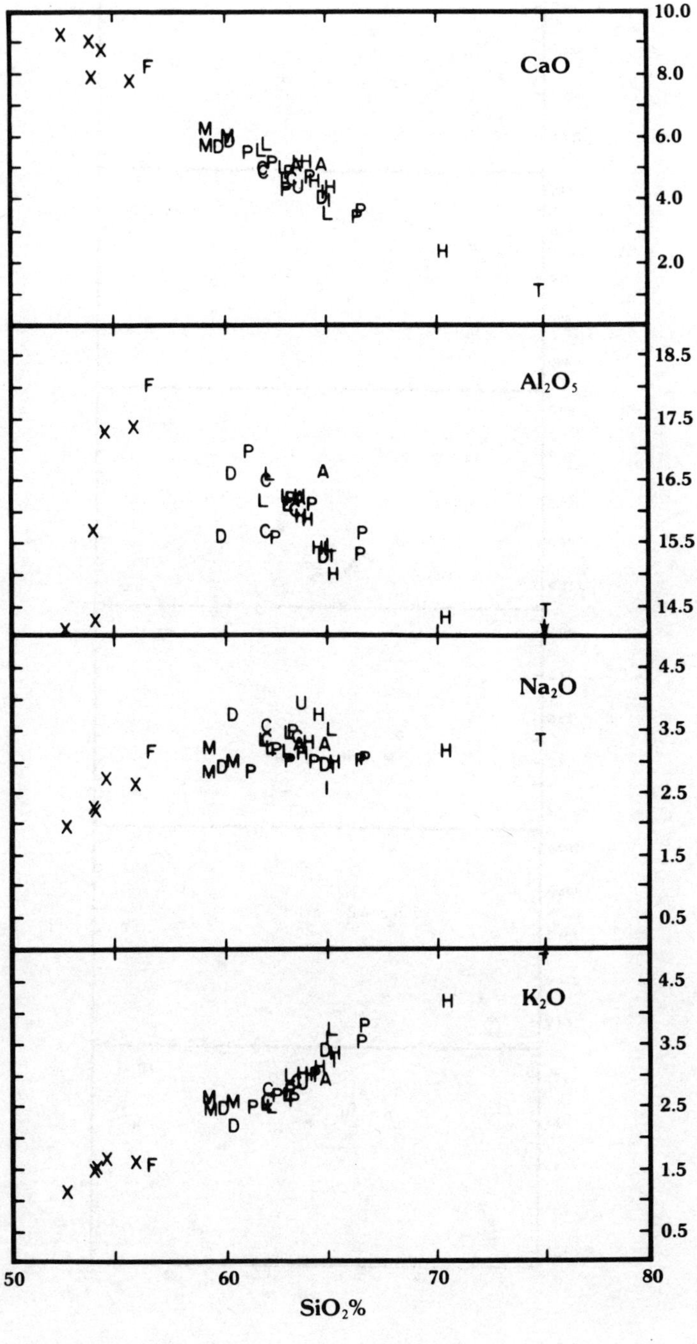

Figure 3 (contd.)

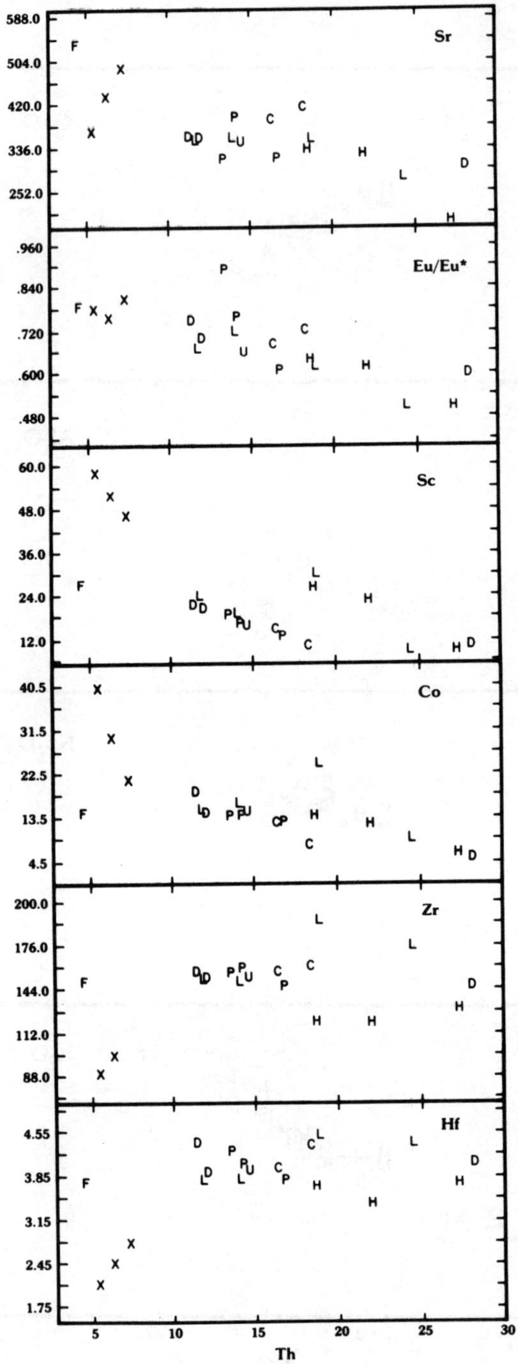

Figure 4 Trace element variation diagrams for rocks of the Cerro Purico complex. Symbols as in Figure 3

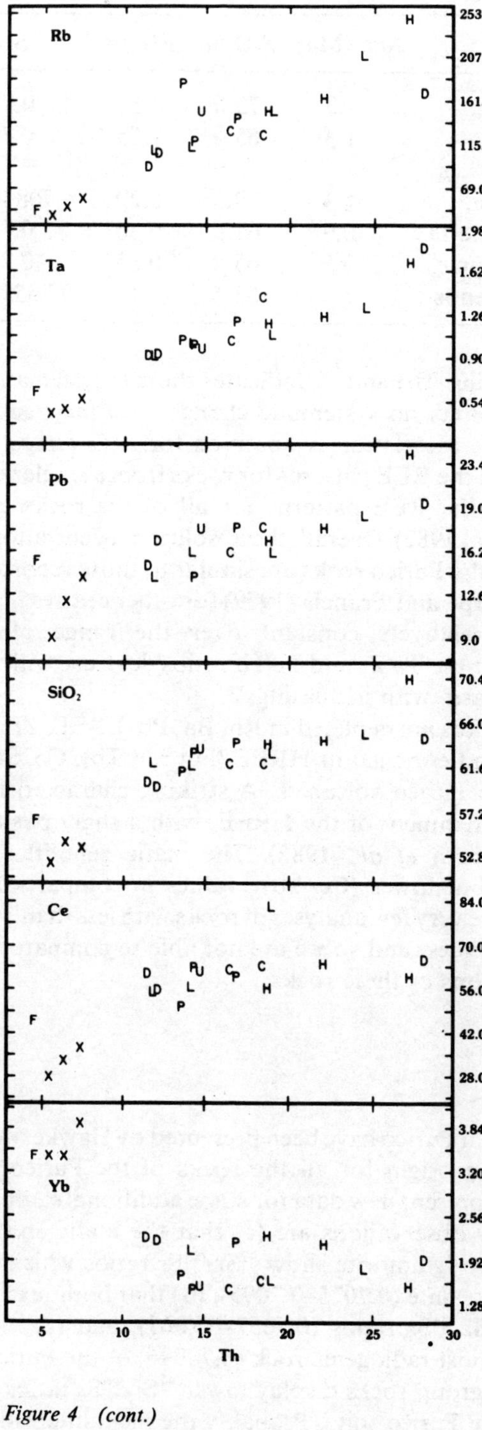

Figure 4 (*cont.*)

Table 2 **Selected geochemical and Sr isotope data**

Sample	Age (Ma)	SiO_2%	Rb/Sr	$^{87}Sr/^{86}Sr$
Toconao ig.	1.3	75.0	4.1	0.7100
Purico ig.	1.3	65.0	0.55	0.7085
Mafic pumices in Purico ig.	1.3	59.5	0.29	0.7084–0.7088
Dacite dome D	0.9	61.5	0.36	0.7088
Chascon dome	>0.5	65.4	0.65	0.7073
Chascon xenos	?	53.4	0.12	0.7057–0.7061

HREE (not including Tb) and Y, indicates there is systematic trace element behaviour. Ba exhibits no systematic change with increasing Th, or SiO_2, although a positive correlation is observed for K_2O versus K/Ba. There is limited variation in the REE patterns for rocks from a single vent, whereas the total variation in the REE patterns for all of the rocks is much greater (Hawkesworth *et al.*, 1982). Overall, the absolute concentration and the slopes of REE curves for the Purico rocks are similar to those reported by Thorpe *et al.* (1976) and Thorpe and Francis (1979) for other centres in north Chile. Zr and Hf remain relatively constant over the range of SiO_2 and Th concentrations, but the Ti/Zr and Ti/Hf ratios decrease with increasing SiO_2 and Th, and increases with increasing V.

The mafic xenoliths are depleted in Rb, Ba, Pb, LREE, Zr, Hf and Ta ('the incompatibles'), and enriched in HREE (but not Tb), Co, Sc, ±Cr and V in comparison to the Purico volcanics. A striking characteristic of the mafic xenoliths is the enrichment of the HREE, with a slight positive slope in the HREE (Hawkesworth *et al.*, 1982). The mafic xenoliths display similar $(Ce/Sm)_N$ ratios, but lower $(Ce/Yb)_N$ values in comparison to the Purico volcanics. There are very few analyses of rocks with less than 55.0% SiO_2 in the Central Chilean Andes, and so we are not able to compare major and trace element compositions of these rocks.

Sr ISOTOPE DATA

Sr isotope data from Purico have been presented by Hawkesworth *et al.* (1982) who argued for an origin for all the rocks of the Purico suite by crustal anatexis. Here we present new data for some additional samples (see Table 2). The principal new observations are (a) that the mafic and banded pumice clasts in the Purico ignimbrite show $^{87}Sr/^{86}Sr$ ratios which are the same as those of the Purico suite (0.7075–0.7095), (b) that both textures of xenoliths have the same $^{87}Sr/^{86}Sr$ ratios (0.7057–0.7061), and (c) that the Toconao ignimbrite is the most radiogenic rock (0.70996) of the Purico complex. The Chascon/Aspero group rocks display lower $^{87}Sr/^{86}Sr$ ratios (0.7069–0.7079) than the rest of the Purico suite. Probably the most interesting aspect of the

Purico complex is its overall isotopic evolution through time, ranging from 0.7100 in the Toconao rhyolite to 0.7073 in the Chascon dacites (Table 2), in which the Sr isotope signature is apparently becoming less radiogenic with time, while the composition of the magmas continues to vary.

Discussion

A PHYSICAL MODEL FOR THE PURICO IGNIMBRITE AND LAVAS

The Toconao ignimbrite, the most evolved rock of the entire complex, presents particular problems of interpretation. It is the earliest member of the complex, and its eruption was separated from that of the Purico ashflow by a minor, but distinct, erosional episode. No other rocks attributable to this phase of activity have yet been identified. Whether the Toconao ignimbrite represents a direct crustal partial melt, or originated as a result of extensive assimilation and fractional crystallization cannot now (and possibly never) be determined. Physically, there is little difference between the two processes; both require large thermal influxes into the sub-Purico crust and result in the creation of a subvolcanic magma chamber that was only partially emptied during the Toconao eruption. Simply from consideration of its volume, we suggest that it is likely that the Toconao ignimbrite was erupted from a compositionally zoned magma chamber, but for reasons which we touch on later, and discuss more fully in a forthcoming petrological paper, it would be difficult to support a chemical model in which the Toconao ignimbrite represents the silica-rich cap to a larger, compositionally zoned magma chamber, which ultimately gave rise to the other younger members of the complex.

Regarding the Purico ignimbrite with its andesitic and banded pumice clasts, the immediate questions that arise are these: first, assuming that there was compositionally zoned magma chamber beneath Purico, what is the first evidence of its existence? Second, if it existed, how did the compositionally zoned system develop initially? Third, given that the chamber was chemically zoned, was it also isotopically zoned, and how?

There appear to be three different mechanisms which could cause clasts of mafic and banded pumices to be erupted in an ignimbrite: (a) partial evacuation of a compositionally zoned magma chamber, (b) extensive evacuation of a compositionally zoned magma chamber, or (c) an eruption of a previously homogenous magma chamber triggered by the injection into it of a basaltic magma. In the latter case, suggested by Sparks *et al.* (1977) and Sigurdsson and Sparks (1981) for the Askja, Iceland eruption of 1875, abundant evidence of wholesale hybridization of magmas, with production of intermediate compositions, would be expected. This is not seen at Purico, and inclines us to discount this possibility.

If a large, stable, compositionally-zoned magma chamber were extensively tapped during an eruption in the manner modelled by Blake (1981), one might expect to see a major change in composition between pumices at top and bottom of the ignimbrite, with a smooth variation in relative abundances of the two types between top and bottom. Exactly such a variation is seen in the Crater Lake ignimbrite (Ritchey, 1980; Noble *et al.*, 1969), but it is not seen at Purico, where mafic and banded pumices are found only in the upper third of the ignimbrite, and there only in minute proportions. We conclude that the eruption which fed the Purico ignimbrite only barely tapped down as far as the andesitic horizon, and that only a small proportion of this magma was entrained during the eruption. Had the eruption been more voluminous, the proportions of andesitic to dacitic clasts would have shown a marked increase.

Regarding the formation of the initial compositional zonation, the volume of the Purico ignimbrite is about 15–25 km^3. Spera and Crisp (1981) have suggested that small magma systems, of the order of 10 km^3 will generate compositional zonation over a short (10^2–10^4 a) time scale. According to Hildreth (1981), compositionally-zoned silicic magma chambers should be density stratified with an upper, volatile enriched, stable rhyolitic layer, overlying a homogenous mafic, convecting layer, with episodic thermal and chemical inputs from diapirs of basaltic or gabbroic material at the base. Spera and Crisp (1981) suggest that minor and trace elements may fractionate as a result of Soret diffusion or double diffusion across thin thermal boundary layers in compositionally zoned magma chambers. Such fractionation may also affect significantly the isotopic composition of such a system.

As far as our present understanding goes, relatively simple magmatic processes (e.g. partial melting and crystal fractionation) cannot fractionate radiogenic isotopes, except possibly during very protracted fractional crystallization (McCarthy and Cawthorn, 1980). Hildreth (1981), however, has argued that isotopic fractionation may take place in highly silicic melts as a consequence of the physical structure of the melt and 'convective driven thermogravitational diffusion', and has shown that the uppermost part of the Bishop Tuff was characterized by a Sr isotope gradient of 0.7079 to 0.7061, from top to bottom.

Whether or not the Purico magma system was isotopically zoned depends largely upon the role played by the Toconao ignimbrite. This may represent the very silicic, volatile enriched, isotopically zoned cap to the *same* magma chamber as that which supplied the younger Purico ignimbrite, or it may represent the top of a distinct, chemically evolved, isotopically enriched magma chamber, which experienced further evolution after the eruption of the Toconao and prior to the eruption of the Purico ignimbrite. The former case would require the Purico and Toconao ignimbrites to be co-magmatic, although now exhibiting different isotopic signatures, whereas the latter would require the magma chamber vacated by the Toconao ignimbrite to be replenished with a new batch of magma. Through subsequent mixing and diffusion processes a new isotopically homogeneous, compositionally zoned magma developed, which gave rise to the Purico ignimbrite.

A number of factors suggest that the second is the more likely alternative. Although the time interval between the two ignimbrite eruptions is uncertain, it was sufficient for minor erosion to take place. Second, the fact that the

Purico ignimbrite erupted as an ignimbrite (as opposed to lavas) indicates that substantial accumulation of volatiles had taken place; sufficient to drive a second major pyroclastic eruption, and from volume-repose time estimates (Spera and Crisp, 1981) we would suggest an elapse time of at least 10^2–10^3 a between eruptions. Third, the presence, in the upper part of the Purico ignimbrite, of mixed andesite-dacite pumice clasts demonstrates that the post-Toconao, sub-Purico magma chamber was compositionally zoned, and no such evidence for such zonation has been found in the Toconao ignimbrite. Fourth, the Sr isotope data for Purico ignimbrite mixed dacite (0.7085) —andesite (0.7087) pumice, indicates that there was a very slight (beyond analytical uncertainty) Sr isotope gradient during the Purico ignimbrite eruption, but this Sr isotope difference is much less than the difference found in the Bishop Tuff, and the difference is in a reverse sense to that discussed by Hildreth (1981).

Although the andesite pumice streaks are more mafic than the dacites, we note that they are not 'primitive' compositions, either isotopically or in their major or trace chemistry. Thus, if the eruption of the Purico ashflow was triggered by injection of hot, basaltic magma into the base of the magma chamber, as aruged by Sparks *et al.* (1977) and Sigurdsson and Sparks (1981) for eruptions such as that of Asjka 1875, then the effects of such an influx were physical rather than chemical. If the Purico magma chamber was stably density stratified prior to the introduction of a fresh basalt, it is possible that the thermal and/or volatile input from the basalt was sufficient to lower the density of the overlying andesite enough to initiate convective overturn. Convective overturn perturbs the compositional zonation, and by transporting volatile rich magmas to shallower levels and lower pressure, may cause vesiculation and thus trigger an eruption, in which streaked pumices are conspicuous in the ejecta. In the absence of any physical evidence of such a basaltic input, however, this must remain speculative.

We conclude that at the time of the Purico ignimbrite eruption the sub-Purico magma was chemically zoned, but most likely not isotopically zoned, and that this zoned magma developed initially according to the model outlined by Hildreth (1981), with the initial introduction of basaltic magmas, providing both a thermal and chemical influx, into the sub-Purico crust.

A PHYSICAL MODEL FOR THE CHASCON SUITE

The Chascon suite has a different set of features requiring a different physical model. The Chascon dacites are compositionally similar to the Purico ignimbrite and post-ignimbrite lavas; the most striking difference being the lower Sr isotope ratios of the Chascon suite, lower concentrations of Zr, ±Hf, Ba and Ti, and flat HREE patterns. On a TiO_2 versus V diagram (Figure 5) the rocks of the Chascon suite plot in a separate field that appears to be distinct from the Purico suite and the xenolith suite. Although no radiometric dates are available for Chascon, the morphology of the dacite dome shows that

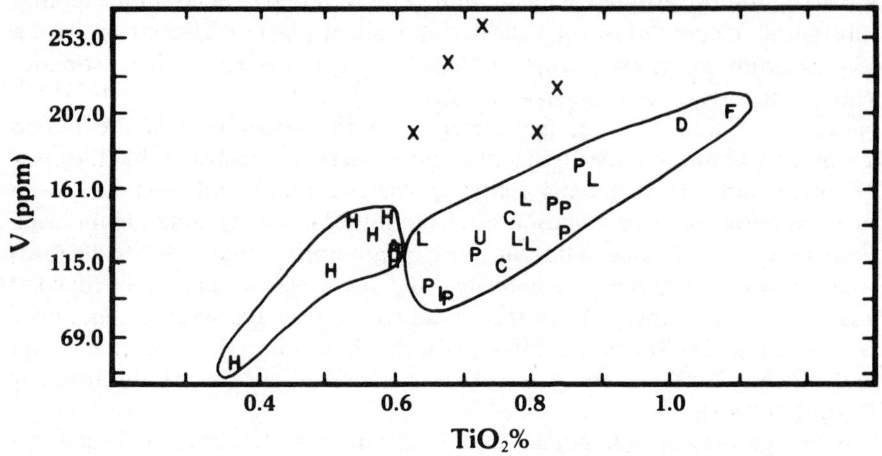

Figure 5 V versus TiO_2 for rocks of the Purico complex. Symbols as in Figure 3

these rocks are much younger than the Purico suite, perhaps by as much as 1 Ma younger.

The most distinctive feature of the Chascon suite is the presence of mafic xenoliths and xenocrystic olivine scattered throughout the dacites. Many of these xenoliths are glassy, markedly vesicular, and their globular nature suggests that they were chilled as amorphous magma blobs on being injected into the molten dacitic magma chamber. The xenocrystic olivines are presumably derived from the same source as the xenoliths. The xenoliths are basaltic *(sensu lato)* in composition, and have much less radiogenic isotopic signatures ($^{87}Sr/^{86}Sr$ = 0.7057–0.7061) than their host lavas ($^{87}Sr/^{86}Sr$ = 0.7069–0.7072) (Hawkesworth *et al.*, 1982). Ratios of incompatible trace elements in the xenoliths are similar to Purico and Chascon ratios. The trace element characteristics suggest that the xenoliths may be related (i.e. cogenetic) to the Chascon dacites. Some degree of crustal contamination is required to explain the Sr isotope ratios of the xenoliths, which are enriched relative to more primitive compositions. By contrast, the less radiogenic character of the Chascon dacites, compared to the other rocks of the Purico complex, is supportive evidence for contamination of their source magma by the mafic xenoliths.

We envision that prior to the eruption of the Chascon dacite (and perhaps a cause of the eruption), the magma chamber was invaded by a fresh intrusion of basaltic magma with 'primitive' compositional characteristics. A limited amount of intermingling occurred which was responsible for the minor xenocrystic olivine and other phenocryst phases seen in the dacites. In the case of the mixed andesite/dacite pumice clasts in the Purico suite, although mixing on a scale of millimetres occurred, there are no xenocrystic minerals, which is indicative of widespread blending (Sigurdsson and Sparks, 1981). In the case of the Chascon xenoliths, the 'mixing' is on a scale of tens of centimetres. Thus, while moderate convective overturn may have taken place in the case of the Purico suite, it seems more probable in the case of the Chascon suite that a denser, basaltic magma intruded into the lower portion of the magma chamber, with parts of this basaltic magma being entrained during the eruption.

The physical properties of the xenoliths and dacites are relevant in this context. The mafic xenoliths are more vesicular than the host dacites, and the fact that no large scale pyroclastic activity appears to have accompanied the extrusion of the Chascon dacites, suggests that the dacites were degassed, and were therefore relatively denser than the mafic xenoliths (Huppert *et al.*, 1982). Eichelberger (1980) proposed that the vesiculation of a mafic magma at the interface between it and a silicic magma may cause partial disruption of the boundary layer, and result in the incorporation of a mafic 'froth' into the silicic magma. Such a process might be responsible for the introduction of scattered mafic xenocrysts into the dacite magma. Some larger, more vesicular chunks of the mafic material may also have 'floated' up into the dacite, to become the xenoliths themselves. We emphasize that the density and viscosity relationships in the Chascon suite must have been such that wholesale turnover and mixing to the point of homogenization did not take place.

Regarding the time relationships, the thermal models of Smith and Shaw (1975) and Norton (1977) showed that for large plutons, the upper portions will cool well below magmatic temperatures in 1–2 Ma, in the absence of the introduction of any fresh magma. However, the lower parts remain at magmatic temperatures for longer periods of time. Thus, it is possible that the lower, more mafic parts of the Purico magma remained liquid and that these liquids gave rise to the Chascon suite (with, most likely, the addition of more isotopically primitive material). In principal, however, it is equally possible for a relatively small volume of dacitic magma, represented by the Chascon suite, to form by fractional crystallization, and/or thermal-convective diffusion from a mafic parent, not requiring any contribution from the earlier Purico suite.

Conclusions

A detailed discussion of the petrological evolution of Cerro Purico is forthcoming. We do not pretend to have fully understood all the processes that have operated in the evolution of the Purico complex, but we hope that we have illustrated some of the problems encountered there. In general terms it is clear that the evidence from Cerro Purico may be useful in the context of chemical and isotopic studies of Andean volcanic rocks. If nothing else, it emphasizes some of the complexities encountered in detailed studies of individual centres, and the need for more such studies. Apart from detailed collecting, we also stress the need for close stratigraphic and geochronological control; without which, attempts to interpret magmatic history at Cerro Purico would not be very meaningful. The work on Cerro Purico also supports the contention of Domenick *et al.* (1983), based on their isotopic

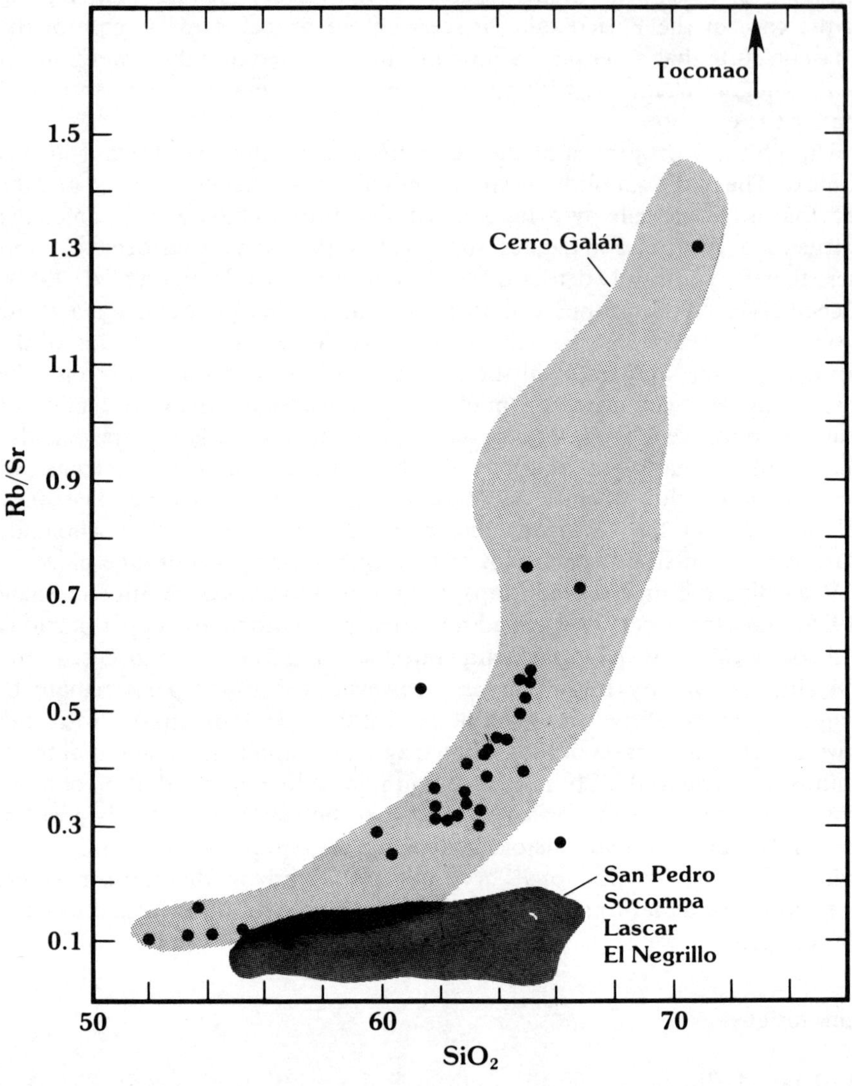

Figure 6 Rb/Sr versus SiO_2 for volcanic rocks of the Central Andes. Data for San Pedro from Francis *et al.* (1977); Cerro Galan from Fancis *et al.* (1980), Socompa, Lascar and El Negrillo from Deruelle (1982)

studies of xenoliths in the Sierra Nevada batholith, that the use of simple models, such as simple two-component mixing, are no longer adequate in detailed petrological studies.

Having said this, how typical is Cerro Purico of volcanic centres in the CVZ as a whole? In terms of its major element chemistry, it is fairly representative. For the most part, its trace element chemistry is closely similar to other studied centres. In detail, however, some important differences emerge. For most of the young volcanic centres of the central Andes, for which a sufficient body of data exists, there is little variation in either Sr isotope ratio or Rb/Sr ratio with increasing silica content (Figure 6). In contrast, both Cerro Purico and Cerro Galan exhibit opposite trends, in particular, showing marked increases in Rb/Sr over a range of silica content. Centres which show little tendency towards increases in these ratios are typically conventional andesitic composite volcanoes, such as San Pedro (Francis *et al.*, 1974) and Lascar (Deruelle, 1978, 1982). These centres are representative of the most common kind of Andean volcanism. Cerro Purico and Cerro Galan, by contrast, have both erupted large volumes of ashflows, and experienced long magmatic histories, greater than five million years in the case of Cerro Galan. Both, therefore, may represent longlasting crustal thermal anomalies.

In modelling the interactions of mantle-derived magmas with the crust, De Paolo (1981) emphasized the importance of the relative rates of assimilation and fractional crystallization (AFC). Varying the rate of assimilation to the rate of fractional crystallization can result in significant changes in the abundances of incompatible trace elements in the melt. Watson (1982) has further shown, through experimental studies and discussions of previous field and experimental studies, that if the assimilation process does not go to completion (crustal materials are only partially melted), then the chemical modelling of the AFC process is greatly complicated. In addition, the rate of assimilation reflects the rate at which heat is input into the system. Thus, this simple physical factor can have important chemical differences. We suggest that this may be one explanation for the differences between centres such as Cerro Purico and others such as San Pedro, and urge that such physical factors be taken into account in attempting to model the petrological evolution of Central Andean volcanics.

ACKNOWLEDGEMENTS

Fieldwork in Chile was financed by the National Environment Research Council. The manuscript was written while P.W.F. and W.F.M. were visiting scientists at the Lunar and Planetary Institute, Houston and the manuscript constitutes LPI contribution No. 581.

Quantitative Models for Crustal Contamination in the Central and Northern Andes

D. E. JAMES

Abstract

Systematic variations in O, Sr, Nd and Pb isotope ratios and related trace element abundances in modern andesitic lavas of the central and northern Andes are examined and compared in terms of crustal contamination models of assimilation and fractional crystallization. Isotopic and trace element compositions of the 'primary' magmas are estimated and used to assess the degree of involvement of crustal material in the deep crust and/or the subduction zone. Andesitic volcanics of south Peru are not markedly different in major element chemistry from those of southwestern Columbia, but they do exhibit large differences in isotopic and trace element composition. Systematic geochemical variations in the south Peruvian lavas are shown to be the result of shallow to intermediate depth assimilation of Precambrian gneiss accompanied by dominantly plagioclase fractional crystallization. Correlated O-Sr-Nd isotope ratios in southwestern Columbia andesites are not as easily described by simple assimilation-fractional crystallization models, but do require significant involvement of crustal material at some stage of their magmatic evolution. Pb isotope ratios, which are strongly diagnostic of crustal contamination in south Peruvian volcanics exhibit minimal variation in southwestern Columbia lavas. Geochemical variations due to fractional crystallization appear to be controlled more by fractionation of mafic minerals, including olivine, than by fractionation of plagioclase. These and other differences in the style of crustal contamination between the CVZ and NVZ are examined in light of dissimilar regional crustal structure and composition.

Introduction

This paper examines some of the isotopic and other geochemical differences between modern volcanics in portions of the NVZ and CVZ, with the objective of characterizing the geochemical imprint produced by crustal contamination and, ultimately, of estimating compositions of the primary magmas. The presentation is summary in nature, with the focus on specific geochemical characteristics that serve to compare and contrast magmatic evolution in the central and northern Andes. A more comprehensive presentation of results for the individual central and northern volcanic provinces may be found in James (1982a, 1983).

The quantitative analysis of crustal contamination presented here is based upon an assimilation-fractional crystallization (AFC) model (Taylor, 1980; James, 1981a; De Paolo, 1981) in which the heat required to melt assimilated rock is produced by magmatic crystallization. In this model, magmatic evolution is a process of assimilation accompanied by fractional crystallization and removal of precipitated phases from the melt. A quantitative treatment of crustal contamination using the AFC model requires knowing (or assuming) the compositions of the primary melt and assimilated material, the bulk partition coefficients of precipitated phases and their O-isotope fractionation factors (i.e. $\delta^{18}O$ value of precipitated phases relative to the $\delta^{18}O$ value of the melt), and the weight proportion of crystal cumulate to assimilated rock. In practice, these variables are adjusted to provide an optimum model fit to the analytical data.

The discussion which follows describes the process of crustal contamination in both CVZ and NVZ lavas. The AFC model provides a good working fit to the observed data in both regions and allows primary magmatic compositions to be estimated. Only in the NVZ, however, can a direct estimate be made of a possible crustal component in the primary melt.

The central Andes

The modern andesitic lavas of southern Peru, in common with volcanics elsewhere in the CVZ are characterized by anomalous isotopic ratios and high abundances of incompatible elements. Over the years, a multitude of varying explanations have been offered to explain the origin of those compositional characteristics. These include derivation from an enriched continental lithospheric mantle, magma mixing, subduction zone melting of oceanic crust and/or subducted continental sediments, melting of lower crust, and crustal contamination of 'normal' calc-alkaline magmas (see James, 1982a, for references).

The first geochemical measurements on central Andean volcanics in south Peru, involving mostly $^{87}Sr/^{86}Sr$ ratios and trace element (including rare earth element) abundances, did little to discriminate between the various theories. Subsequent measurements of O isotope ratios (Magaritz *et al.*, 1978; James, 1981b, 1982a) and especially the Pb isotope measurements of Tilton and Barreiro (1980) demonstrated conclusively, however, that crustal contamination was a dominant factor in producing the distinctive geochemical characteristics of the CVZ lavas.

The two principal volcanic groups of southern Peru are the Arequipa and Barroso volcanics (Figure 1). It is not a trivial exercise to obtain a quantitative estimate of crustal contamination in these lavas. The southern Peru volcanics exhibit both regional geochemical variations (between volcanic centres in south Peru) and local geochemical variations within comagmatic volcanic series. Figure 2, in which $^{87}Sr/^{86}Sr$ ratios are plotted versus Rb, Sr and Ba abundances, illustrates the nature of both the local and regional geochemical variations. Regionally, the Barroso volcanics of southernmost Peru have lower average $^{87}Sr/^{86}Sr$ ratios and Sr abundances and higher average Rb and Ba abundances than do the Arequipa volcanics. The variation of trace element abundances relative to $^{87}Sr/^{86}Sr$ ratios is similar for both the Arequipa and Barroso volcanics.

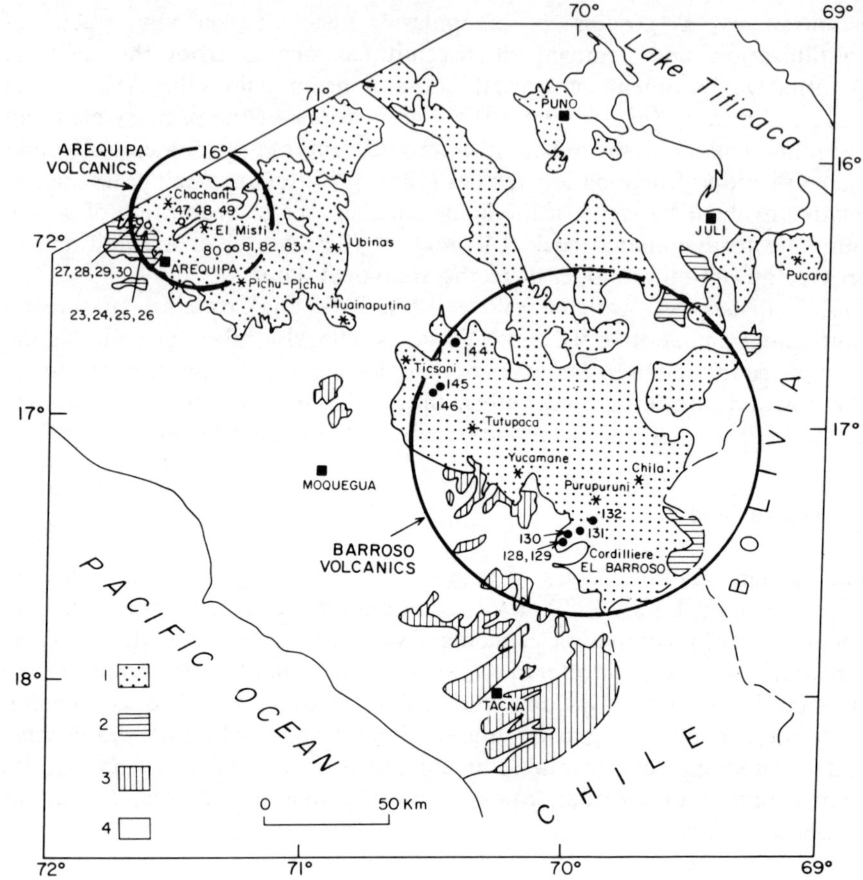

Figure 1 Sample locations in south Peru for the Arequipa and Barroso volcanics. 1—late Cenozoic lavas; 2—late Cenozoic ignimbrites; 3—Miocene/Pliocene ignimbrites; 4—other formations. Andesitic stratovolcanoes indicated by asterisks

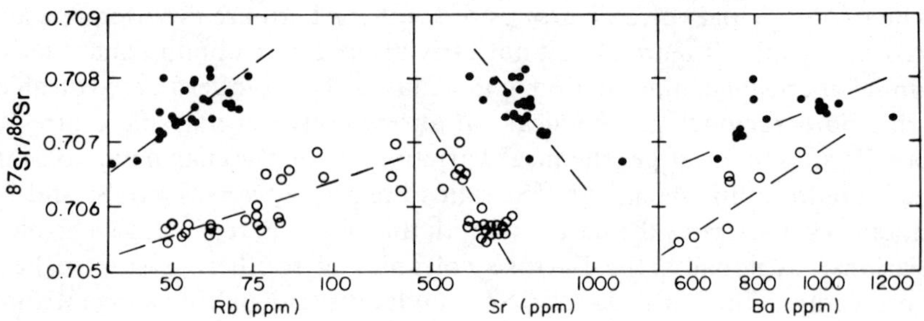

Figure 2 $^{87}Sr/^{86}Sr$ versus Rb, Sr and Ba in Arequipa (closed circles) and Barroso (open circles) volcanics

The data shown in Figure 2 show that the Arequipa and Barroso volcanics are derived from different parental magmas. The contrast in the parental magmas, as well as their comparatively 'continental' character, strongly suggest that the magmas experienced a major prior episode of crustal contamination and crystal fractionation, possibly in the lower crust. The model calculations shown here for crustal contamination pertain only to the local geochemical variations. While these calculations give one measure of crustal contamination and are essential for determining appropriate parental magma compositions, the full measure of crustal contamination cannot be determined until more is known about crustal involvement in the initial stages of magmatic evolution.

RESULTS AND DISCUSSION

Sr, Nd, Pb and O isotope ratios and Sr, Nd, Pb and Rb elemental abundances in the Arequipa and Barroso volcanics constitute the primary data set for the modelling calculations (see James, 1982a, for tabular data summary). Of these data, the Sr and Pb isotope results and the Sr and Rb abundances proved to be the most diagnostic of the AFC process.

The Pb isotope results of Tilton and Barreiro (1980) indicate that the Charcani gneiss, the basement rock of the southern Peru region, is a likely contaminant. In modelling calculations, contaminant compositions close to average Charcani gneiss give a good overall numerical fit to the analytical data (see James, 1982a, for details and a table of Charcani gneiss compositions used in modelling calculations). It is important to note that mixing trajectories predicted by the AFC model will not in general trend toward the contaminant end-member. Figure 3, for example, shows the computed Sr–Pb mixing trajectories for the Arequipa volcanics. For small degrees of assimilation, the strongly convex downward shape of the mixing lines (produced by a high $(Sr/Pb)_{magma}/(Sr/Pb)_{contaminant}$ ratio) disguises the high $^{87}Sr/^{86}Sr$ of the Charcani gneiss. In this instance, the mixing trajectory defined by the data yields little direct information about the isotopic composition of the contaminant.

The anticorrelation between $^{87}Sr/^{86}Sr$ ratios and Sr concentrations suggests that plagioclase is the dominant cumulate phase. Plagioclase is the major phenocryst phase and the only common phase with a partition coefficient for Sr greater than 1.0.

The radiogenic isotopic and trace element data shown in Figures 4 and 5 have been fit with AFC mixing trajectories. Assumed primary magmatic compositions for the Arequipa and Barroso volcanics are indicated by A and B, respectively. The contaminating end-member composition (Charcani gneiss) is out of the field for all the plots shown. Model fits to the analytical data are somewhat better for the Arequipa volcanics than for the Barroso volcanics, probably a reflection of the more limited sampling area and younger age of the Arequipa series.

Measured $\delta^{18}O$ values for both the Arequipa and Barroso volcanics exhibit

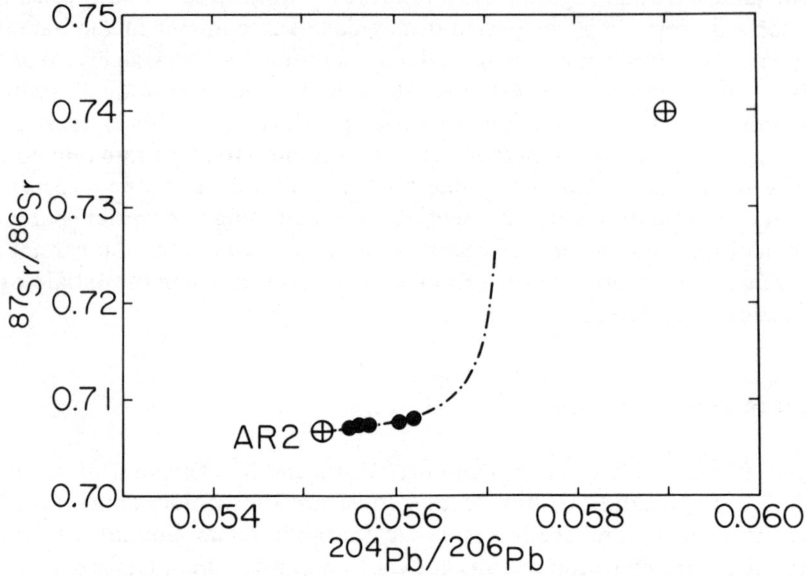

Figure 3 AFC mixing trajectory for $^{87}Sr/^{86}Sr$ versus $^{204}Pb/^{206}Pb$, Arequipa vòlcanics. Circles with crosses denote primary magma (A) and contaminant end-members. Dash-dot line is predicted mixing trajectory. Data shown as solid circles. Model parameters given in James (1982a)

little systematic behaviour relative to the radiogenic isotope ratios. Accordingly, O isotope results are not included in the data plots here. The scatter in $\delta^{18}O$ values is apparently the result of large variations in the O isotope composition of the Charcani gneiss. The lack of coherency in $^{18}O/^{16}O$ ratios effectively precludes the use of O isotopes for quantitative modelling of the south Peru volcanics.

The best interisotope correlation is between Sr and Pb. The strongly convex downward shape of the Sr–Pb isotope mixing trajectory has two important consequences. First, it tightly constrains the $^{87}Sr/^{86}Sr$ ratios for the primary magmas to lie between about 0.7062 and 0.7067 for the Arequipa volcanics and between about 0.7050 and 0.7052 for the Barroso volcanics over the entire range of normally permissible Pb isotope ratios for island arc volcanics. While the flatness of the Pb–Sr isotope mixing curve near the primary magma end-member closely constrains the possible $^{87}Sr/^{86}Sr$ of the parental melt, the Pb isotope ratios of the parent are essentially indeterminant. Secondly, the Sr–Pb mixing trajectory (see Figure 3) shows that for minor degrees of assimilation and perhaps up to 20%, the $^{87}Sr/^{86}Sr$ ratio of the contaminant only weakly controls the shape of the mixing curve. Thus, the comparatively large variations in $^{87}Sr/^{86}Sr$ ratios measured for the Charcani gneiss are not sufficient to produce significant scatter on the Sr–Pb mixing curve.

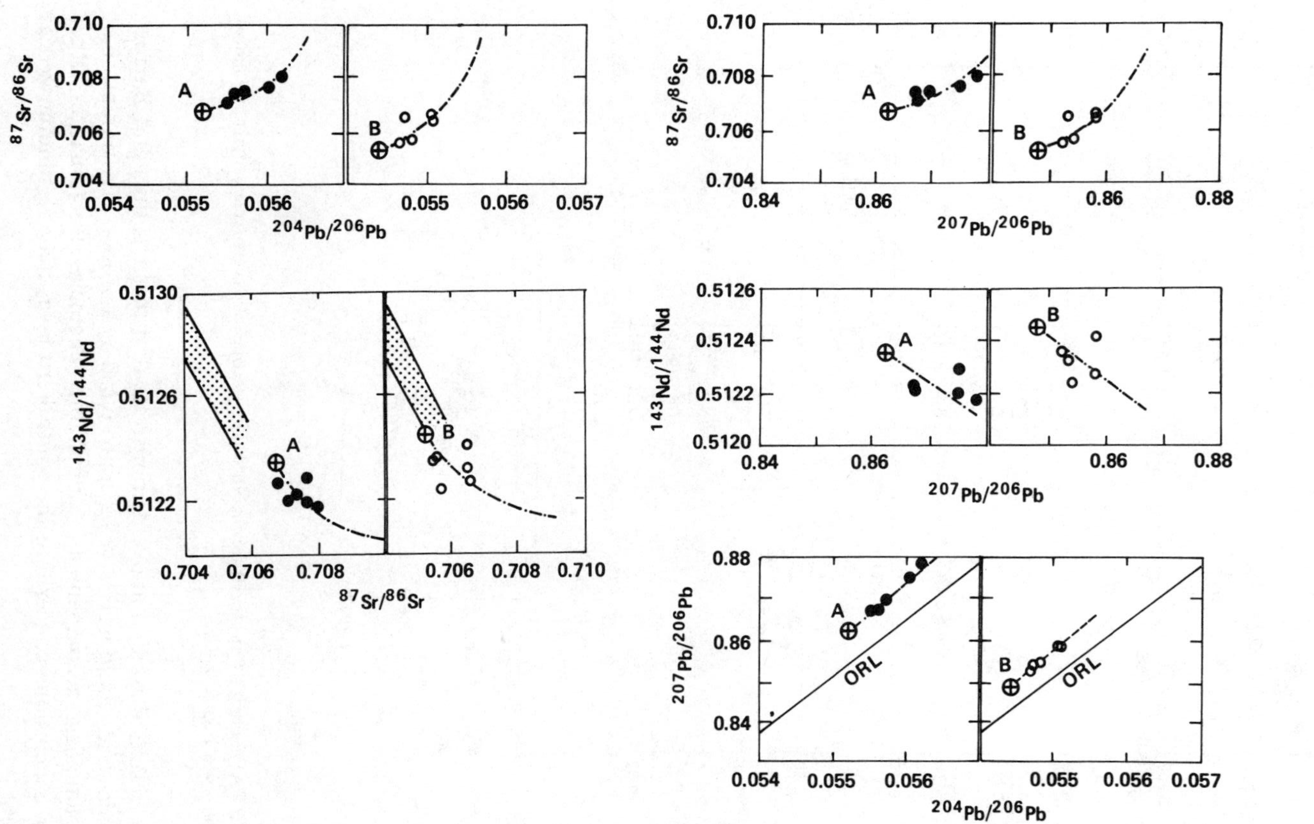

Figure 4 AFC mixing trajectories for Sr, Nd and Pb isotope ratios. Computed model mixing curves shown by solid lines. Primary magma compositions indicated by A for Arequipa and B for Barroso. Contaminant compositions are out of the field of the diagrams shown. Other symbols same as in Figure 2

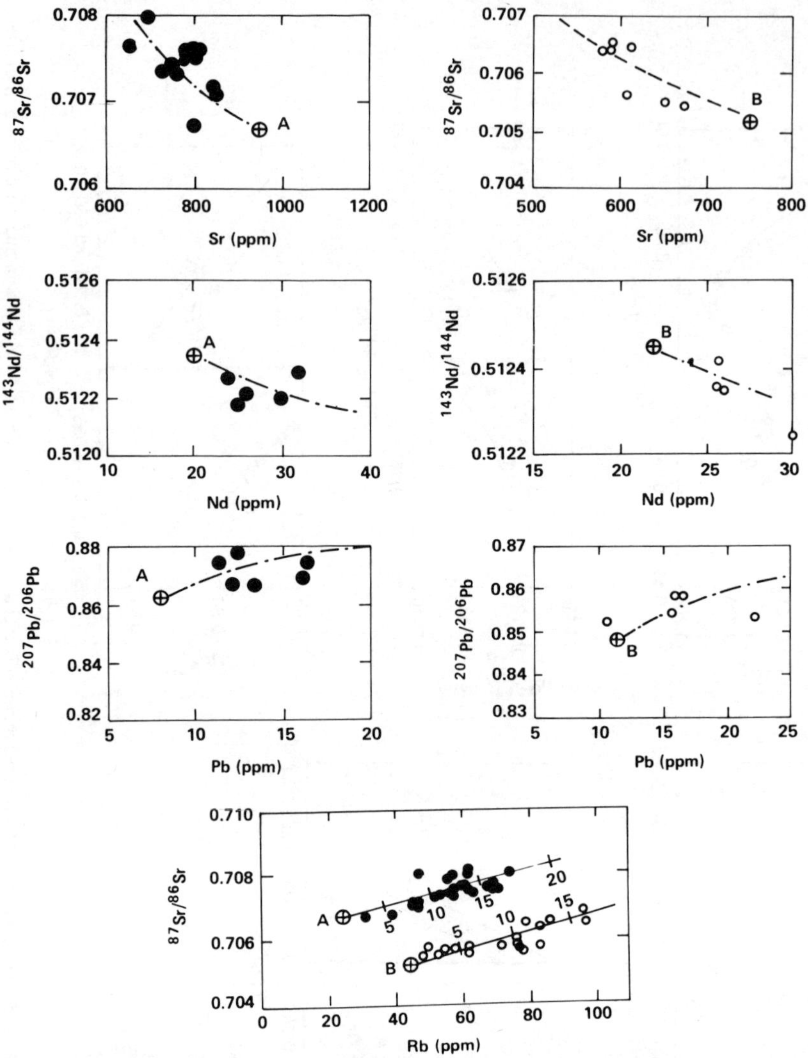

Figure 5 AFC mixing trajectories for Sr, Nd and Pb isotope ratios versus elemental abundances. Symbols as in Figure 4

The results described above demonstrate that the Arequipa and Barroso volcanic series could well have been produced by a process of assimilation of Charcani gneiss accompanied by dominantly plagioclase fractionation. Other major conclusions concerning the south Peru lavas may be summarized as follows:

1. The parental magmas for the Arequipa and Barroso volcanics had $^{87}Sr/^{86}Sr$ ratios of at least 0.7062 and 0.7050, respectively, and

$^{143}Nd/^{144}Nd$ ratios of probably no greater than about 0.51255. These primary magma compositions are well constrained by the observed isotopic correlations and are clearly outside the range of compositions for normal island-arc magmas.

2. The Sr and Nd isotope compositions of the primary magmas and their high Sr concentrations, together with a pronounced interisotopic colinearity for the Arequipa and Barroso series (see James, 1982a) suggests prior contamination of the magmas by Precambrian crustal material similar to Charcani gneiss. That contamination did *not* involve plagioclase fractionation, either because the process took place in the lower Andean crust outside the plagioclase stability field, or because precipitated plagioclase did not separate from the melt.
3. Computed parental magma compositions are characterized by high Sr concentrations which are not reflected in uniformly high concentrations of other incompatible elements. As a result, Sr/Nd ratios are high (34–80) relative to 'normal' mantle or chondritic ratios of about 18 (De Paolo and Johnson, 1979) and Rb/Sr ratios are low, especially for the Arequipa volcanics, relative to oceanic calc-alkaline volcanics (Jakes and White, 1972). Sr/Pb ratios are not high relative to those measured in the Mariana lavas, but they are higher than those measured for many other island arcs (Whitford, 1975).
4. Maximum contamination levels for both the Arequipa and Barroso volcanics are less than 20% (i.e. total weight of assimilated material is less than 20% of the initial weight of the parental magma). The bulk of the volcanic samples involve 5–15% assimilation. At these levels of assimilation, the major element compositions of the lavas are not markedly altered.

The south Peru data thus constitute a well-defined example of pervasive crustal contamination. The contamination is particularly evident because the parental magmas for the Arequipa and Barroso volcanics *already* had anomalous radiogenic isotopic signatures. As shown by the southwestern Columbia examples described in the following sections, even quite significant contamination may produce what appear to be only subtle effects in the radiogenic isotopic ratios. Entire contaminated volcanic series may exhibit isotopic values well within the 'normal' range of 'uncontaminated' island arc lavas.

Northern Andes

Modern andesitic lavas in southwestern Colombia exhibit isotopic compositions that lie well within the range of 'normal' island-arc andesites. On that basis alone they would therefore appear to be largely unaffected by the kind of pervasive crustal contamination so characteristic of andesites in south Peru. Detailed analysis of O, Sr, Nd and Pb isotope ratios reveals,

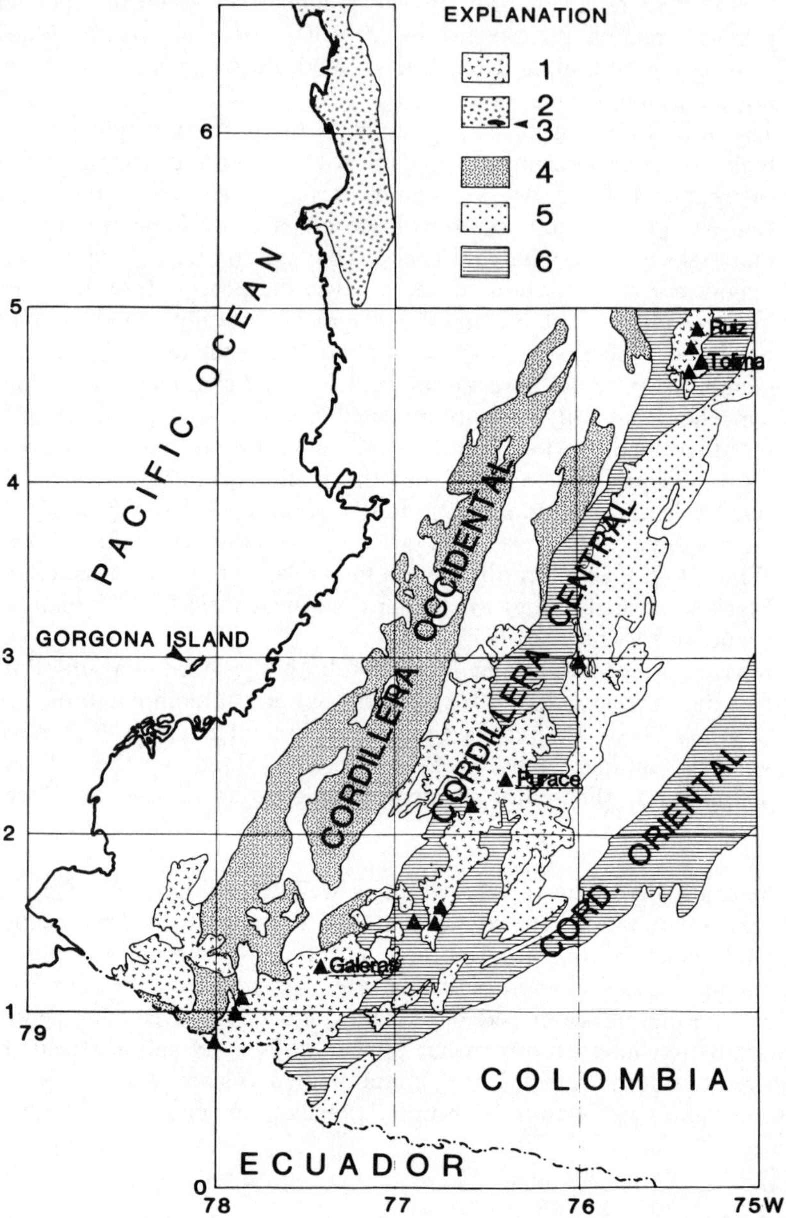

Figure 6 Generalized geologic map of southwestern Colombia showing area of late Cenozoic volcanism. Active volcanoes shown by solid triangles. 1—late Cenozoic calc-alkaline volcanic rocks; 2—Tertiary mafic volcanic rocks; 3—Tertiary ultramafic rocks; 4—Mesozoic/Tertiary mafic rocks (Diabase Group); 5—Mesozoic intrusive rocks; 6—Precambrian/Palaeozoic (?) metamorphic complex

however, that the magmas have experienced a substantial degree of crustal contamination, perhaps nearly as great as that affecting the south Peru volcanics.

The analytical data summarized here pertain entirely to lavas of two volcanic centres: Galeras to the south and Ruiz to the north (Figure 6). These two volcanoes, separated by about 450 km, are representative of the most southerly and most northerly extent of volcanism in the Colombian Andes. Both volcanoes have had a number of historic eruptions and the andesites collected exhibited few signs of weathering. Ruiz (elevation 5400 m), situated near the northern terminus of the active Andean chain, appears to be underlain by thin crust as judged by comparatively small (100 mgal) Bouguer anomalies (Case *et al.*, 1971). Galeras (elevation 4500 m), in the southwest corner of Colombia, is situated on the structural 'knot' formed by the convergence of western, central, and eastern Andean ranges. Bouguer anomalies around Galeras exceed 250 mgal, consistent with a moderate crustal thickness of 40–50 km.

RESULTS

Figure 7 shows $\delta^{18}O$ values for samples from Galeras and Ruiz plotted against $^{87}Sr/^{86}Sr$ and $^{143}Nd/^{144}Nd$ ratios, respectively. Figure 8 shows $^{87}Sr/^{86}Sr$ ratios plotted versus $^{143}Nd/^{144}Nd$ ratios. Calculated mixing trajectories based on the AFC crustal contamination model described below are shown as solid lines.

The plots of $\delta^{18}O$ versus $^{87}Sr/^{86}Sr$ for both Galeras and Ruiz exhibit a good overall isotopic correlation. There is also a significant amount of overlap between the Galeras and Ruiz data, although the latter tend toward distinctly higher overall $\delta^{18}O$ values. For $\delta^{18}O$ versus $^{143}Nd/^{144}Nd$, the Ruiz data set produces a reasonably coherent array, while the Galeras data are somewhat ambiguous. Two Galeras samples have relatively high $\delta^{18}O$ values that plot well above the O–Nd isotope trend defined by the remaining samples. Both high-^{18}O samples exhibit some slight petrographic evidence for groundmass alteration, suggesting that the high $\delta^{18}O$ values may be the result of secondary alteration produced by low temperature interaction with groundwater. Because the high-^{18}O samples plot well to the right of both the Ruiz and remaining Galeras data points (Figure 8), it is also possible that the relatively high $^{87}Sr/^{86}Sr$ ratios and $\delta^{18}O$ values relative to $^{143}Nd/^{144}Nd$ could be produced by the introduction during petrogenesis of a large component of altered oceanic crust from the slab. It is unlikely, in any event, that the high-^{18}O, high-$^{87}Sr/^{86}Sr$ isotope signature is the result of normal crustal contamination.

Pb isotope data are shown in Figure 9, but are not included in the model calculations because, as of this writing, Pb concentrations had not been determined. Again, however, as for Sr and Nd, Pb isotope compositions are virtually identical for Galeras and Ruiz. There is one anomalously radiogenic sample from Galeras ($^{206}Pb/^{204}Pb = 20.0$), but both Sr and Nd isotope ratios in

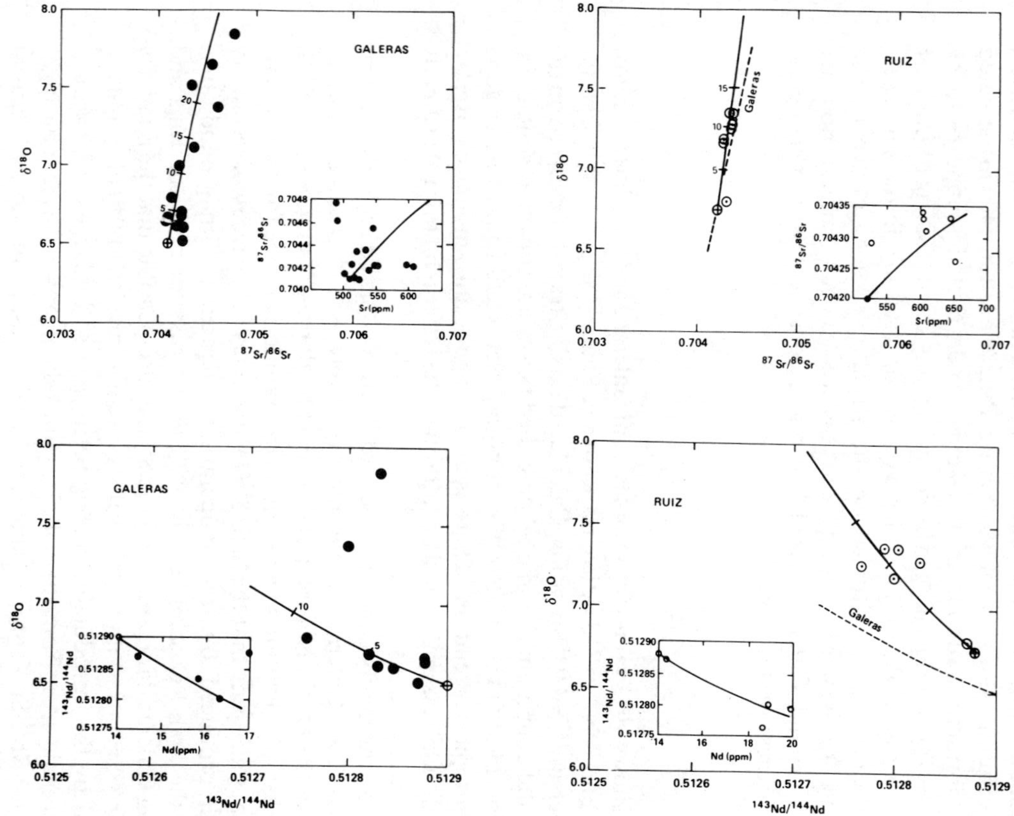

Figure 7 $\delta^{18}O$ versus $^{87}Sr/^{86}Sr$ and $^{143}Nd/^{144}Nd$ for andesitic samples from Galeras and Ruiz. Solid circles—Galeras; open circles—Ruiz. Circles with crosses are assumed primary magma compositions. Computed mixing trajectories shown by solid lines. Tic marks denote weight % assimilated material. Isotopic ratios versus elemental abundances shown on inset plots

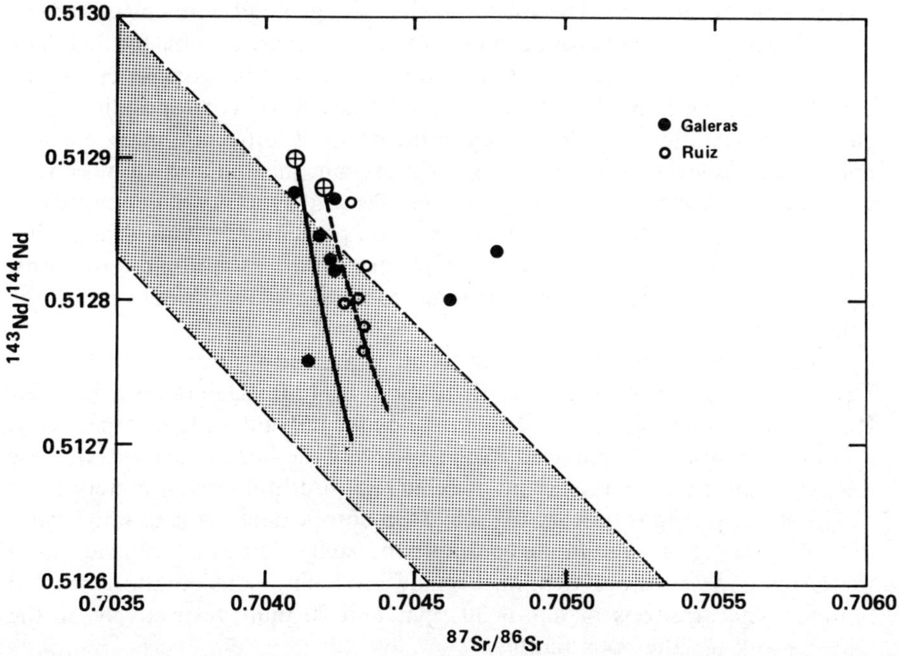

Figure 8 Sr–Nd isotope plot for northern Andean volcanics. Symbols as in Figure 7. Computed mixing trajectories denoted by solid line for Galeras, dashed line for Ruiz. Shaded area shows mantle array

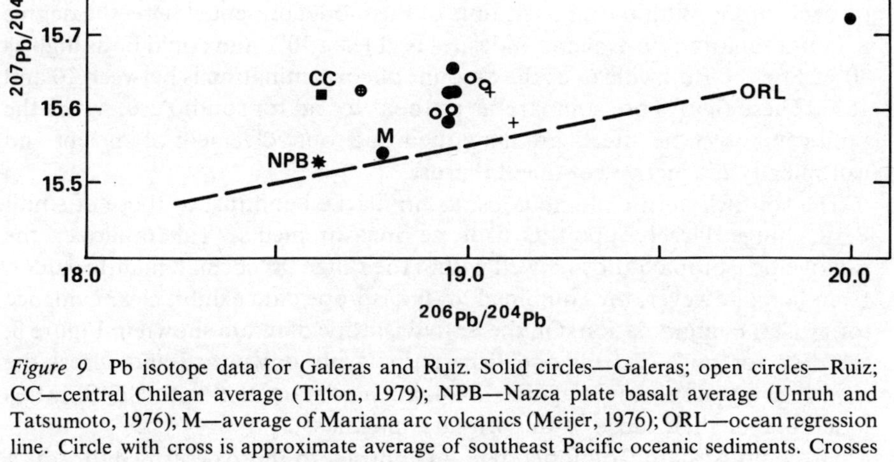

Figure 9 Pb isotope data for Galeras and Ruiz. Solid circles—Galeras; open circles—Ruiz; CC—central Chilean average (Tilton, 1979); NPB—Nazca plate basalt average (Unruh and Tatsumoto, 1976); M—average of Mariana arc volcanics (Meijer, 1976); ORL—ocean regression line. Circle with cross is approximate average of southeast Pacific oceanic sediments. Crosses indicate country rocks near Galeras and Ruiz

this sample are normal. The source of the radiogenic Pb remains unknown.

The trace element data for both Galeras and Ruiz are notable for their lack of coherence with the isotopic data. Given the reasonably good interisotopic correlations exhibited by the lavas, the absence of corresponding trace element correlations could be an indication of differing trace element concentration levels in the primary magmas, but this remains an unknown.

Unlike south Peru, none of the samples of country rock we have analysed to date from southwestern Colombia has a composition that would make it a possible contaminant. On the basis of the observed isotopic variations, however, we can place some general constraints on the nature of the contaminant.

For a contaminant with an assumed but plausible $\delta^{18}O$ value of 10–11‰, $^{87}Sr/^{86}Sr$ ratios cannot be greater than about 0.712 for Galeras and 0.710 for Ruiz. Corresponding $^{143}Nd/^{144}Nd$ ratios are 0.5120 and 0.5122, respectively. As might be expected from the tectonic history of the Cordillera Central, these isotopic values are characteristic of relatively youthful crustal material.

The steep mixing trajectory for the O–Sr isotope data requires a high ratio of Sr in the magma to Sr in the contaminant, suggesting Sr concentrations in the latter of less than 100 ppm. Similar modelling considerations for Nd require concentrations of about 30 ppm and 20 ppm, respectively, in the Galeras and Ruiz contaminants.

Based upon phenocryst populations in the volcanics, pyroxene (both cpx and opx) and olivine are taken to be the dominant cumulate phases, but significant amounts of plagioclase may also be involved, particularly for the later differentiates. Values assumed in the model calculations are $K_{Sr} = 0.4$, $K_{Nd} = 0.2$, and $A - B = 0.5‰$. R is assumed arbitrarily to be a value appropriate for shallow to moderate depth (3–10 km) in an active volcanic terrain. For Galeras, R = 2.0 and Ruiz, R = 3.0.

Magmatic evolution curves based upon the above assumptions and parameters are shown as solid lines in Figures 7 and 8. The percentage of assimilated material relative to the starting weight of the magma is indicated on each curve. Within the constraints of the model presented here, the degree of contamination for Galeras andesites is at least 10% and could be as high as 20%. For the Ruiz volcanics the amount of contamination is between 10 and 15%. These figures are comparable to those found for south Peru, where the radiogenic isotopic effects are magnified due to involvement of ancient and isotopically distinctive continental crust.

The southwestern Colombia results, in marked contrast to those of south Peru, hinge largely upon O isotope measurements. Taken alone, the radiogenic isotopic ratios lie well within the range of 'normal' mantle values. Even here, however, the combined Sr–Nd isotope data exhibit clear evidence for crustal contamination. On the Sr–Nd isotope diagram shown in Figure 8, the data for both Galeras and Ruiz cut sharply downward and across the mantle array along a mixing line involving a contaminant with Nd/Sr ratios much higher than those of the primary magmas.

The southwestern Colombia data, in contrast to those described for south

Peru, could reflect a single episode of crustal contamination. If this possibility is made to be an assumption, the *maximum* degree of crustal involvement at the magma source (i.e. source contamination) can be estimated. Two factors are important here: First, the $\delta^{18}O$ data cluster heavily about 6.6‰ for Galeras and, somewhat less obviously, about 7.1‰ for Ruiz. Second, the primary magmas for both the Galeras and Ruiz series have Sr–Nd isotope ratios that plot to the right of the mantle array. Taken together these two factors suggest that the primary magma already contained a significant crustal component, possibly derived from the descending slab. If the $\delta^{18}O$ value of that crustal component is taken to be 11‰ (see James, 1981a) and normal mantle is considered to be 5.8‰ (see Harmon and Hoefs, this volume) then a $\delta^{18}O$ value of 6.5‰ for the primary magma requires a crustal contribution of about 10–15% by weight. Similarly, for a $^{87}Sr/^{86}Sr$ ratio of 0.705 in the descending slab and 0.703 in the overlying mantle wedge, and given $Sr_{slab}/Sr_{mantle} = 10$, then a $^{87}Sr/^{86}Sr$ ratio of 0.7041 for the primary magma also requires a 10–15% crustal contribution. While such unconstrained calculations plainly do not yield unique results, the estimates do account satisfactorily for all of the O, Sr, Nd and Pb isotope characteristics of the primary magmas, including the offset to higher $^{87}Sr/^{86}Sr$ ratios on the Sr–Nd diagram.

The isotopic signatures of the southwestern Colombia volcanics may thus reflect both source contamination and crustal contamination. The source contamination produced primary magmas with $\delta^{18}O$ values close to 6.5‰ and with Sr–Nd isotope compositions shifted to the right of the mantle array. Subsequent crustal contamination involved crustal material rich in ^{18}O but not very radiogenic, possibly indicating comparatively young (post-Palaeozoic?) crust. The crustal contamination produced magma mixing trajectories trending steeply across the field of the mantle array. The marked contrast in correlated isotopic variations produced by source contamination as compared to crustal contamination allows the overprint of the latter to be distinguished from the effects of crustal involvement in subduction zone magmagenesis.

Discussion and conclusions

Andesitic volcanics in both the CVZ and NVZ exhibit evidence of pervasive crustal contamination. The degree of assimilation required to produce the *local* geochemical variation in the volcanics of southern Peru is about the same as that required to produce the observed range of isotopic variations in the southwestern Colombia volcanics. In both instances, calculated weights of assimilated material constitute less than 20% of the magma.

The geochemical character of the assimilated rock differs profoundly between northern and central Andes. In south Peru the crustal contaminant is ancient Precambrian gneiss with appropriately distinctive radiogenic isotope characteristics. In southwestern Colombia the contaminant is representative of much younger continental crust, probably metasedimentary in nature, exhibiting only slightly evolved radiogenic isotopic ratios, but characterized by marked ^{18}O enrichment.

The contrasting nature of the isotopic evidence for crustal contamination in the two regions illustrates the importance of utilizing as many isotopic systems as possible. In south Peru, Pb isotopes provided conclusive proof of crustal contamination and lead to a clear identification of the contaminating material. By contrast, O isotopes, while indicative of contamination, gave no clue as to the degree of assimilation or the nature of the contaminant. In southwestern Colombia, Pb isotopes are singularly uninstructive, and it is the variation in O isotopes relative to Sr and Nd which provides the essential data for quantitative modelling.

The probability that the primary magmas of the south Peru volcanics experienced a crustal contamination prior to evolving into the Arequipa and Barroso series effectively precludes any evaluation of the amount of crustal component that may have entered the melt in the subduction zone. If the southwestern Colombia volcanics experienced only one episode of crustal contamination, an *upper* bound of 15% can be placed on the contribution of crustal material from the slab.

For both southwestern Colombia and south Peru volcanics, magmatic evolution appears to occur at shallow to moderate depths, as judged by a relatively high ratio of cumulate to assimilated material (~3). It may be, therefore, that volcano elevation more than crustal thickness may be a critical factor leading to the sort of magmatic evolution observed. While elevations are not markedly different between northern and central Andes, crustal thicknesses in the central Andes are 10–30 km greater than in the northern Andes. The greater crustal thickness in the central Andes may account for the earlier episodes of contamination.

Correlated Sr Isotope and Geochemical Variations in Basalts and Basaltic Andesites from Guatemala

N. K. GRANT, W. I. ROSE, Jr and L. A. FULTZ

Abstract

The incompatible elements in the high alumina basalts and basaltic andesites of six composite volcanoes in Guatemala are most abundant in the volcanoes of San Pedro, Atitlán and Tolimán which lie in the 84 000 year old cauldron of Lake Atitlán. Modelling of the major and trace elements, using the high alumina basalts of Fuego as the parental magma, indicates that the five other volcanoes contain excess incompatible elements but are deficient in Na. We infer the excesses to mean that the magmas have evolved through cycles of magma replenishment and fractional crystallization prior to their delivery to the individual volcanoes, probably in magma ponds at the base of the crust (Carr, 1983). The Na deficiency may reflect either the crystallization of sodic plagioclase as would be expected at high pressures, or the diffusional loss of Na as the magmas rose through continental crust.

The average $^{87}Sr/^{86}Sr$ ratios, which range from 0.70385 to 0.70405, correlate positively with average incompatible element abundances and Rb/Sr and incompatible element ratios, and negatively with Sr abundances for each volcano. Because the isotope ratios also correlate with the incompatible element excesses, it would appear the process adding ^{87}Sr to the magmas was linked to the process leading to incompatible element enrichments. We conclude that the explanation lies in the size and longevity of the magma ponds at the base of the crust, where longevity controls the length of time through which the assimilative addition of ^{87}Sr, and incompatible element enrichment through magma replenishment-fractionation cycles, have taken place. The data suggest that the chemical composition of the lower crust lies within the compositional spectrum of the erupted lavas and consists of gabbroic cumulates.

The chemical evolution of the magma ponds in or at the base of the crust may change abruptly after the volcanoes supplied by the ponds become dormant, because the magmas (previously buffered to uniform compositions by magma replenishment triggered by the removal of eruptive magma batches) would be free to evolve through fractionation and assimilation to silicic compositions. Silicic magmas may also have been produced by the partial melting of the lower crust, where the magmas ponded at the base have contributed the thermal energies for melting.

Introduction

The isotope systematics of the convergent margin magmas of Central America have received less attention than the systematics of the same magmas in the Andes. Although the pre-Pleistocene record in Central America is not known in detail and plutonic igneous rocks are uncommon, the Pleistocene volcanoes and volcanic centres are known in great detail (see Carr *et al.,* 1982 for a summary of recent work). Unlike the Andes where the isotopic variations of O, Sr, Nd and Pb have been studied, in Central America the published work on basalts and basaltic andesites has been primarily restricted to accounts of the isotopic variations of Sr only. This work ranges from the early studies of $^{87}Sr/^{86}Sr$ ratios by Pushkar (1968) to the more recent and more precise ratios reported by Whitford and Bloomfield (1976), Moorbath *et al.* (1978), Thorpe *et al.* (1979b), Rose *et al.* (1977) and a single combined Nd-Sr isotope study by Langmuir *et al.* (1981) which has not been published in full. These results show that the range of $^{87}Sr/^{86}Sr$ ratios of basalts and andesites is small, and for the most part restricted to 0.7035–0.7040. A notable feature of this work is that, unlike the Andes, correlated isotopic and geochemical variations have not been reported, with the exception of ten andesite flows from the immediate vicinity of the volcano Nevado de Toluca in Mexico for which the $^{87}Sr/^{86}Sr$ ratios correlate with Rb/Sr (Whitford and Bloomfield, 1976).

In this paper we show that such correlations can be recognized elsewhere in Central America. We base our work on the least evolved lavas in a given volcanic centre, and present our data in the form of averages for each volcano. In contrast to the earlier isotopic surveys made by most other workers, our stress on mafic lavas minimizes the decoupling of the isotopic and geochemical characteristics of the lavas caused by low pressure fractionation (dominated by plagioclase crystallization). Furthermore, we emphasize only the isotopic and chemical differences between volcanoes rather than the differences between lavas of a single volcano. These differences in approach allow us to better constrain the processes which control the high pressure evolution of the magmas before they were supplied to each volcano.

The importance of high pressure processes in controlling the chemical composition of Central American lavas has been recently stressed by Carr (1983), who has showed that magma and crustal densities together with volcanic edifice heights imply that the volcanoes have been supplied by magma bodies ponded at the base of the crust. Carr has also shown that the average abundance of several incompatible elements at a fixed silica concentration correlates positively with the volume of the volcanic edifices. Because Carr concluded that the edifice volumes are controlled by the sizes of the magma ponds at the base of the crust, this suggests that the size of the ponds also controls the degree of incompatible element enrichment.

The present study

Our work is based on lavas from six Recent composite volcanoes in Guatemala, Santa Mariá, San Pedro, Tolimán, Atitlán, Fuego and Pacaya, which occur over a 115 km section of the Central American arc (Figure 1), and have erupted a range of magma compositions from high alumina basalt to andesite. Of these volcanoes Fuego and Pacaya are currently active, although Santa Mariá is linked to the satellite dome of Santiaguito which has discontinuously erupted dacite since 1922 (Rose, 1972). The volcanoes San Pedro, Tolimán and Atitlán lie in the Atitlán cauldron which erupted the Los Chocoyos ash flow tuff, a large (> 100 km^3) silicic (>72% SiO_2) unit, 84 000 years ago. While the association of composite volcanoes with older silicic cauldrons is not uncommon in Central America, the Atitlán cauldron is unique in terms of the volume of the silicic magmas erupted from it, some 300 km^3 of dense rock in the last 300 000 years (Rose *et al.*, in prep.). The Atitlán cauldron has been the focus of repeated silicic magmatism over the past 11 Ma (Newhall, 1981).

In one respect our approach differs from recent studies of Andean volcanism in which the isotopic composition of two or more elements (O, Sr,

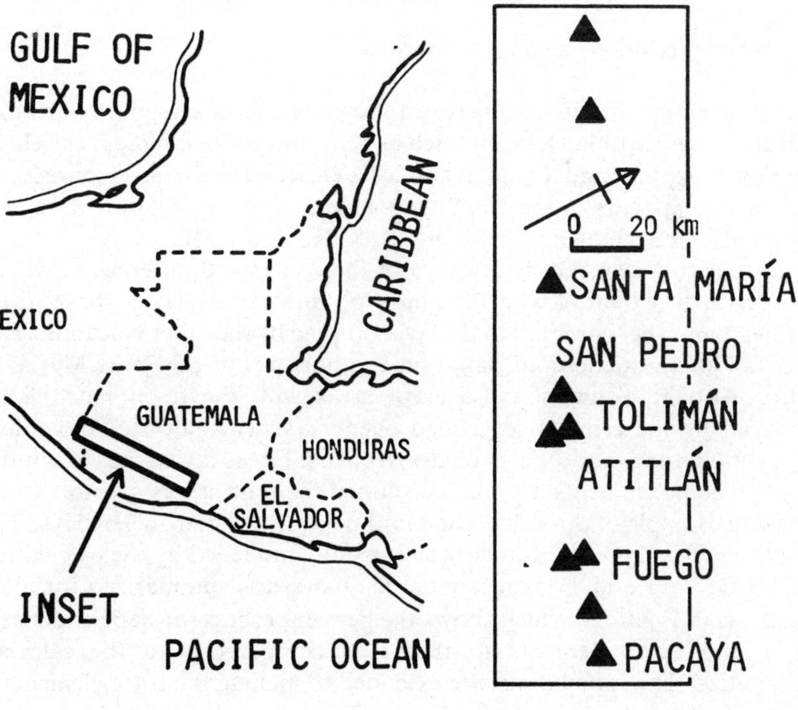

Figure 1 Location of the volcanoes discussed in this paper. Modified from Rose *et al.* (1977)

Nd, Pb) have been combined with a small number of trace elements to constrain the petrogenesis of the lavas (e.g. James, 1982a). We discuss the isotopic composition of Sr only, but complement it with the major and trace element geochemistry of the lavas as well as microprobe measurements of mineral compositions. Because the major element chemistry of the lavas places constraints on their chemical evolution, it is likely that the petrogenesis of the lavas will not be completely understood until all aspects of their chemical composition are simultaneously considered.

The geochemical and isotope data we discuss has been partly reported elsewhere. The major element data, the Sr and Rb abundances and $^{87}Sr/^{86}Sr$ ratios for Santa Mariá are taken from Rose *et al.* (1977), and the same data for the remaining volcanoes come from Fultz (1979). With the exception of the isotopic data for Santa Mariá, the $^{87}Sr/^{86}Sr$ ratios were measured at Miami University. All the ratios are normalized to a value of 0.70800 for the Eimer and Amend interlaboratory standard. The microprobe analyses of the Fuego phenocrysts are reported in Rose *et al.* (1978). The Ba, Zr and Y abundances are new. A total of 24 lavas have been analysed, with between two (Pacaya) and five samples (Atitilán) of the least evolved lavas from each volcano. The average SiO_2 abundances of the analysed lavas are as follows: Pacaya 49.6%, Fuego 50.1%, Santa Mariá 52.2%, Atitlán 52.3%, Tolimán 53.8% and San Pedro 54.1%.

Geochemical relations among the volcanoes

The geochemical differences between the volcanoes are regularly arranged, with the incompatible elements such as K_2O and Rb being most enriched in San Pedro, Atitlán and Tolimán and least enriched in Fuego (Figure 2). The pattern also appears in the Rb/Sr ratios, and more importantly in ratios of incompatibles such as Zr/Y (Figure 2), K/Rb and K/Ba.

In order to assess the significance of these chemical differences, we have considered the extent to which the most primitive of the lavas, those erupted by Fuego, may be 'parental' to the lavas erupted by the other volcanoes. Least squares mixing calculations based on comparisons of the Santa Mariá, San Pedro, Atitlán, Tolimán and Pacaya lavas with the fractionated liquids produced by the removal of Fuego phenocrysts (plagioclase An_{92}, olivine Fo_{72}, clinopyroxene and Fe-Ti oxides) from the Fuego lavas leads to solutions in which some elements are well accounted for, but others are not. Table 1 shows an example of this calculation for Fuego and Atitlán, where SiO_2, TiO_2, Al_2O_3, FeO*, MgO and CaO are successfully modelled by the calculations, but Na_2O, K_2O and P_2O_5 are not. This pattern is summarized for all the volcanoes in Figure 3, which shows the per cent excess (or deficiency) of the major and trace elements in the lavas compared with the calculated abundances. The calculations are extended to include the trace elements Ba, Rb and Zr through the use of the disequilibrium model of crystallization. This model, chosen because it leads to the greatest enrichment in incompatible elements and not because we think it provides the best description of the fractionation processes modifying the magmas, was constrained by the weight fractions of the fractionated mineral phases derived from the major element calculations (e.g. Table 1), published partition coefficients and the abundances of Ba, Rb and Zr in the Fuego lavas. Both sets of calculations, for the major and the trace elements, assume *closed system* fractionation of the

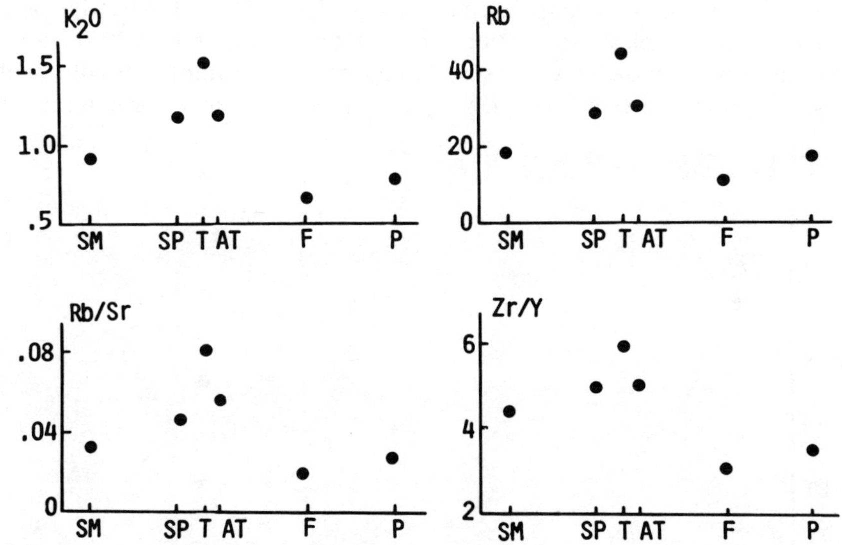

Figure 2. Variations of the average K_2O and Rb abundances and Rb/Sr and Zr/Y ratios in the least evolved lavas sampled along WSW–ESE traverse parallel to the volcanic arc. The volcanoes are labelled in the same order as shown in Figure 1

Table 1 **Fractionation of Fuego to Atitlán lavas**

	Fuego	Atitlán	Predicted composition	Fractionated phases wt. %	
SiO_2	50.31	53.06	53.05	Olivine	4.94
TiO_2	0.94	0.92	0.92	Clinopyroxene	7.94
Al_2O_3	19.11	20.58	20.58	Plagioclase	9.91
FeO*	10.19	8.14	8.14	Ilmenite	0.38
MgO	5.68	3.44	3.45	Magnetite	2.19
CaO	9.72	8.50	8.51	Total	25.36%
Na_2O	3.27	3.95	4.12		
K_2O	0.66	1.21	0.88	Sum of squares = 0.08	
P_2O_5	0.13	0.20	0.17		

Fuego magmas, and that all the volcanoes have been supplied by magmas derived from high alumina basalts of Fuego-type. The extent to which the calculations do not lead to successful solutions is an indication of the extent to which these assumptions break down. Na_2O is deficient in all the volcanoes relative to Fuego and all the incompatibles (K_2O, P_2O_5, Rb, Ba and Zr) are present in excess (Figure 3). If we justify the assumption that all the volcanoes have been supplied by magmas derived from high alumina basalts of Fuego-type on the grounds that the least squares mixing calculations give satisfactory solutions for most of the major elements (SiO_2, TiO_2, Al_2O_3, FeO*, MgO and CaO), then the simplest explanation for the poor solutions is that the magmas have not evolved under closed-system conditions. For example, the pattern of excess incompatible elements can be qualitatively accounted for in terms of

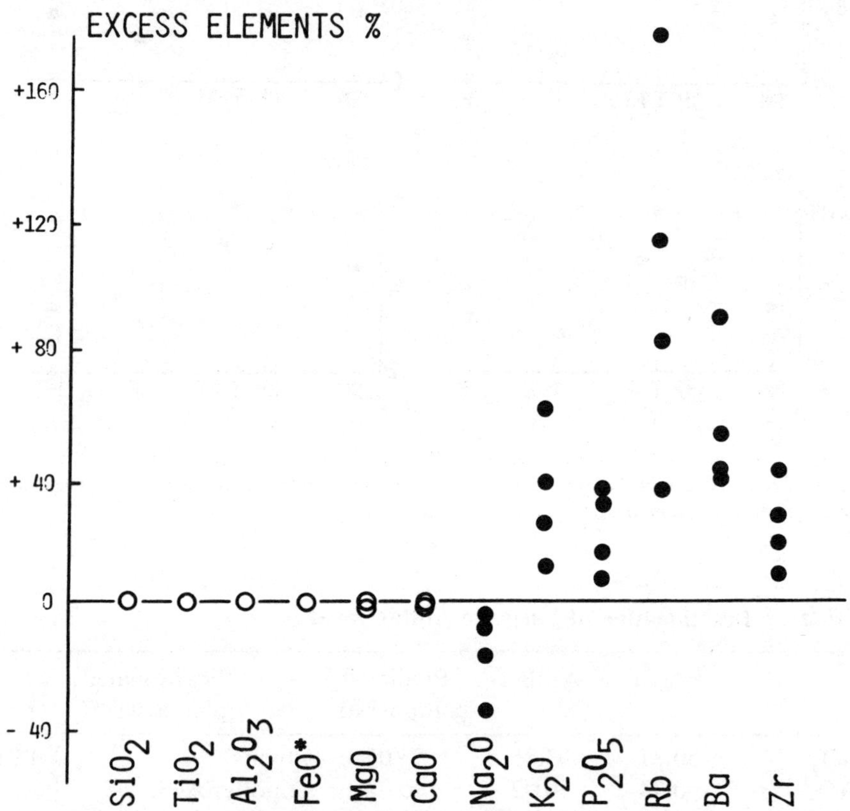

Figure 3 A summary of the least squares mixing calculations showing the excess (or deficiency) of the major and trace elements in the lavas compared with the calculated concentrations. In all cases the Fuego high alumina basalts were taken as the parental magma and the compositions of the fractionated phases was based on microprobe analyses of Fuego phenocrysts. An example of the major element calculations is shown in Table 1. The calculated trace element abundances used the disequilibrium model of fractional crystallization, in order to place maximum limits on the calculated abundances

the evolution of the magmas in systems where cycles of magma replenishment, crystal fractionation and eruption have taken place (O'Hara, 1977; O'Hara and Mathews, 1981). Under these conditions the replenishment-fractionation cycles can lead to the buffering of the major element chemistry to some composition which varies only within narrow limits, while the incompatible elements increase continuously in abundance. The magnitude of the increase depends on a number of factors, but includes the number of replenishment-fractionation cycles. We therefore conclude that the pattern of excess incompatible elements reflects replenishment-fractionation cycles, possibly operating in the magma ponds at the base of the crust (Carr, 1983), and which have modified the magmas supplied to each volcano.

A different explanation must be invoked to account for the Na_2O deficiency in the lavas. At its simplest this must involve a Na sink which removed Na from the magmas faster than occurred through the crystal fractionation of the calcic plagioclase in the Fuego magmas. We cannot confidently constrain the nature of the Na sink, but it might be a more sodic plagioclase than present in the Fuego lavas, as would be expected if the open-system fractionation took place at high pressure at the base of the crust (Green, 1969), or it might be the loss of Na through wall rock reaction during the rise of basaltic magmas through continental crust (Watson and Jurewicz, 1983).

Correlated geochemical and Sr isotope variations

The average $^{87}Sr/^{86}Sr$ ratios for the volcanoes range from 0.70385 to 0.70405, being lowest for Fuego and highest for Tolimán. These isotopic variations are linked to the geochemistry of the lavas in a distinctive way in that they correlate (Figure 4) positively with K_2O, Rb, Ba and Zr, and negatively with Sr (excepting Fuego which has an anomalously low Sr abundance compared with the other volcanoes). These correlations are similar to those reported by James (1982a) in the Barroso and Arequipa volcanics of Peru. In contrast, the $^{87}Sr/^{86}Sr$ ratios do not correlate with the major element variations, for example MgO (Figure 4) or FeO*/MgO, which are often taken as measures of magmatic fractionation. The $^{87}Sr/^{86}Sr$ ratios also correlate with Rb/Sr and Zr/Y ratios (Figure 5). These links between the isotopic and geochemical variations bear on the interpretations which have been proposed for correlated isotopic and geochemical variations in lavas. Two component mixing, between magmas or between a magma and a contaminant, can be ruled out because not all elements correlate with the $^{87}Sr/^{86}Sr$ ratios (i.e. MgO, Figure 4). The view that the correlations between $^{87}Sr/^{86}Sr$ and Rb/Sr ratios are 'mantle' or 'pseudo-isochrons' (Brooks *et al.,* 1976) appears unlikely because the isotopic variations also correlate with Zr/Y (Figure 5) and other incompatible element ratios which are not linked to time-dependent changes to the $^{87}Sr/^{86}Sr$ ratios. Correlations between $^{87}Sr/^{86}Sr$ and Rb/Sr ratios would appear to be more realistically interpreted in terms of assimilative processes accompanying fractional crystallization as De Paolo (1981) has discussed.

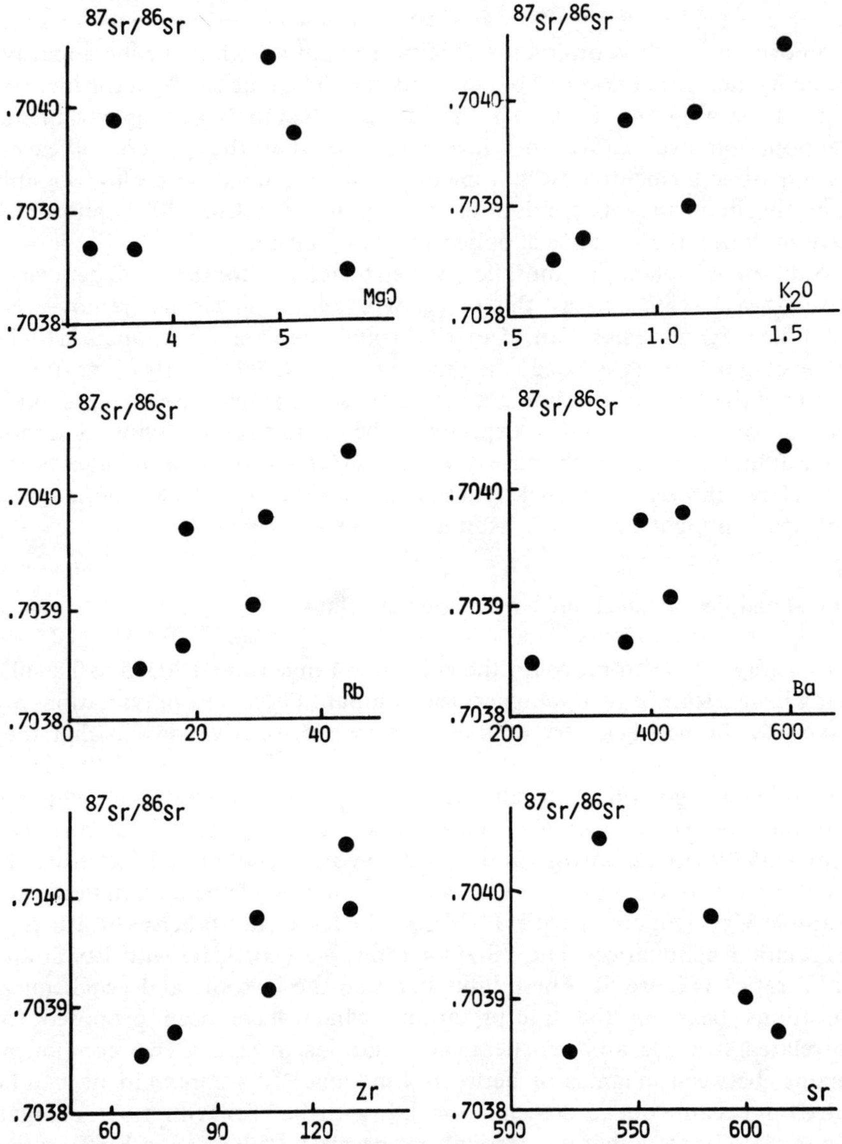

Figure 4. The relations between average $^{87}Sr/^{86}Sr$ ratios and average MgO, K_2O, Rb, Ba, Zr and Sr abundances for the six volcanoes. Note that correlations exist for all elements except MgO, and that Fuego is Sr deficient compared to the other five volcanoes

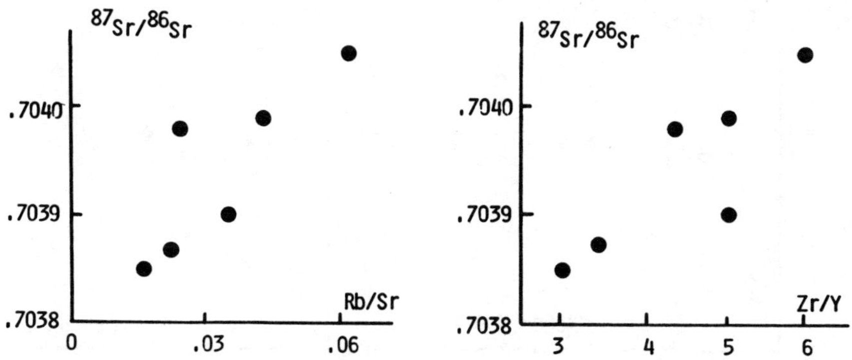

Figure 5 The positive correlations between the average $^{87}Sr/^{86}Sr$ versus Rb/Sr and Zr/Y ratios for the six volcanoes

Discussion

We have shown two features of the geochemical and isotopic variations between the six volcanoes. One is that the magmas evolved in open-system magma chambers subjected to repeated cycles of magma replenishment and crystal fractionation before they were delivered to each volcano. We presume this open-system evolution took place at high pressure in the magma ponds inferred by Carr (1983) to exist at or near the base of the crust. The second is that the $^{87}Sr/^{86}Sr$ ratios are linked to the chemical variations between the volcanoes, notably in the positive correlations with the incompatible element abundances and ratios (Figures 4 and 5). These correlations imply that mixing or assimilative processes (but not simple two-component mixing) have played a part in the high pressure evolution of the magmas.

We now show that these two features are connected, in that the $^{87}Sr/^{86}Sr$ ratios correlate with the per cent excesses of the incompatible elements (Figure 6). This means that the magnitude of the assimilative processes adding ^{87}Sr to the magmas correlates with the degree of incompatible element enrichment in the magmas. A simple explanation for this connection relates both features to the longevity of the magma ponds in which these magmas evolved prior to their delivery to the individual volcanoes. Longevity would control the importance of the assimilative processes adding ^{87}Sr to the magmas, and also the number of magma replenishment-crystal fractionation cycles producing the incompatible element enrichments. This explanation leads to three further considerations. The explanation is consistent with both the correlations of the $^{87}Sr/^{86}Sr$ ratios with the incompatible elements, and the absence of such correlations with the major elements. This is so because, while the behaviour of the incompatible elements is insensitive to the composition of the bulk extracts fractionated from the magmas, the opposite is true for the major elements. Thus it is possible for the magmas supplied to the individual volcanoes to have been previously buffered to different compositions by open-

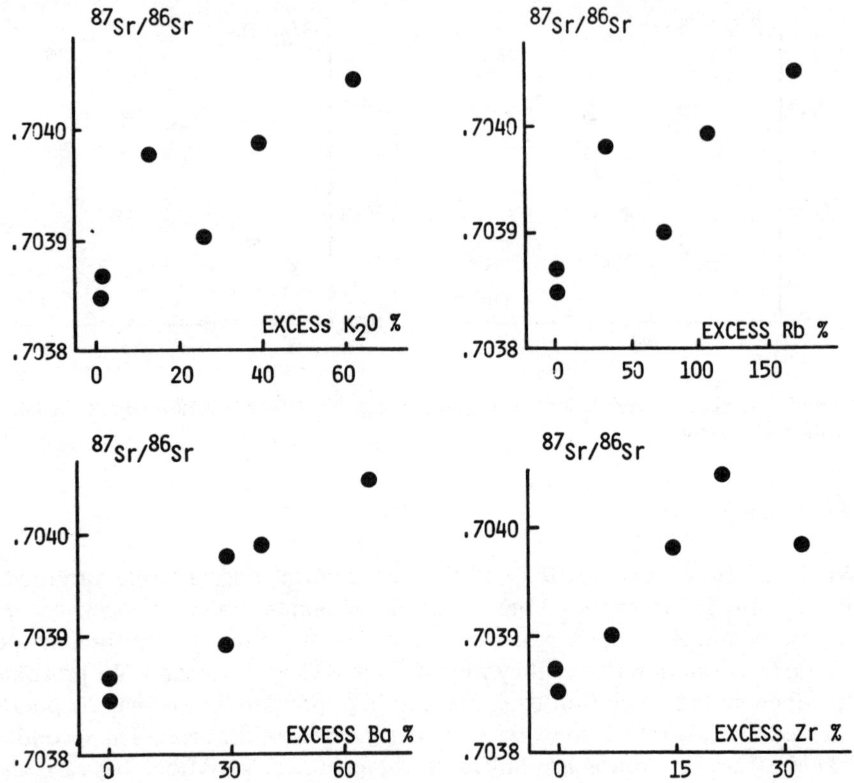

Figure 6 The positive correlations between average $^{87}Sr/^{86}Sr$ ratios and the excess per cent K_2O, Rb, Ba and Zr in the six volcanoes. The per cent element excess are also shown in Figure 3

system fractionation in such a way that the assimilative addition of ^{87}Sr was not correlated with the major element variations of the magmas, but was correlated with the enrichment of the incompatible elements.

The second point is that if we are correct in suggesting that the longevity of the magma ponds supplying the volcanoes is an important control on the extents to which the assimilative addition of ^{87}Sr and the degree of incompatible element enrichment, then it would appear reasonable to think that longevity is a function of the size of the magma pond. In Figure 7 we show that the incompatible element excesses (e.g. excess Rb %) are focussed on the San Pedro, Atitlán and Tolimán volcanoes, and that these volcanoes have the greatest edifice volumes of the volcanoes discussed here. Thus edifice volume is related to the maximum enrichments in the incompatible elements and ^{87}Sr. We think that the linkage between all of these features is the size and longevity of the magma ponds supplying each volcano, ponds which Carr (1983) has concluded are located at the base of the crust.

The third point is that the successful solutions to the least squares mixing calculations for most major elements imply that the chemical evolution of the

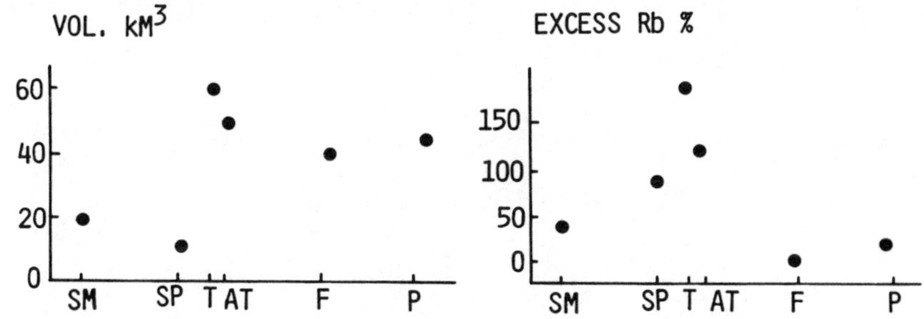

Figure 7 The edifice volumes of the six volcanoes (from Carr, 1983), and the excess per cent Rb in WNW–ESE traverse parallel to the volcanic arc. Note that the greatest edifice volume and excess Rb occurs in Tolimán

magmas has not included the addition of components with chemical compositions which are distinct from the high alumina basalt-basaltic andesite spectrum, or the linear extrapolation of the spectrum. This indicates that the assimilant, which we infer was responsible for the addition of ^{87}Sr to the magmas, must be restricted to the same compositional range, and suggests that the lower crust consists of the crystallization products of high alumina basalt to basaltic andesite, or their fractionated derivatives.

Finally, we note that the open-system processes are focussed on the San Pedro, Atitlán and Tolimán volcanoes which lie inside the Atitlán cauldron, suggesting that there may be a connection between the episodic production of large volumes of silicic magmas and the large open-system magma ponds at or in the base of the crust. The nature of this connection is not clear, but we speculate that it might involve the production of silicic magmas after the volcanoes supplied by the magma ponds became dormant on reaching the critical cone heights above which magma buoyancy was unable to continue eruptions from the summit vents (Rose *et al.,* 1977). With the onset of dormancy, eruption-triggered magma replenishment would cease and the magma ponds, with earlier histories of open-system fractionation and assimilation which buffered them about some constant composition, would be free to evolve through fractionation and assimilation to more evolved or silicic compositions. Silicic magmas may also have been produced by the partial melting of the lower crust, where the major contribution of the magmas ponded at the base of the crust has been the thermal energy required for melting.

ACKNOWLEDGEMENTS

We thank M. J. Carr for his comments on an early draft of this paper.

Section 2: Plutonic Andes

Phanerozoic Plutonism in the Peruvian Andes

W. S. PITCHER

Abstract

A synthesis is presented of the geological settings of the several granite types of the Peruvian Andes, viz. intraplate reactivation, interplate collision, intraplate rifting, marginal ensialic spreading and plate-edge thickening. A Lower Palaeozoic, gneiss-granite series is located within an Arequipa crustal block; a syntectonic S-type granite series occurs within an Upper Palaeozoic flyschoid slate-belt; a group of high-K, relatively low-Ca, I-type granites is associated with a Permo-Triassic rift-zone and its molasse; a low-K, high-Ca, I-type granodiorite batholith formed the core of a marginal arc during the Cretaceous, and a belt of crust-contaminated, I-type granodioritic stocks is associated with the Tertiary volcanism. The possible sources of each type are discussed.

Introduction: the overall geological context

Granites are important in each of the several geological environments revealed in the Peruvian Andes. Here, as elsewhere, the type of granite differs with the context (Pitcher, 1983) in parallel with contrasts in the sedimentary facies and deformation.

In outlining the magma-tectonic styles in Peru it is important to emphasize that the Andean mobile-belt has been encratonic in character throughout much of the Phanerozoic. Thus an ancient crystalline basement underlies all Phanerozoic sequences and the essential control of Andean geology lies in the vertical and horizontal movements of separate blocks of this Precambrian basement.

During the Palaeozoic the formation of linear, extensional basins, and their eventual high-strain compression, resulted from the interaction between two forelands, an Arequipa massif in the west and a Bolivian massif in the east. In the Mesozoic, in great contrast, a largely tensional and certainly low-strain regime was expressed as vertical movements on deep-reaching faults in the Arequipa massif in response to long-standing ocean-continent subduction along the Pacific margin.

Of these two contrasted environments, that of the Lower Palaeozoic, located within the Eastern Cordillera, involved a clastic, flysch trough that was deformed, though little metamorphosed, and then uplifted during a mid-Devonian *Eohercynian* orogenic phase. This trough was re-established during the Carboniferous and Lower Permian with the deposition of a cover sequence of a continental molasse passing up into shallow-water clastics. In its turn this sequence was gently folded, faulted and uplifted, and these *Tardihercynian* movements were accompanied by the coeval deposition of a Permo-Triassic, red-bed molasse identified as the Mitu Group (Dalmayrac *et al.*, 1980).

In the Mesozoic, and within the Western Cordillera, fault-controlled troughs were established within the margin of the South American craton in response to ensialic spreading-subsidence (cf. Levi and Aguirre, 1981). The most westerly of these, that of Huarmey-Cañete, represented, during the mid-Cretaceous, a marginal basin that failed to develop into an oceanic rift (Dalziel, 1981; Cobbing and Pitcher, 1983; Atherton *et al.*, 1983). This was closed at about 100 Ma during a *Mochica (Subhercynian)* compression, after which followed a number of short periods of compression in which the movements were taken up on Andean-oblique wrench-faults and in localized belts of Andean-trending folds. These progressively waned in strength into the Cenozoic at the same time as the continental edge was buttressed by the granitoids, thickened and uplifted, with the formation of a red-bed molasse during the Eocene. Thus the scene was set for the plateau-type volcanicity of the Cenozoic Andes, representing another and different environment.

In Peru the plutonic episodes which occupy these several tectonic niches are separated in space, time and type and provide a pattern for the Central Andes as a whole.

The Palaeozoic granites of the foreland

Lower Palaeozoic plutonic activity is revealed in the Arequipa massif in two forms, viz. the syntectonic, Camana-Atico gneiss complex (CA in Figure 1), dated at 440 Ma (Rb–Sr, Shackleton *et al.*, 1979) and the late-tectonic granodiorite batholith of San Nicholas (SN in Figure 1) (Pitcher, 1974), dated at both 392 Ma (Rb–Sr, Shackleton *et al.*, 1979) and 440–430 Ma (K–Ar, Wilson, 1975). The occurrence of a granite series of this type, representing a late stage of a 'Caledonian' reworking of a foreland massif, would not be remarkable if it were not for the paucity of evidence of such a magma-tectonic event in the Lower Palaeozoic trough itself.

Mid-Palaeozoic plutonism within the Palaeozoic slate-belt

Within the Lower Palaeozoic rocks of the Eastern Cordillera plutonism is first in evidence associated with the mid-Devonian (Eohercynian) deformation that produced the existing slate-belt. Dalmayrac *et al.* (1980) and Bard *et al.* (1974) report on a two-mica gneissose granite lying within a substantial contact aureole at Zongo-Yani in Bolivia, and this might be grouped with a similar granite gneiss complex at Amparaes, near Cuzco in Peru, and

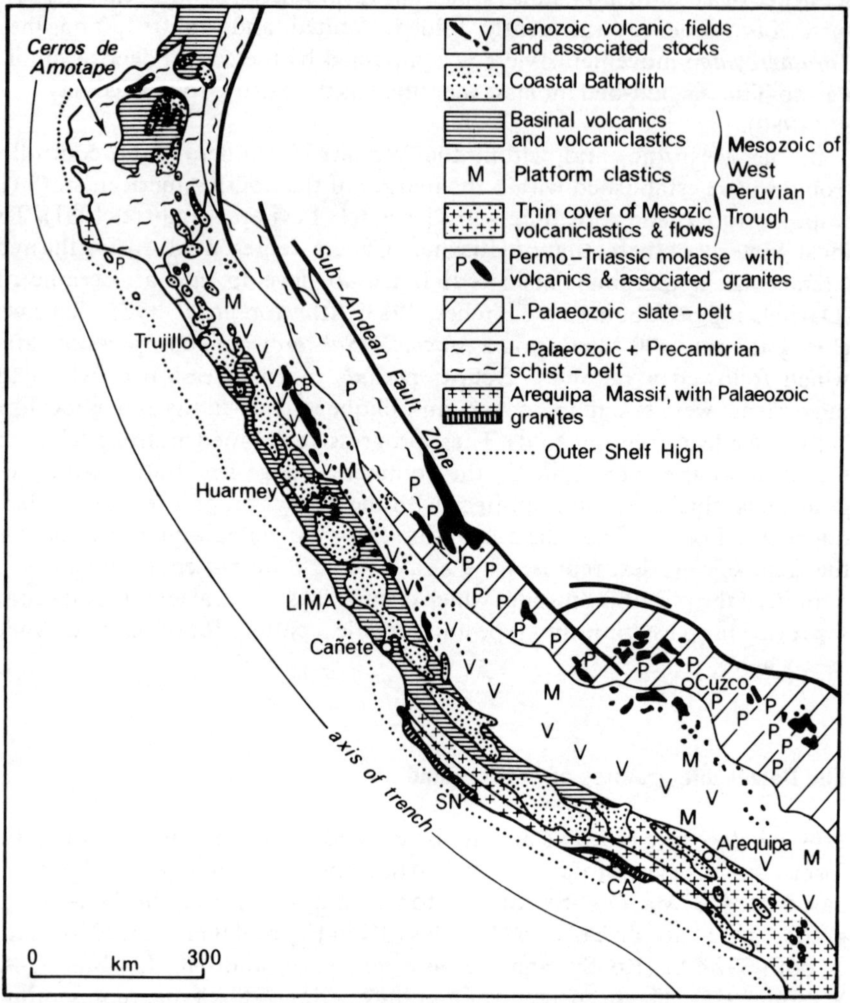

Figure 1 The main belts of granitoids in Peru in relation to their overall geological context. SN—San Nicholas batholith; CA—Camana-Atico gneiss complex; CB—Cordillera Blanca batholith

represent a syntectonic plutonism dating at around 330–346 Ma (Marocco in Dalmayrac *et al.*, 1980, p. 202; Mégard, 1978). Within this same zone there are also granites with late-syntectonic and post-tectonic relationships to the Palaeozoic envelope (e.g. Mégard, 1973) and, although these granites are not all peraluminous types, it is likely that an entire S-type granite series is represented, within its natural slate-belt setting (Pitcher, 1983), and of crustal source.

Permo-Triassic magmatism in the Eastern Cordillera

Also within the Eastern Cordillera of Peru volcanism is associated with the rifting and the deposition of the Permo-Triassic red beds (Jenks, 1951). Some of the high-K rhyolitic and dacitic ash-flows and lavas are of distinctly peralkaline character and there is even an association in space with a peralkaline syenite complex (Francis, 1956). Such an alkali-magma associated rift environment fits well with a thesis of crustal extension (Fricker and Weibel, 1960; Mégard, 1973; Noble *et al.*, 1978; Kontak *et al.*, this volume).

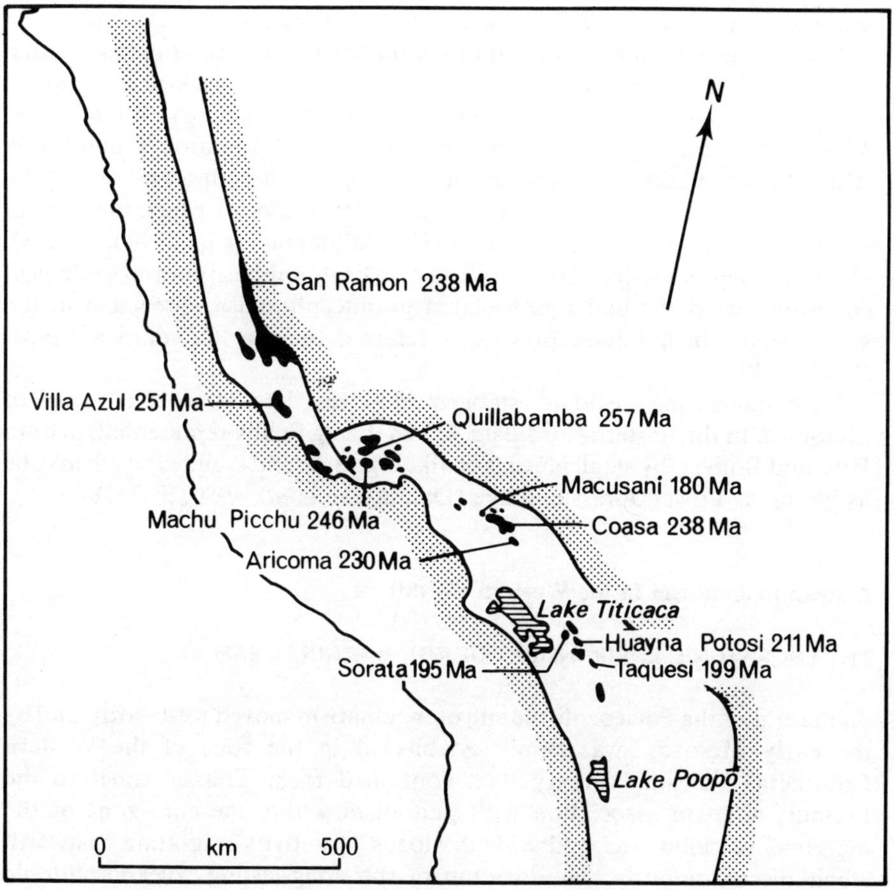

Figure 2 Distribution of Permo Triassic granite plutons in the Eastern Cordillera of Peru and Bolivia, after Lancelot *et al.* (1978), Mégard *et al.* (1971). Radiometric ages quoted: *San Ramon* 238 ± 10 Ma Rb/Sr, Capdevila *et al.* (1977); *Villa Azul* 251 Ma K/Ar, Stewart *et al.* (1974); *Quillabamba* 257 ± 5 Ma U/Pb, Lancelot *et al.* (1978); *Macchu Picchu* 246 ± 10 Ma Rb/Sr, Priem *et al.* unpublished, quoted in Lancelot *et al.* (1978); *Macusani* 180 Ma, Stewart *et al.* (1974); *Coasa* 238 ± 11 Ma U/Pb, Lancelot *et al.* (1978); *Aricoma* 230 ± 10 Ma U/Pb, Dalmayrac *et al.* (1980); *Sorata* 195 Ma K/Ar, *Huayna Potosi* 211 Ma, *Taquesi* 199 Ma, Clark and Farrar (1973)

This incipient break-up of the old sialic crust may have marked the beginning of the Pacific subduction regime, but within a zone far removed, at least 500 km, from any possible Permo-Triassic ocean-continent margin.

Closely associated in space and time with this volcanicity and rifting are several arrays of large granite plutons (Figure 2), clearly cross-cutting, and with narrow yet strongly developed contact aureoles. These plutons show a clear spatial relationship to major faults, and within the graben the red beds of the Mitu Group rest unconformably on certain granites whilst they are cut by others, illustrating the coevality of uplift and intrusion (Dalmayrac *et al.*, 1980; Lancelot *et al.*, 1978; Capdevila *et al.*, 1977). In Peru radiometric determinations yield ages spanning 260–234 Ma (but see Kontak *et al.*, this volume; Carlier *et al.*, 1982), whilst in Bolivia they range to 195 Ma.

Within these Permo-Triassic plutons biotite granites predominate, often coarse grained, megacrystic in K-feldspar, and generally lacking in xenoliths except in marginal situations. Such granites are calc-alkaline to sub-alkaline, with K/Na ratios greater than unity and A/CNK ratios around 1.05 (Cobbing, in press), factors which certainly contrast with granitoids of the Andean, less certainly with those of the Eohercynian for which there is but scanty data (cf. Capdevila *et al.*, 1977; Dalmayrac *et al.*, 1980, p. 233). However there is a remarkable parallelism in both composition and geological environment to rift and uplift-related granite plutonism elsewhere in the world, and which I have previously referred to as a 'Caledonian I-type' (Pitcher, 1983).

With increasing tectonic stabilization this late-Palaeozoic cycle of plutonism in the Eastern Cordillera waned, being finally represented, in both Peru and Bolivia, by small plutons of alkali-granite and syenites which may be as young as Lower Jurassic in age (Dalmayrac *et al.*, 1980, p. 232).

Andean magmatism in the Western Cordillera

THE ENCRATONIC ENVIRONMENT OF THE MARGINAL BASINS

At the end of the Palaeozoic the site of magmatism moved westwards, and by the early Mesozoic was firmly established in the zone of the Western Cordillera. Volcanic activity then continued from Triassic times to the Present, often in association with plutonism within the core zone of the marginal volcanic 'arc' and with the locus of activity migrating eastward, albeit discontinuously. The location of this long-lasting volcano-plutonic lineament within the mobile edge of the continental plate is the principal evidence for the equally long-standing operation of a subduction mechanism.

For much of the Mesozoic the coastal zone of Peru was the site of marine transgressions on to the block-faulted lip of the craton, resulting in a ribboned development of narrow sedimentary troughs parallel to the continental edge (Cobbing *et al.*, 1981). There is ample evidence that a crystalline basement everywhere fringed, and probably underlay, these Mesozoic troughs, even that of Huarmey-Cañete. An unconformable relationship can be observed at margins (Bellido, 1969; Cobbing, 1978), also in faulted inliers (Garcia, 1968; Vargas, 1970) and in a deep bore (Lehmann, 1978) and, although these findings concern mainly the southern limit of the basin, there is the more general supporting evidence in the form of the derivation of material from a westerly block of metamorphic rocks (Wilson, 1963; Webb, 1976; Aguirre, *pers. comm.* 1980; Rivera *et al.*, 1975, p. 175). The latter has persisted until the Present in the form of the coastal cordilleras of Arequipa and Amotape (Cobbing and Pitcher, 1972a; Cobbing *et al.*, 1977b; Shackleton *et al.*, 1979; Martinez, 1970; Morris and Aleman, 1975; Caldas, 1979) and their submarine interconnection (Thornberg and Kulm, 1981). Furthermore there is the geophysical evidence of a thick crust under the Andes at present (James, 1971a).

There is, however, the important finding by Couch and his co-workers (Couch *et al.*, 1981; Jones, 1981) that a welt of high density material rises into the lighter crust beneath the basinal deposits lying east of the coastal massif, specifically beneath that part of the Huarmey-Cañete basin to the north of Lima. Within the latter an abundance of pillow-lavas, hyaloclastites, coeval basic sills and dykes, interbedded with a volcaniclastic flysch and intruded by near coeval gabbros, provides evidence of submarine, intrabasinal volcanicity. The primitive nature of these magmas (Atherton *et al.*, 1983) is suggestive of a mantle source without crustal contamination. This is a view consonant with the geophysical evidence for a thinned upper-crust directly under the Huarmey basin and with the petrological evidence for high geothermal gradients during an essentially burial-type metamorphism (Offler *et al.*, 1980). It is envisaged that the stretching of the continental crust in the hinterland of a destructive plate margin (Cobbing and Pitcher, 1983) led to intense dyking of the thinned crust by basic magma.

DEFORMATION STYLE AND IGNEOUS UNDERPLATING

This early stretching phase of tectonics is increasingly interrupted by short periods of compression from the mid-Cretaceous onwards, with both the timing and the effect of the deformational phases differing across the Andean belt. Nevertheless the overall style of this deformation is simple, at least within the zone of the Huarmey-Cañete basin. Upright, rarely tight folds are associated with high-angle faults of great extent and with long histories of repeated movement (Myers, 1975b; Mégard, 1978).

The finding that strong cleavage is restricted to fold cores and to relatively narrow belts confirms the overall estimate of the shortening as just reaching 40% (Mégard, 1978), though it is much less than this in openly folded volcanic rocks of the coastal belt (cf. Myers, 1974, p. 484). This low degree of shortening across the belt, coupled with the general observation that vertical movements predominate in the central Andes, suggests that the present great thickness of the Andean crust is not wholly due to compression but might have resulted in part from igneous underplating, whether this be represented by an underridden oceanic slab or by magma directly segregated from a mantle wedge.

During the long history of Upper Cretaceous magmatism several cycles of magma-tectonic activity can be recognized (Moore, 1979; Bussell, 1983), viz. extension with the injection of a basic dyke swarm, then mild compression with the rise of granitoid magma, followed by uplift. Such cycles may eventually be related in some complex way to cyclic changes in subduction rates (Frutos, 1981).

THE GABBROS: A PRECURSOR TO A GRANITOID BATHOLITH?

A string of elongate basic plutons and associated basaltic dykes occupies the axial zone of the Huarmey-Cañete basin (Regan, 1976; Cobbing *et al.*, 1981). Intrusion occupied the period between the ending of sedimentation and the emplacement of the early members of the granitoid batholith and so spanned the mid-Cretaceous phase of folding. Multiple plutons, often sheet-like, can be as large as 1000 km^2 in outcrop, though this is presently much reduced by intersection with the younger, granitic intrusions.

Gabbros are important within a range extending from ultrabasic rocks to meladiorite, though the average major element composition approximates to that of an olivine tholeiite. Hornblende is ubiquitous with a long-extended growth history (Mullan and Bussell, 1977). Geochemically these rocks are primitive in character and clearly separate from the granitoids of the superposed batholith (McCourt, 1981).

These basic plutons seem to represent shallow crustal, subvolcanic magma reservoirs relating more to the mid-Cretaceous volcanicity than to the granitoid batholith, yet their precursor relationship to the latter shows that they must play an important role in the evolution of the granitoids. Once again the occurrence of these gabbros is in accord with the model already presented of an extensional marginal basin above a thinned, continental-type crust and its underlying welt of basic, mantle-derived material.

THE COASTAL BATHOLITH

The immense batholith forming the core of the Western Cordillera of Peru is the dominant plutonic feature of the Mesozoic magmatism. This highly composite and multiple batholith consists of a 1600 km-long array of upwards of 1000 plutons occupying a 60 km-wide band along the axis of the Huarmey-Cañete marginal basin (Figures 1 and 3). Its constituent intrusions were assembled, espisodically, between 97–35 Ma (Pitcher, 1978), and in association with the thickening of the crust.

The emplacement was controlled by growth-fractures on all scales (Bussell, 1976; Pitcher and Bussell, 1977) and, at depth, magmas are likely to have been

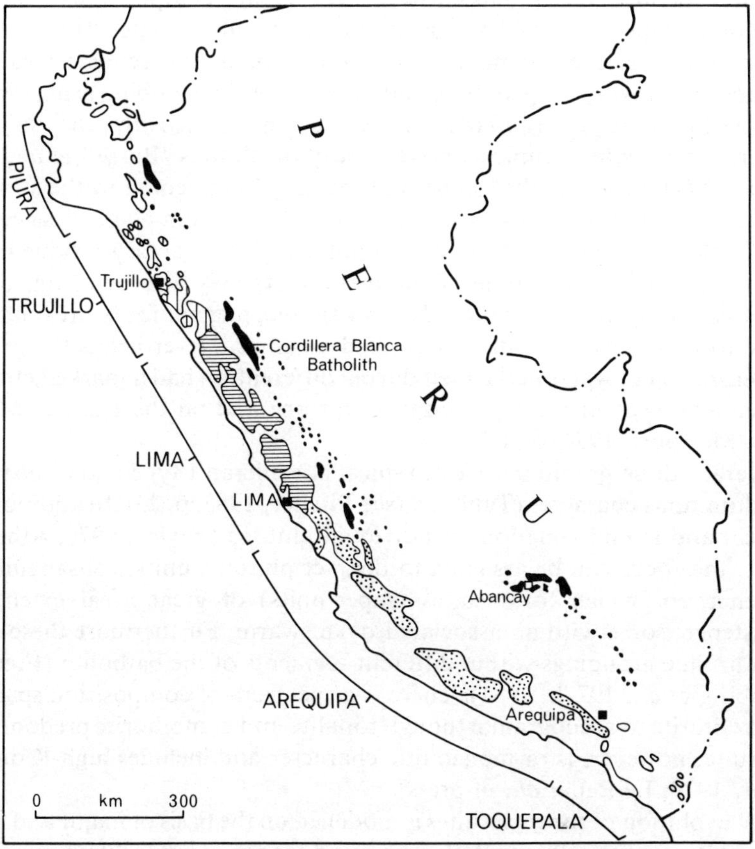

Figure 3 The segmentation of the Cretaceous Coastal Batholith, illustrated by various stipples or blank, and the belt of Cenozoic stocks and associated batholiths, shown in black

channelled along a single, deep-reaching megalineament. At the present high crustal level of exposure the form of all the intrusives, both the plutons with their steep walls and flat roofs, and the vertical dykes, was controlled by the attitude of a resurgent fracture system. The plutons were emplaced by permissive cauldron subsidence and associated piecemeal stoping, both involving mechanical replacement without lateral opening or distortion of the country rock except for some minor doming (Taylor, 1976). Such a brittle fracture mechanism is consonant with the localized, short-lived, contact metamorphism (Atherton and Brenchley, 1972), the comparative lack of mineral alignments and the occurrence of ring-dykes and tuffisitic pipes (Bussell and McCourt, 1977). Taken together with the non-deformative character of the burial metamorphism (Offler *et al.*, 1980), such features indicate the permissive intrusion of magma at a shallow level in the crust (Myers, 1975a), perhaps to within 3–4 km of the surface.

The tonalites and granodiorites form great, compositionally zoned, lenticular-shaped complexes, whilst the later, subordinate, 'true' granites occur either as large homogeneous plutons or as nested assemblages of plutons and ring-dykes (Cobbing and Pitcher, 1972b, Cobbing *et al.*, 1981). It is possible that a denudation series is represented by the sequence caldera—ring-dyke complex—nested array of plutons (Bussell *et al.*, 1976; Knox, 1971), implying that some plutons may have vented to the surface.

Within the tonalitic rocks there are basic cores within plagioclase crystals that could well represent restitic material, i.e. the refractory residue from a remelting process, and some of the mafic clots may also be of this origin. Otherwise, however, the granitoids show simple, magmatic textures and seem likely to represent water-undersaturated magmas. Nevertheless the gradual concentration of water in the melt during differention had a marked effect on the mineralogy, on the crystallization history and on the texture (Mason, 1982; McCourt, 1981, p. 417).

Overall, these granitoids are of typical cordilleran I-type and of distinctly metaluminous character (Table 1), (see Pitcher, 1983, p. 23). In addition to a vertical and lateral zonation in individual plutons (Taylor, 1976; Atherton, 1981), the rocks can be assigned to distinct plutonic units, consanguineous sequences of which form suites (super-units) of great areal extent and persistence, some with an associated dyke swarm. Furthermore these suites form unique groupings within different segments of the batholith (Figure 3) (Cobbing *et al.*, 1977b, 1981). There is a spectrum of composition spanning quartz diorite to syenogranite though tonalite and granodiorite predominate; one suite, however, is of monzonitic character and includes high-K diorites (Agar, 1979; Le Bel *et al.*, in press).

The evolution of the rock suites is modelled on the basis of major and minor element data, especially the REE, as being due to the crystal-differentiation of crops of plagioclase and hornblende starting at depth and continuing until the final emplacement of the magmas at the present level (McCourt and Taylor, 1979; Atherton *et al.*, 1979; McCourt, 1981; Cobbing *et al.*, 1981; Atherton, 1981). Problems remain, however, in identifying the composition of the parent magma and explaining the virtual absence of coeval cumulate phases.

Within the Lima segment of this linear, multiple batholith there is a remarkable space-time uniformity of geochemical character, with low initial $^{87}Sr/^{86}Sr$ ratios of $\sim$0.7042 (Sanchez, 1982; W. P. Taylor and R. Beckinsale, *pers. comm.*, 1979), low $\delta^{18}O$ values (Beckinsale, *pers. comm.*, 1983) and ϵ_{Nd} values near zero (Le Bel *et al.*, in press), together suggesting a homogeneous, deep-seated source. Furthermore the Pb isotope systematics (Mukasa and Tilton, this volume) confirm that there is virtually no contamination by the Precambrian basement, at least in this central segment of the batholith. This again is in accord with a thinned sialic crust beneath the batholith.

These magmas were possibly derived from the remelting of the earlier-formed underplate of basaltic material segregated under the greatly thinned crust of the back-arc basin. This is consonant with the geological evidence for the beginning of a general thickening of the crust at about the same time as the

Table 1 **Summary of contrasting features, environments and possible sources of the granitoids of the Peruvian Andes**

Lower Palaeozoic	Upper Palaeozoic	Permo-Trias	Upper Cretaceous	Mid-Cenozoic
Syn- and late-tectonic granites of Arequipa massif	Syn- and late-tectonic granites of Eastern Cordillera	Rift-associated granites of Eastern Cordillera	Granodioritic batholith along axis of marginal basin in Western Cordillera	Fault-associated granodiorite stocks and aphyric lava domes of the High Plateau
Migmatitic acid and basic gneisses in a possible *series* with hornblende-biotite granodiorite	Two-mica granites pre-dominate with narrow composition range; possibly from a *granite series*	Biotite granite (± hornblende) with narrow composition range	Biotite-hornblende tonalites and granodiorites with wide composition range: associated with gabbro	Quartz diorites and monzonites of fine grain and porphyritic in biotite
Early phases very strongly deformed	Early phases strongly deformed	Largely undeformed	Largely undeformed except in early stages of plutonic cycles	Undeformed
Metamorphic reworking of ancient basement	High-strain slate-belt. Green schist facies metamorphism	Rifting and uplift. No coeval regional metamorphism	Alternate tension and low-strain compression. Burial-type metamorphism	Faulting. Weak burial-type meta-morphism
Late members cut platform clastics	Post-date a clastic flysch; coeval volcanism rare	Coeval with red-bed molasse and sub-aerial, sub-alkaline, siliceous volcanism	Post-dates volcaniclastic flysch and basaltic volcan-icity of a marginal basin	Coeval with a molasse and sub-aerial, siliceous volcanism
I-type of ?meta-igenous source	S-type granite series of ?metasedimentary source	Ca-poor I-type of ?meta-igneous source	Ca-rich I-type of ?mantle source	I-type of ?subducted oceanic crust source

initiation of the granitoid batholith, also by the presence of the relatively high-density welt at depth, and the availability of basic magma in the form of coeval dyke swarms during the whole period of its assembly. Finally it is suggested that the episodic remelting of the first-stage source rocks is likely to have been triggered by the ascent of volatiles and heat generated from the subducted oceanic crust in response to periods of extra-rapid sea-floor spreading. However, without involving a crustal source the origin and efficient separation of so great a volume of silicic magma remains a central problem.

THE METALLOGENETIC CONNECTION

According to Cardozo and Vidal (1981) and Vidal (1980) there are three stages in the metallogenetic evolution of the batholithic belt of the Western Cordillera. The first is pre-batholithic and associated with the formation of the mid-Cretaceous, marginal basin, taking the form of either strata-bound barite-Zn deposits or Cu-Fe metasomatic deposits genetically attached to the gabbroic intrusives.

During the succeeding batholithic event, Cu–Fe veins; Cu–Mo–W skarns, Cu–Fe–Mo disseminations or stockworks, and Au-quartz veins, were all generated as a result of late magmatic to hydrothermal activity, often related to specific plutonic suites (Hudson, 1979; Agar, 1979). This mineralization is relatively more intense in the two southern segments of the batholith, especially that of Arequipa (Figure 3), and this despite the intrusives being everywhere eroded to a similar level. An explanation lies in the possibility that the greater abundance of salts within the deposits of the southern sector of the Huarmey-Cañete basin enhanced the salinity of the metal-scavenging brines. More fundamentally variations in the abundance of both halogens and metals can be expected to relate to the segmental changes in crustal structure which are a particular feature of the central Andes.

The third stage of mineralization involved the formation of the cluster of porphyry-Cu deposits of Cerro Verde, Toquepala, Cuajone and Quellaveco, the stocks and tourmaline breccia pipes of which are aligned along a 170 km-long lineament to the south of Arequipa. Despite their spatial relationship to the batholithic rocks of the Toquepala segment, they are of a significantly younger, late-Palaeocene age (see Laughlin *et al.*, 1968; Estrada, 1978) though very near to that of the central complexes of the Lima segment in age.

Cenozoic plateau volcanicity and the eastern stocks

Throughout the Western Cordillera of Peru an erosion surface is cut from the main intrusives of the Batholith and their envelope. This surface is overlain by a thick pile of subaerial volcanic rocks, the Calipuy Group, which form a broad field along the cordillera, some 40 km wide and with an axis well to the east of the Coastal Batholith (V, Figure 1). North of Lima the group is of the order of 3000 m thickness, a lower part consists of basalt and latite-andesite flows and pyroclastics capped by a sequence of largely rhyodacite ash-flows. In fact two separate formations can sometimes be recognized, especially in northern Peru (Cobbing *et al.*, 1981), separated by an important phase of faulting, and this hiatus is possibly widespread. The age span of these volcanic rocks is not yet closely controlled but radiometric data extend from 52.5–12 Ma, with preferred values, from near unconformities, of 41 Ma and 18 Ma defining two sequences in south-central Peru (Noble *et al.*, 1978; McKee and Noble, 1982).

Approximately corresponding to the outcrop area of this Cenozoic volcanic field is a belt of small stock-like plutons 2–10 km^2 in outcrop, associated with andesitic domes, breccia pipes and dykes, the ages grouping as 32–29 Ma and 23–12 Ma (Figure 4) (Cobbing *et al.*, 1981, p. 86–87; Petersen and Vidal, in press). The intrusions often diapirically punch through and dome the volcanic rocks and the underlying Mesozoic sediments (e.g. Mégard, 1978, p. 232), and many are associated with major fault lines (Figure 4).

Quartz micromonzonite and microdiorite, porphyritic in biotite and hornblende, are common types in these volcanic centres, representing strongly calc-alkaline magmas which were the hypabyssal material feeding coeval volcanic rocks.

Such relatively small intrusions and their associated breccia pipes form the principal loci of mineralization so that they are extensively hydrothermally altered, with contact skarns, mineralized haloes and marginal vein systems carrying Cu, Pb, Zn, Ag and, sporadically, W (Cobbing *et al.*, 1981). In southern Bolivia similar domes of rhyodacite are associated with the major tin field of the Altiplano (Grant *et al.*, 1977).

Some large batholiths of the same general age also occur within this belt: that of Abancay-Cotabombas intrudes strata of Miocene age whilst that of the Cordillera Blanca has been dated radiometrically as between 12–9 Ma.

The former lies within the east-west deflexion zone of Abancay (see Figure 1), where some considerable plutons of coarse-grained, biotite-hornblende granodiorite are stoped high into the Tertiary red-bed molasse, cutting rocks as young as the Miocene (Marocco, 1971) but also intersecting much older deformed diorites (Carlier *et al.*, 1982). Doubtless the structural control of the major deflexion was important in localizing this east-west belt of intrusives.

Located within the Cordillera Blanca, near Huaraz (Figures 1 and 4) (Egeler and De Booy, 1956; Wilson *et al.*, 1967; Cobbing *et al.*, 1981), is a great, steep-sided, flat-roofed, multiple intrusion, with outlying plutons, cut out of dark mudstones of Upper Jurassic age, in which is produced a considerable aureole of andalusite and garnet-bearing schists. The dominantly leucogranodioritic rocks of this Cordillera Blanca Batholith contrast strongly with those of its coastal analogue being coarser in grain, rich in K-feldspar megacrysts, more siliceous overall, much less mafic and with muscovite commonly present; features that are reflected in a relative peraluminicity (A/CNK = 1.05). There is a strong analogy here with the change in composition recorded in plutons lying to the east of the Sierra Nevada Batholith of California, a feature there ascribed to crustal

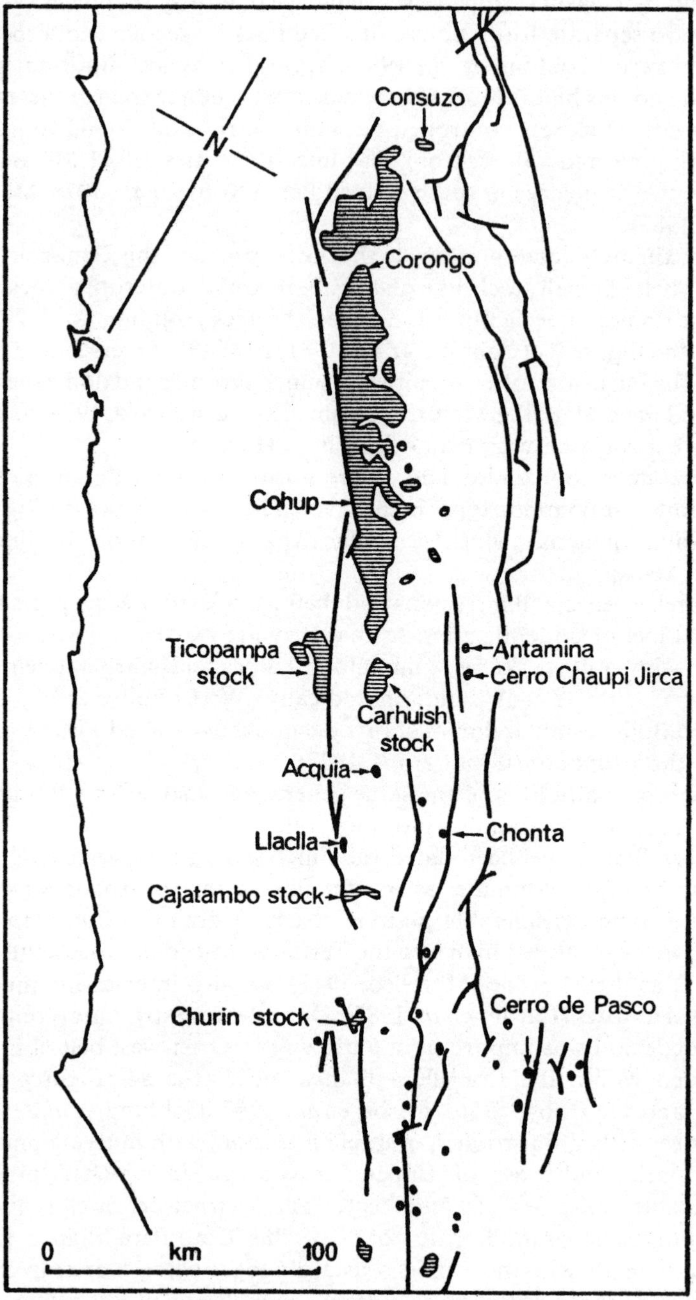

Figure 4 Plutons of the Cordillera Blanca Batholith and the associated high-level stocks in relation to major faults (after Cobbing *et al.*, 1981, Figure 57)

contamination in locations well within the continental crustal block.

This group of intrusions is again clearly associated with a major fault line (Figure 4). The presence of contact schists, a peripheral foliation and an internal mineral alignment oblique to the margins, when coupled with evidence for a degree of forceful emplacement, is suggestive of intrusion into an active strike-slip fault-zone.

Such structural features are remarkable in view of the very high level character of this young batholith, so closely associated in time with a major, multistage peneplanation event (Puna) that its magmas must have risen very near to the surface. Indeed the overall similarity in age and chemistry of the batholithic rocks, the accompanying high-level stocks and late-Cenozoic volcanic rocks, might further suggest the most direct connection of a feeder chamber to a volcanic field. If so we approach a model of magma forcing its way to the surface along an active, deep-reaching fault-system.

Lacking significant geochemical parameters it is not profitable to discuss the origin of the magmas of these Cenozoic stocks and batholiths except to note that, at the time and in the place of their intrusion, the crust must have been relatively thick and rigid, so that a crustal contribution is to be expected; as indeed is indicated by the crustal signatures carried by the Neogene and Quaternary volcanic rocks, the geochemistry of which is the subject of several papers in this present volume. Furthermore, in what is presently the thickest belt of the Andean crust, the Eastern Cordillera of Peru and Bolivia, there are some small monzonitic plutons and ash-flow tuffs of mid-Tertiary age that are so peraluminous as to be cordierite-bearing (Kontak *et al.*, this volume). These must represent the ultimate term in the crustal derivation of magma within a thickened crust, and their presence heightens the strong contrast between the crust-related magmatic environment to which such Cenozoic intrusives belong and the mantle-related magmatism associated with the thinned crust of the Mesozoic times.

A comparison of granite types and their environments

During the Phanerozoic, magmatism in the Peruvian Andes was always restricted to specific, NNW-trending belts, differently located at different times. Only from the beginning of the Mesozoic is there any evidence of a progressive migration of the locus of magmatism (see Cobbing and Pitcher, 1983, Figure 5), but even so there are so many overlaps and lacunae that this apparent progression is probably an illusion, the magmatism being more realistically viewed as a reflection of different stages in the structural evolution of the central Andes, viz. interplate collision, intraplate rifting, marginal spreading-subsidence, and plate-edge thickening. It is worthy of mention, however, that the overall degree of stratigraphic cohesion is too great to permit a thesis involving displaced terrains, implying that these several environments must have evolved in the same part of the South American Plate.

In all but the most ancient of these magma-tectonic environments there is a close relationship between the coeval sets of plutonic and volcanic rocks. Nevertheless this is not to be generally modelled as direct supply from a magma-chamber, for it is likely that the ever-present dyke swarms represent a mechanism of fissure supply that predominated over extrusion from plutonic central complexes.

Concerning the comparison of the various age-environmental groups of Peruvian granites, both Cobbing (in press) and Dalmayrac *et al.* (1980) have pointed to differences, and these are summarized in Table 1, with the proviso that the data are very incomplete.

The most obvious contrast is between the granitoids of the Eastern and Western Cordilleras, yet within both of these a further two groupings are apparent on compositional as well as environmental grounds. Thus in the former location the early, two-mica peraluminous granites, mesozonally emplaced into the Palaeozoic slate-belt, differ from the late, biotite-hornblende granites, epizonally emplaced within the rift-molasse environment of the Permo-Trias. The former are likely to represent S-types whilst the latter are sufficiently metaluminous to be included within the broad grouping of I-type granites. Even so they are very different from the more distinctly metaluminous I-type granodiorites of the Coastal Batholith that form the sub-volcanic core to the marginal volcanic arc of the Cretaceous.

This regional contrast is revealed in the very different compositional spectrum—granite as distinct from a range of tonalite-granodiorite-granite with a basic association—and also in a reversal of the role of the alkalis, in significant differences in both the Ca content and the peraluminicity, and also in the Sr-isotope systematics—as much as this data is presently available in the papers cited (and see Kontak *et al.*, this volume). Such differences between types of metaluminous granitoids, apparently dependent on whether they occur in subduction or post-subduction uplift environments, are found the world over (Pitcher, 1983) and are well exemplified in the South American Andean and the European Caledonian mobile belts respectively.

The rocks of the small, hypabyssal intrusions lying to the east of the Coastal Batholith are also compositionally of I-type and not dissimilar chemically to the quartz-diorites and granodiorites of the batholith. However, bearing in mind that their magmas must have had to pass through a much thicker and more heterogeneous crust than those of the latter it is not surprising that the larger intrusions among them show some evidence of crustal contamination, to the extent that the rocks of the Cordillera Blanca Batholith are mildly peraluminous.

In Peru these differences between several groups of granitic rocks offer a prime target for geochemical studies on the varied source of granitic magma, particularly because different protoliths can be expected where different environments are involved.

Presently most data exist for the Coastal Batholith where a consideration of major and trace element compositions and isotope systematics suggests an essentially mantle source; possibly represented by a magma underplate secreted from a mantle wedge. In contrast the granites of the Eastern Cordillera lack detailed study (but see Kontak *et al.*, this volume) though there are indications that they are both of crustal origin, the S-types of the Palaeozoic possibly having a sedimentary source and the I-types of the Permo-Trias possibly a lower-crustal meta-igneous parentage. A rewarding discussion of such origins awaits the results of researches, such as those reported in this volume. What is important to understand is that not all magmatism, especially that involving granites, is directly related to conventional subduction mechanisms, so that other models of heat generation and transfer in mobile belts have to be devised or resurrected.

The Coastal Batholith of Peru

M. P. ATHERTON

Abstract

The 1600 km long Coastal Batholith of Peru is emplaced, in the main, within an 'ensialic' marginal basin, of lower Cretaceous age, made up of volcaniclastics, pillow lavas, sills and gabbros. Major and trace element analysis indicate much of the variation, from quartz diorite to granite *sensu stricto*, is the result of fractionation in place, though some may have occurred on ascent or in intermediate magma chambers. The fact that the batholith crosses from the marginal basin in the north, with no underlying old crustal component, on to old Precambrian thick crust in southern Peru, without any major change in chemistry or petrology indicates that crustal contamination, if present, is of minor importance in the genesis of the batholith rocks.

The Coastal Batholith of Peru belongs to the circum-Pacific, Mesozoic-Cenozoic volcanic-plutonic province, and has been considered to be one of the clearest examples of Cordilleran calc-alkaline magmatism related to ocean crust subduction beneath an old continental mass (Pitcher, 1978). It stretches from Piura in the north, near the border with Ecuador, to south of Arequipa (Toquepala) close to the southern border with Chile, a distance of nearly 2000 km and it is generally less than 65 km wide (Figure 1). The linear form of the batholith has been emphasized by Pitcher (1978), who thought that magma intrusion was focussed up a deep linear structure throughout the whole life of the batholith (~ 40 Ma), to freeze at subvolcanic levels just short of the surface (Atherton and Brenchley, 1972). Intriguingly, this structure lies parallel to the present trench, at a distance of about 200 km to the east (Figure 1).

The batholith is made up of about 1000 plutons which vary in shape from large, elongate or near circular bodies to ring dyke or sill-like bodies. During the original mapping it was found that the plutons could be grouped into super-units, uniquely identifiable in the field, individual units of which vary in composition from quartz-diorite to leucomonzonite. Super-units show, often over long distances, 'consistent spatial and temporal association of all the constituent units . . . progressive evolution of both the mineralogy and textural interrelations, especially of quartz and K-feldspar and the maintenance of similar granularity and aphyric character' (Cobbing *et al.,* 1977b). Super-units, which usually include more than one rock type and may often encompass the whole compositional range, were considered by Cobbing *et al.* (1977b) to be consanguineous cf. Southern California Batholith (Larsen, 1948) and the Sierra Nevada Batholith (Bateman and Dodge, 1970) for similar groupings.

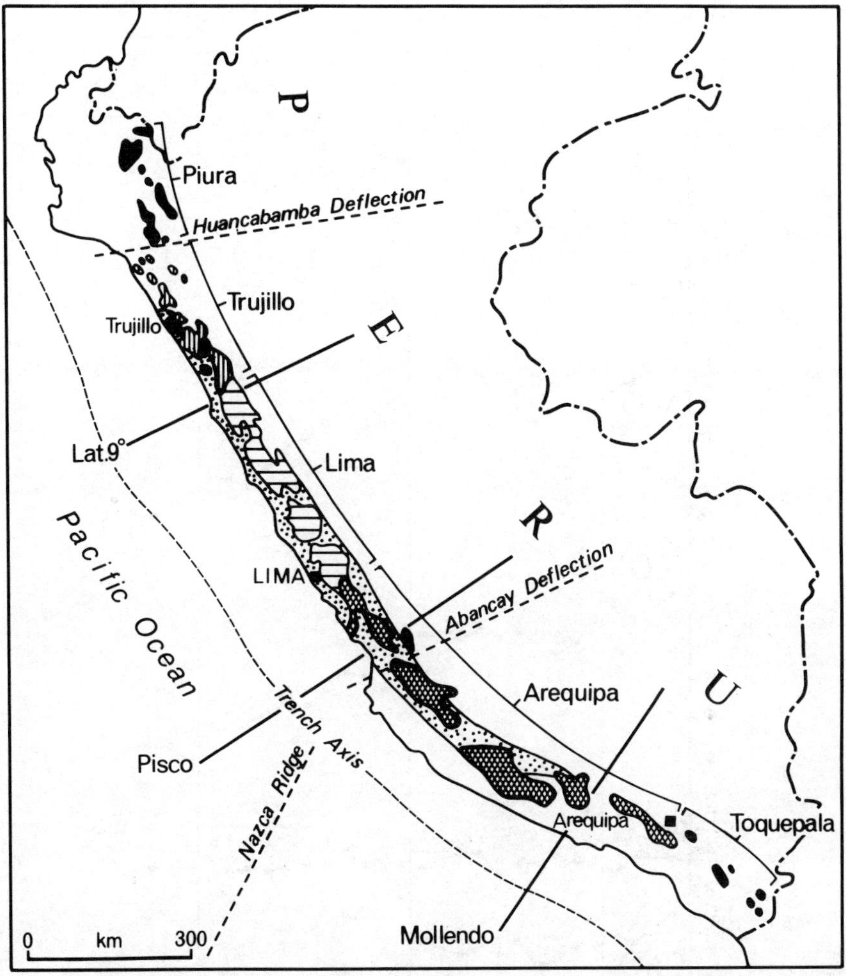

Figure 1 Coastal Batholith of Peru (black + lined ornament), showing the segmentation and the marginal basin ('stippled'—eugeosyncline of Cobbing, 1978)

Five segments of the batholith have been defined by specific super-unit groupings, each segment having about 4–5 super-units characterizing it (Cobbing *et al.*, 1977b; Cobbing, in press; Figure 1). It is important to appreciate that the setting up of super-units was based entirely on field criteria, and more recent chemical and isotopic studies have aimed to try and understand the deep physical processes producing them, as well as their validity in terms of age and chemical uniqueness. Detailed aspects of the batholith are reviewed by Pitcher (1978), Atherton *et al.* (1979) and McCourt (1981).

The plutonic rocks of the batholith may be divided into two groups: (a) an older group of large, often elongate plutons mainly tonalitic in composition;

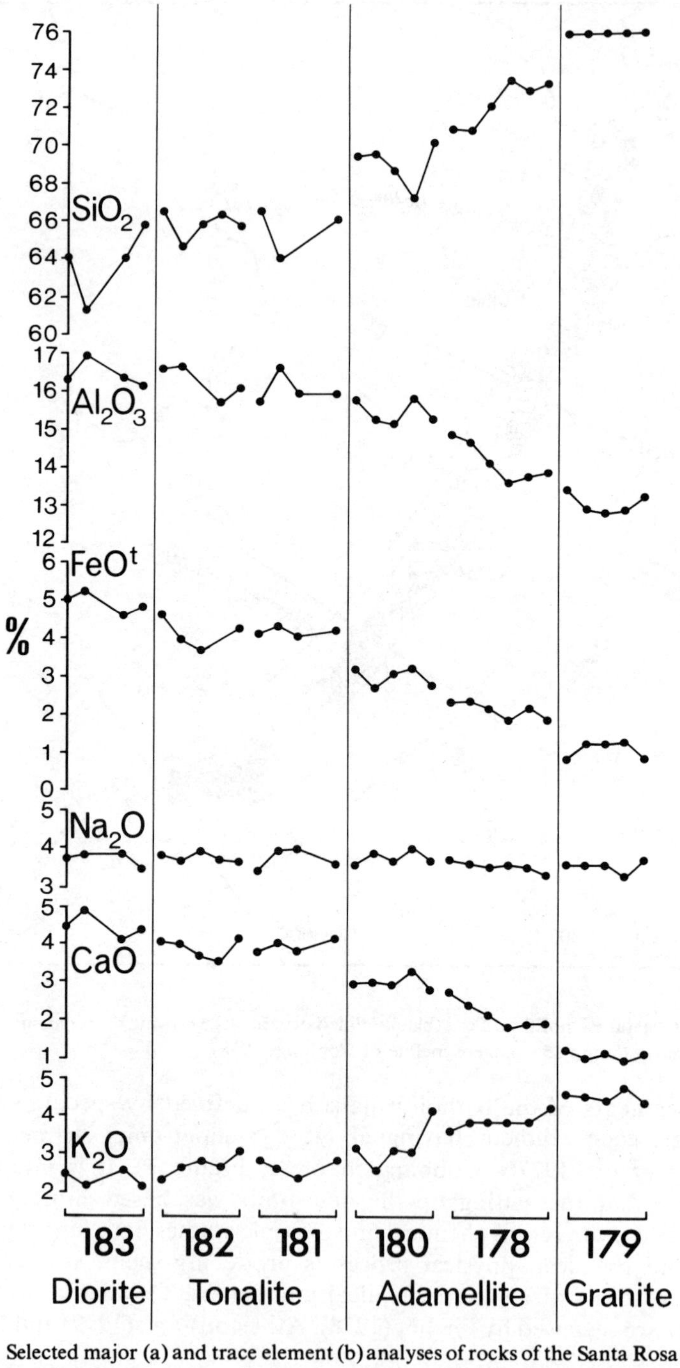

Figure 2 Selected major (a) and trace element (b) analyses of rocks of the Santa Rosa super-unit, in the Huaura area. Samples were collected along traverses at right angles, and relatively close, to the contacts. Sample points are not equidistant and do not belong to a single linear traverse across the super-unit, which is normally zoned (basic rocks succeeded inwards by more acid rocks)

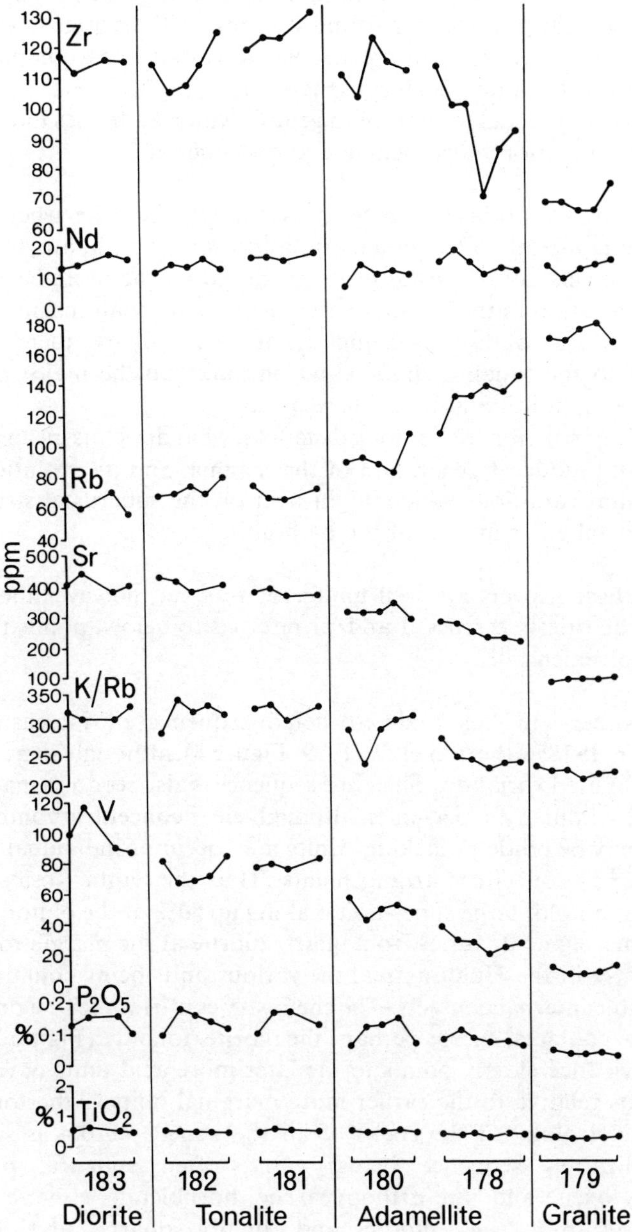

Figure 2 (contd.)

forming the major part of the batholith and (b) a younger group of plutons, commonly circular or arcuate in form with compositions often near to granite *sensu stricto*; these occur in, or are associated with, the ring complexes which are confined to the Lima Segment (Figure 1).

Any consistent model for batholith genesis which addresses the problem of crustal contamination must include a knowledge of:

1. The pluton variation at the present level and its cause, i.e. whether it is due to magma mixing or fractional crystallization or both; are, for instance, the more acid rocks crustal melts produced by the more basic units?
2. Axial variation in the batholith and the possible relationship to different levels in the batholith magma chamber, i.e. are there cumulates related to the tonalitic 'liquids' which make up the major part of the exposed batholith and if so where are they?
3. Super-unit coherence over long distances; what does this mean in terms of place and mode of generation of the magma, and its evolution?
4. Segmental variations which might bear on the subcrustal structure.
5. The crustal environment of the batholith.

Not all these aspects are well understood as yet, however the evidence, which will be briefly discussed and/or referred to below, points to minimal crustal involvement.

1. Each super-unit has a time-ordered sequence of increasing acidity (Pitcher, 1978; Atherton *et al.*, 1979, Figure 3), although some may show rather limited variation. This time sequence is also seen as a marked zonal structure in many super-units, although clear concentric zoning in many cases may be crude or lacking. Units may occur as individual plutons or together as concentric arrangements. Thus the Santa Rosa super-unit, which is an old, large super-unit making up 50% of the plutonic rocks of the Lima segment, varies from quartz diorite at the margin to granite at the centre in the Huaura area, the various units being contained within mappable internal contacts. The chemistry confirms the first order nature of these contacts, except perhaps the diorite/tonalite (Figure 2), and the field evidence clearly points to the later more acid units having moved upwards relative to the earlier more marginal units (Atherton, 1981).

   The petrology of the rocks of all the zoned plutons is similar, the crystallization sequence being: early, often euhedral plagioclase, clinopyroxene and rare orthopyroxene, hornblende, either euhedral or replacing pyroxene, biotite and finally quartz and K-feldspar (microperthite). Often pyroxene is absent or replaced by hornblende which is then the major ferromagnesian phase. Apatite, ilmenite and sphene are common minor phases (for details see Atherton, 1981; McCourt, 1981; Sanderson and Atherton, in press). The normative mineral variation across the Santa Rosa super-unit is shown in Figure 3, and corresponds closely with the modal mineralogy except that

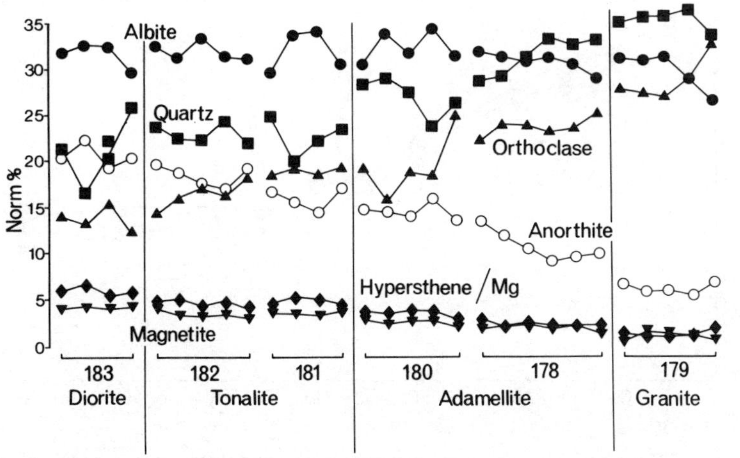

Figure 3 Normative values for the rocks of the Santa Rosa super-unit in the Huaura area. Details as in Figure 2

hornblende ± pyroxene are the actual ferromagnesian phases rather than hypersthene. In general, the textures indicate clear liquidus equilibria with respect to the early phases (Atherton, 1981), and occasionally interlocking crystal configurations in the marginal rocks, suggesting crystal accumulation, are seen (Pitcher, 1978). However, this does not mean that the more evolved units were entirely liquid, although the final granitic (*sensu stricto*) unit almost certainly was so (Atherton, 1981). This latter magma was the most mobile and reached the highest structural level of all the earlier, less evolved units. The large Eu anomaly seen in the granite REE pattern (Figure 4) testifies to the essential liquid nature of the magma, which evolved by dominant plagioclase precipitation. The ternary mineral assemblage of quartz, K-feldspar and plagioclase with interstitial granophyre is compatible with the liquid character of the granite.

Major, trace and rare-earth element modelling (Sanderson and Atherton, in press) all indicate fractionation by plagioclase (dominant), hornblende, clinopyroxene, with biotite and magnetite coming in at more acid compositions (Figure 5). Although variations occur in detail between the different models generated for a given super-unit the observed mineralogy and liquidus phases are consistent with all the modelling. Selected major and trace element data across the Santa Rosa pluton emphasize the continuity across the whole composition range, although abrupt changes in composition of individual elements occur across the

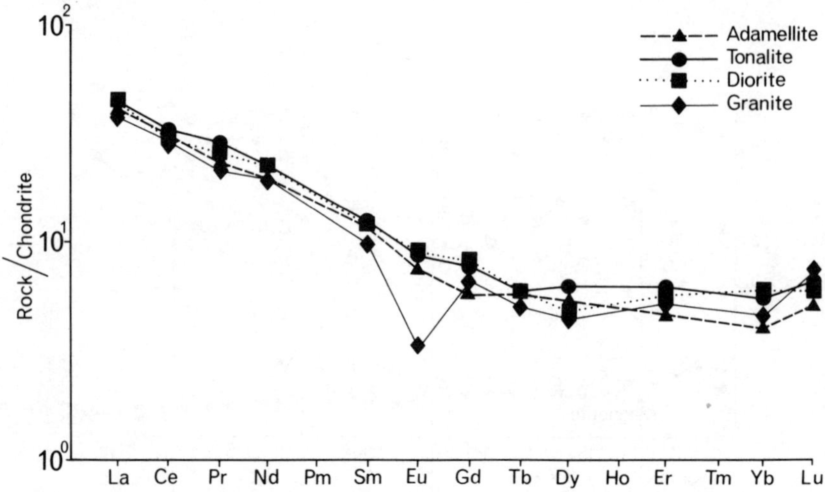

Figure 4 REE plots of four representative rocks from Santa Rosa super-unit in the Huaura area, showing a very strong coherence and implied consanguinity and a very clear Eu anomaly in the granite. Note K_D^{REE} must have been very close to 1 for all elements, apart from Eu, during crystallization

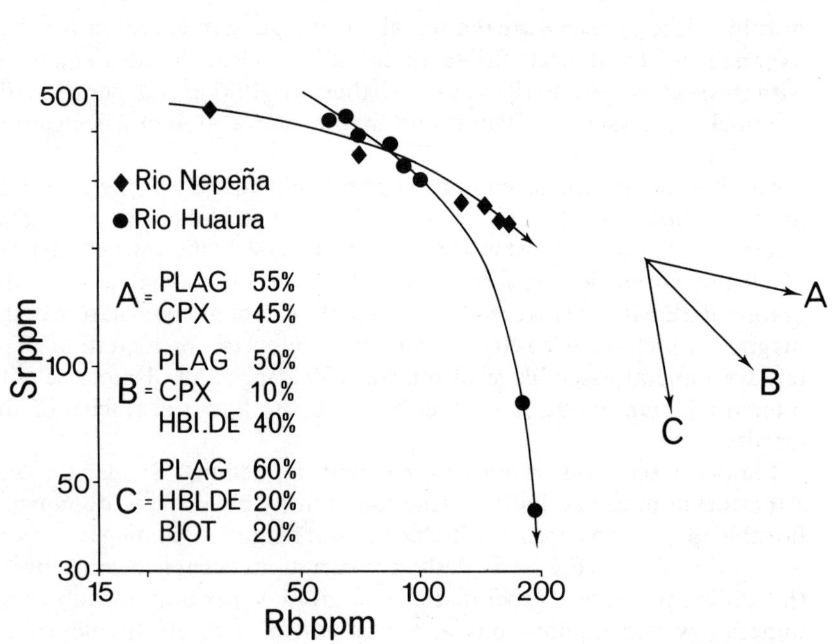

Figure 5 Graphical plot of Sr (ppm) versus Rb (ppm) for rocks of the Santa Rosa super-unit in the Nepeña and Huaura traverses, with percentages of precipitating phases determined numerically from the modelling

internal contacts. However, ratios, such as K/Rb, remain unchanged across the most significant breaks, for example between the granite and the adamellite (see Figure 2), indicating liquid compositions are continuous and simply related. The changes across this contact are compatible with liquid separation from the earlier precipitate of plagioclase, hornblende, clinopyroxene and buoyancy concentration of the liquid at higher levels of the magma chamber (Atherton, 1981).

It could, and has been argued, that the chemical variation seen in calc-alkaline Cordilleran batholiths may result not so much from high-level differentiation, as argued here and elsewhere (Atherton *et al.*, 1979; McCourt, 1981), but may be explained by partial melting, the trace element variations, particularly the HFS element behaviour, relating to their retention at source in minerals such as ilmenite, rutile, apatite, sphene and zircon (Tarney and Saunders, 1979). Certainly Zr, Hf, Ta, Nb, Ti and P do not show systematic incompatible behaviour in the Antarctic Peninsula Mesozoic volcanic and plutonic rocks studied by Tarney and Saunders (1979). In the super-units studied here, HFS element abundances, particularly Ti, P and Zr often decrease systematically (see Figure 2), although in the Linga super-unit Zr content increases dramatically with SiO_2 content. In the volcanic rocks associated with the batholith some of these elements show increases then marked depletions when late minor phases precipitate, e.g. TiO_2, while others show pronounced incompatible behaviour, e.g. Zr (Atherton *et al.*, 1983). It is perhaps unwise to group plutonics and volcanics of different ages, i.e. the Antarctic Peninsula rocks studied by Tarney and Saunders (1979), for the discrete data from Peru indicate HFS element behaviour is different in plutonic and volcanic rocks and can often be related to precipitation in the minor phases seen in specific rock suites.

In the Arequipa segment the Linga super-unit was considered by Stewart to be the result of gabbro-granite mixing, however more recent work has indicated that high level fractionation is more likely, involving plagioclase (dominant), hornblende, biotite, magnetite ± pyroxene and the high-K nature of this super-unit is a primary feature related to depth of generation (Sanderson and Atherton, in press; Le Bel *et al.*, in press).

In conclusion, although magma variation in the batholithic rocks may result from crystallization over a considerable time and may have started deep in the crust, the compositional variation is explicable in terms of fractionation at levels above which plagioclase is stable (Atherton, 1981), the bulk of which occurred at very high levels in the crust, in the magma chambers we see exposed today.

2. High-level differentiation as outlined above implies the presence of cumulates, if genesis from mantle-derived basalt is accepted (Atherton, 1979; McCourt, 1981). Over the whole length of the batholith it is only in the south, in the Linga super-unit, that cumulate type rocks occur. The most basic rock in the Linga super-unit is an olivine, two-pyroxene gabbro showing adcumulate textures which also contains K-feldspar. The

high K_2O content of all the rocks of the super-unit indicate it belongs to a high-K suite (Agar, 1978). Trace and major element modelling of the Linga super-unit clearly indicates the cumulate nature of the most basic unit (Sanderson and Atherton, in press; Le Bel *et al.*, in press) while the later, more evolved, members can be derived by the extraction of plagioclase, hornblende, biotite and magnetite ± clinopyroxene in very similar proportions to that seen in the rest of the super-units (Sanderson and Atherton, in press; Le Bel *et al.*, in press).

There is no reason to believe that this super-unit is more deeply eroded than the others although it is certainly more basic than the other super-units. This being so, the equivalent 'cumulates' of the other super-units of the batholith may lie at depth either as underplated material or part of the deep basic arch structure (see section 5). Thus 'cumulates' do not appear to lie at the bottom of the high level magma chambers of the Coastal Batholith.

3. Although super-units are coherent over long distances in the field (400 km, Cobbing *et al.,* 1977b), and have unique trace element chemistries, in some cases they have significantly different compositions from place to place (Atherton *et al.,* 1979; Figure 5). Thus the Santa Rosa super-unit in the Lima segment is very different in trace element composition in the Nepena section to that in the Huaura section, some 300 km to the south (Atherton *et al.,* 1979; Figure 5). Similar, but less extreme, cases are seen in some of the super-units in the Arequipa segment and it may well be that the fine-structure in individual super-units is more variable than has been yet recognized. Although this implies that parts of a super-unit are not consanguineous, there is insufficient data, at present, to ascertain whether the changes from one section to another are abrupt or continuous. If the latter is the case, the differences might relate to subduction/mantle wedge geometry and composition varying continuously axially, so that magmas generated at the same time over long sections of the Andes may not have homogenized at depth, even though they may have formed at the same time. Alternatively, if the changes are abrupt, individual parts of a super-unit may not be consanguineous and the field appearance may merely be a superficial high-level imprint indicating a similar final history.
4. Segmental differences in the plutonic rocks have been discussed elsewhere (Atherton *et al.*, 1979). Briefly, considering only the Arequipa and the Lima segments (the other segments are very poorly known) the former has no ring complexes, is mineralized, and is generally K-rich compared to the Lima segment. These differences must relate to the origin of the magmas at depth, as well as to the crustal structure, and are best understood in relation to the marginal basin structure (see section 5).
5. Perhaps the most important single feature when considering possible crustal involvement relates to recent work indicating that the Coastal Batholith was emplaced mainly into a marginal basin (Atherton *et al.*, 1983) very similar in terms of age and components to that seen in southern

Chile (for a review see Dalziel, 1981). Up to 10 km of material was deposited in the basin during the Mesozoic and included pillow lavas, hyaloclastites, turbiditic volaniclastics, cherts, oozes, etc. The basin shows a classic marginal basin asymmetry with the pillow lavas and volcaniclastics dominating on the trench or sea (western) side and aerial pyroclastics and continental platform sediments coming in towards the continental (eastern) side of the basin. Within the basin are sill and dyke complexes plus large gabbro masses (for details see Atherton *et al.*, 1983), with rock compositions similar to equivalent lavas, dykes, sills and gabbros from Sarmiento and Tortuga, southern Chile (Saunders *et al.*, 1979), and lavas from the Bransfield Straight and East Scotia Sea spreading centre (Weaver *et al.*, 1979; Saunders and Tarney, 1979).

The marginal basin is underlain by a huge arch structure, coincident with the basin, which extends from just north of Trujillo to just short of near Arequipa (see Figure 1). Details of this structure are given in Couch *et al.* (1981), and Figure 6.

In a gross sense the structure of this dense wedge (3.0 $g\,m^{-3}$) widens to the north and is centred near the coast at 14°S, but is mainly offshore at 9°S. The western margin is formed by a thin sliver of continental crust trending out to sea northwards, but connected to the Precambrian Arequipa massif to the south, where crustal rupture ceases. Jones (1981) thought the dense material could be imbricate upthrust oceanic crust, or intrusion at depth of upper mantle material, or even lower crust mafic material. Whatever, the nature of this new crust is not granulite facies lower crust, indeed the Arequipa massif rocks have densities of 2.75 $g\,m^{-3}$ and lie above this dense material, which is likely to have mantle-type characteristics assuming it is not actually mantle. Crustal contamination in such a situation would be either negligible or unrecognizable. The Coastal Batholith of Peru is particularly suited to test the effect of crustal contamination, as the plutons of the segment to the north of Trujillo lie in presumed Palaeozoic basement and not within the marginal basin, which in that region lies out to sea (west). To the south in the Lima and the northern part of the Arequipa segments the batholith lies within the marginal basin (more correctly within the two basins making up the marginal basin viz. Huarmey and Cañete basins, see Cobbing, 1978). Further south, in the Arequipa area where there is no arch-like structure and the crust has not split, the Coastal Batholith has intruded the thick Precambrian Arequipa massif (Figure 6). We have very little evidence from the north, near Trujillo, relating to differences in the plutonic rocks of the Coastal Batholith as they move from a marginal basin (new crust) into an older crustal environment. In the south near Arequipa interaction of the mantle-derived tonalite (Atherton *et al.*, 1979) with crust has certainly not involved major melting of the Precambrian massif as acid rocks (possible 'S' type melts) are absent. This is in sharp contrast to the Lima segment, in which the magmas did not pass through Palaeozoic or Precambrian crust, and granite *(sensu stricto)* is quite common. These

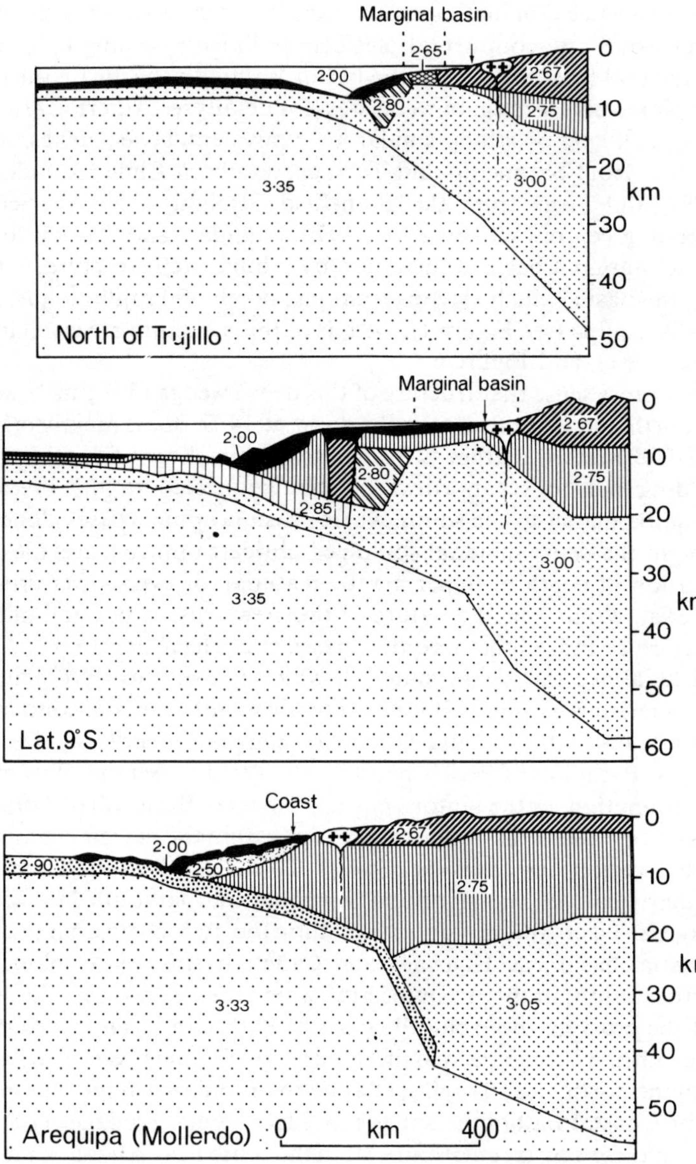

Figure 6 • Sections across the Andean margin in Peru along traverses shown in Figure 1. Gravity data from Crouch *et al.* (1981) and Jones (1981). Position of the batholith (crosses) and marginal basin indicated. Note the section north of Trujillo is drawn by inference from surface exposure and extrapolation from the southern sections. There is no geophysical evidence from this area

granites, we maintain, are produced by liquid-crystal fractionation (Atherton *et al.*, 1979). Neither does it seem likely that crustal interaction via some mixing process was important in the Arequipa segment, as the overall chemical variation can be modelled in a very similar manner, albeit the rocks may be richer in K_2O, to those in the Lima segment. But more significantly the Sr and Nd isotope studies (Le Bel *et al.*, in press) indicate the magmas, even in the mineralized zone near Arequipa, are mantle-derived with little or no crustal component. Given the geological setting of the Lima segment plutonic rocks it is even less likely that crustal contamination would occur there. Low values of $(^{87}Sr/^{86}Sr)_i$, all in the region of 0.704 (Atherton *et al.*, 1979; McCourt, 1981) positive ϵ_{Nd} values (McCulloch, *pers. comm.*) and Pb isotope data (Mukasa and Tilton, this volume) all indicate crustal involvement in the genesis and crystallization history of the plutonic rocks of the Lima Segment was negligible. In conclusion, it is apparent that although the Peruvian Coastal Batholith was intruded into two, possibly three, very different environments, i.e. the thick Precambrian crustal wedge in the south, the new crust in the central zone and Palaeozoic crust in the north, it shows no evidence of crustal contamination in either of the two environments studied so far.

Lead Isotope Systematics in Batholithic Rocks of the Western and Coastal Cordilleras, Peru

S. B. MUKASA and G. R. TILTON

Abstract

Lead isotope ratios have been determined on plutonic rocks of various compositions and ages from the San Nicolás, Coastal and Cordillera Blanca Batholiths in the Western and Coastal Cordilleras, Peru. The data have been evaluated in the Pb correlation diagram framework established by previous workers from Pb isotope determinations on the Precambrian Mollendo and Charcani granulitic gneisses, volcanic and plutonic rocks, and ores of southern Peru and central Chile.

Results indicate that the dominant Pb component in the plutonic rocks came from an isotopically homogeneous reservoir very similar to that identified by Tilton (1979) and Tilton and Barreiro (1980) for Quaternary ores, and volcanic and plutonic rocks in central Chile. The homogeneous reservoir might be enriched subcontinental mantle.

Additionally, the data also show that there is spatial association between the occurrence of a Precambrian basement Pb component in the Phanerozoic batholithic rocks and exposures of the Precambrian granulitic gneisses. Precambrian basement Pb is detected in all samples from the San Niçolás Batholith and in those from the Arequipa segment of the Coastal Batholith. In contrast, samples from the Cordillera Blanca Batholith and associated stocks, and those from the Lima segment of the Coastal Batholith, which occur where Precambrian exposures are totally lacking, exhibit no detectable old crustal Pb component. However, the more felsic rocks in this group seem to have undergone high-level mixing with crustal materials such as those found in the Jurassic to Upper Cretaceous marginal basin of Peru.

Introduction

The application of isotope compositions of Pb, Sr and Nd in the study of magmatic processes along convergent plate margins in which continental crust is involved has proven significant in numerous studies worldwide. The success of these radiogenic isotopic systems lies in their ability to detect various components which may contribute to magma generated in this tectonic environment. Perhaps the most thoroughly studied area using these systems is the Western Cordillera of North America (e.g. Doe, 1967; Doe *et al.*, 1968; Doe and Delevaux, 1973; Zartman, 1974; Farmer and De Paolo,

1983). Here, it has been shown that Pb isotope compositions in plutonic rocks of the Mesozoic batholiths are highly influenced by the nature of the underlying basement rocks. For the Western Cordillera of North America and for a few other convergent margins (e.g. Zartman, 1974; Kay *et al.*, 1978; and Hawkesworth *et al.*, 1979b), subducted sediments, subducted basaltic slab and overriding mantle wedge have also been proposed as potential contributors of materials to magmas generated in this tectonic environment. In principle, using variations in the radiogenic components of isotope tracers such as Pb, Sr, Nd and Hf may yield information indicating, or placing limitations on, the possible source materials. For this work, the U–Th–Pb isotopic system has been applied on plutonic rocks of the continental margin of Peru.

Source Pb isotope signatures

Four zones that can potentially undergo melting with addition of water from phase dehydration reactions, in an Andean-type margin, have been identified (e.g. the reviews by Green, 1980 and Wyllie, 1982). These zones are illustrated in Figure 1a, and their total Pb concentrations and Pb isotope compositions given by Chow and Patterson (1962); Reynolds and Dasch (1971); Unruh and Tatsumoto (1976); Meijer (1976); and Tilton and Barreiro (1980) are summarized in Figure 1b. Zone 1 represents a veneer of subducted Pacific Ocean pelagic sediments which are characterized by relatively high $^{207}Pb/^{204}Pb$, $^{206}Pb/^{204}Pb$ and $^{208}Pb/^{204}Pb$ ratios, and high total Pb abundances. The basaltic slab (zone 2 in Figure 1a), with which the sediments are subducted, is here represented by the Nazca plate basalt analyses of Unruh and Tatsumoto (1976). The basalts are reasonably close to the ocean regression line (ORL) which characterizes depleted sources, and on the average, they have <1 ppm total Pb.

Zone 3 in Figure 1a is the overriding mantle wedge. Phases from this zone have very low (<1 ppm) Pb abundances (Zartman and Tera, 1973; Morioka and Kigoshi, 1978), but it has been demonstrated in various parts of the world that they can have either highly radiogenic or highly depleted isotopic signatures (Doe, 1967; Doe *et al.*, 1982; Mahoney *et al.*, 1982; Church, 1983; Menzies *et al.*, 1983). For the Andes, Tilton (1979), and Tilton and Barreiro (1980) observed that ores and both plutonic and volcanic rocks from central Chile have Pb isotope compositions of a very limited range that plotted well above the ORL in Pb correlation diagrams; and Halpern (1979) reported that even their $^{87}Sr/^{86}Sr$ ratios are uniformly 0.704, a value usually associated with mantle derivatives. Subsequent Pb isotope analyses by Barreiro and Stern (1982), and Barreiro (this volume), on both basalts and andesites, have further shown that central Chilean rocks appear to be derived from an isotopically homogeneous source. Such observations encourage the hypothesis that these samples represent sub-Andean mantle. Evidently, it lacks a depleted character and may, therefore, be labelled enriched subcontinental mantle.

Finally, zone 4 in Figure 1a is ancient lower crust which, in the Peruvian Andes, may be represented by the Mollendo and Charcani granulitic gneisses

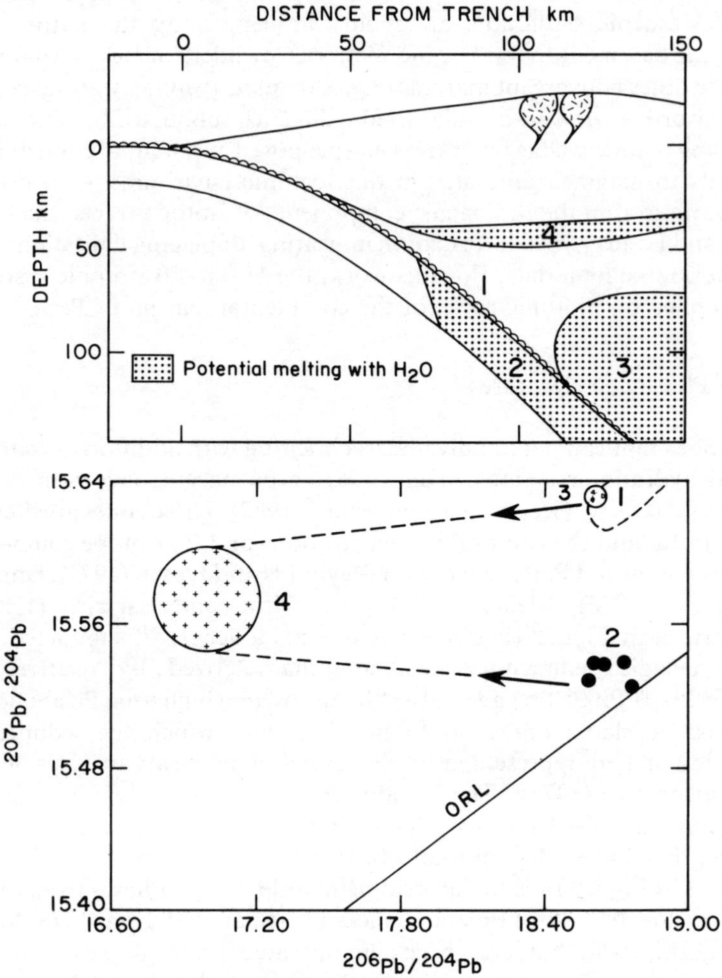

ZONE	MATERIAL		PPM Pb
1	Pacific ocean sediments	:	20
2	Subducted slab	:	<1
3	Mantle phases	:	<1
4	Precambrian basement	:	20

Figure 1 (a) Cross section through a subduction complex showing the four zones (1–4) that could potentially undergo melting with release of water by dehydration reactions (modified after Wyllie, 1982). (b) Pb isotope signatures and total Pb concentrations of the four zones shown in Figure 1a. Data are from Chow and Patterson, 1962; Reynolds and Dasch, 1971; Zartman and Tera, 1973; Unruh and Tatsumoto, 1976; Morioka and Kigoshi, 1978; and Tilton and Barreiro, 1980

or their isotopic equivalents (Tilton and Barreiro, 1980). These rocks have had low μ values since the time of high grade metamorphism at ~2 Ga (Dalmayrac *et al.*, 1977), and they therefore have low $^{206}Pb/^{204}Pb$ but normal $^{208}Pb/^{204}Pb$ ratios, which are typical of Precambrian granulite terrains.

It follows that if a Pb extraction process such as melting or dehydration were to occur in zones 1 and 2 of Figure 1a in conservatively estimated proportions of 20:1 or 95% to 5% basaltic slab to pelagic sediment, the resulting liquid would have a Pb isotope composition half way between these two end members. Then, as it rises and interacts with material in the mantle wedge, definite partial melting should be induced. The Pb isotope composition of the product would approach the ocean regression line if the mantle wedge has been depleted with respect to incompatible elements for a time period in the range of 10^9 years, but would approach the central Chile homogeneous ratios if the mantle wedge is enriched or undepleted subcontinental mantle. Any subsequent interactions between the magma and ancient lower crustal materials should then lower $^{206}Pb/^{204}Pb$ ratios of the magma as shown by the arrows in Figure 1b.

Geologic components at the continental margin of Peru

The continental margin of Peru has been the site of repeated igneous activity through much of the Phanerozoic. This activity is attributable to interactions between the Nazca and American plates in a convergent boundary tectonic setting; and its products are manifested in three major batholiths: San Nicolás, Coastal and Cordillera Blanca, as well as in isolated stocks and various volcanics (Figure 2). The San Nicolás Batholith which intrudes Precambrian granulites and gneisses, is oldest and closest to the boundary between the continental and oceanic crusts. The Coastal Batholith which intrudes a Lower Jurassic to Upper Cretaceous deep-water sedimentary sequence and its own volcanic ejecta in the north, but which cuts through Precambrian basement in southern Peru, is intermediate in both position and age; the Cordillera Blanca Batholith, which intrudes a Lower Jurassic to Upper Cretaceous sequence of platform deposits, is youngest and farthest from the lithospheric boundary. The three batholiths, the sedimentary and metamorphic rocks they intrude, and the Triassic to Plio-Pleistocene volcanics in the area constitute the Western and Coastal Cordilleras of Peru.

This discussion is restricted to the three batholiths, and greater emphasis is placed on the Lima and Arequipa segments of the Coastal Batholith. These segments outcrop in the 400 km north of and 1000 km south of Lima respectively. The other two batholiths are dealt with here only in reconnaissance fashion.

In the course of the systematic mapping of the Lima and Arequipa segments (see, for example, Cobbing and Pitcher, 1972b; Cobbing *et al.*, 1977b; Cobbing *et al.*, 1981), it became apparent that the batholith is comprised of many separate plutons which could be grouped on the basis of similarities in

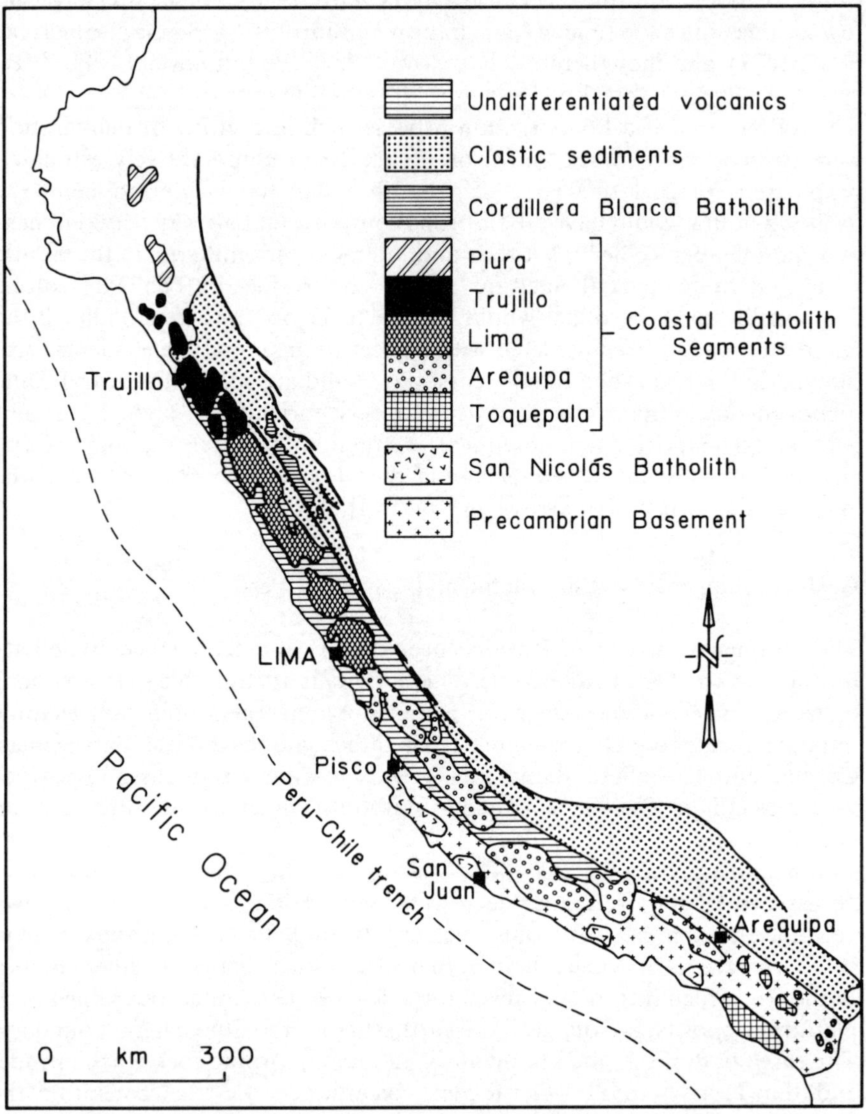

Figure 2 The San Nicolás, Coastal and Cordillera Blanca Batholiths, and other tectonic elements of the Western and Coastal Cordilleras, Peru (modified after Cobbing, 1978)

lithology, temporal relationships, textural and modal variations, xenolith morphology composition and content, dyke associations, consanguinity as exhibited by mineralogy and chemistry, and close spatial association within the same segment of the batholith. Such an approach yielded groups of plutons that are essentially lithostratigraphic units which have been called super-units (Cobbing *et al.,* 1977b). Sampling for the isotopic evaluations presented here closely followed the super-unit concept. In all, Cobbing *et al.* (1977b) have identified nine super-units in the 400 km-long Lima segment and five super-units in the 1000 km-long Arequipa segment. These super-units and their approximate emplacement ages are given in Table 1.

Lack of access has prevented thorough studies in both the San Nicolás and Cordillera Blanca Batholiths. However, it has been shown that the former comprises a group of disconnected yet apparently related plutons of basic to intermediate composition (Pitcher, 1974; Instituto de Geologia y Mineria, unpublished maps). These plutons are characterized by the presence of abundant euhedral hornblende phenocrysts which attain lengths of up to 2 cm and by high contents of magnetite. The San Nicolás Batholith has an age of 350–450 Ma based on reconnaissance K–Ar work by Wilson (1975), Rb–Sr isochrons by Shackleton *et al.* (1979) and U–Pb zircon ages on discordant zircon populations from several plutons (Mukasa and Tilton, in prep.).

Table 1 **Super-units in the Lima and Arequipa segments of the Coastal Batholith and their approximate emplacement ages***

Super-unit	Age (Ma)
Lima segment	
Pativilca monzogranite	34
Cañas-Sayán monzogranite	60
Puscao granodiorite-monzogranite	60
San Jerónimo monzogranite	60
La Mina tonalite-granodiorite	60
Humaya granodiorite-granite (unit)	70
Santa Rosa tonalite-granodiorite	80–90
Paccho diorite-tonalite	95
Jecuan tonalite-granodiorite	102
Patap gabbro	102
Arequipa segment	
Tiabaya tonalite-granodiorite	80
Incahuasi monzodiorite-monzotonalite	80–95
Pampahuasi tonalite	80–95
Linga monzodiorite-granodiorite	97
Patap gabbro	102

*All ages based on the K–Ar work of Stewart *et al.* (1974), Wilson (1975) and Moore (1979).

The Cordillera Blanca Batholith consists principally of a leucogranodiorite named the Cohup Granodiorite by Egeler and De Booy (1956). Within this rock, in some areas, biotite is the only mica present. In others, biotite coexists with muscovite, and yet in others, muscovite replaces biotite altogether (Cobbing *et al.,* 1981). Locally, the leucogranodiorite is cut by monzogranites, and along its margins, dioritic and tonalitic contamination is common. K–Ar age determinations on these rocks range between 13 and 2 Ma with a cluster at 9 Ma (Stewart *et al.,* 1974). Our preliminary U–Pb zircon ages for this batholith fall between 9.9–12.2 Ma (Mukasa and Tilton, in prep.).

Petrogenetic models for the Coastal Batholith

Several petrogenetic studies (Regan, 1976; Taylor, 1976; Cobbing *et al.,* 1977b; Atherton, 1981; McCourt, 1978, 1981), have been carried out in various parts of the Coastal Batholith. For the gabbroic members of the batholith, Regan (1976) proposed diapirically initiated plumes of peridotitic mantle resulting from seismic disturbances related to the subduction process. Partial melting of the diapirs during ascent should eventually result in the segregation of an olivine-rich and low-Al tholeiitic liquid which experiences fractional crystallization of olivine and orthopyroxene at the base of the crust yielding a high-Al tholeiitic magma whose chemical character is identical with that of the batholithic gabbros in general.

Cobbing *et al.* (1977b) favour a crustal melting hypothesis whereby, as a result of the emplacement of large volumes of mantle-derived mafic magma into the base of the crust, melting processes generate the early granitoids. Their hypothesis calls for the progressive contraction of the melt zones so that areal extent of the resulting plutons is diminished with time. The Lima segment is presented as an illustration of this phenomenon. Here, for example, the early gabbroic, meladioritic and tonalitic super-units occupy the whole segment. Granodioritic units occupy almost the same length, but in much smaller volumes, and the youngest monzogranites occur only in small and isolated centred complexes.

Using major oxide, trace element and Sr isotope data, McCourt (1978, 1981) argued for two magma sources. $^{87}Sr/^{86}Sr$ ratios of 0.7042, suggesting a mantle origin, appear to be universal for all units in the Lima segment. However, because trace elements in granodiorites and granites contradict arguments for a single source model, McCourt (1981) suggested that perhaps partial melting of crustal underplated andesitic material, initially of mantle origin, is the source of these more silicic rocks. The gabbros, diorites and tonalites have been attributed to partial melting in the wedge above the subducted slab.

The purpose of our work is to further identify the nature of the mantle source postulated by the models mentioned above, and to estimate the degree of possible contamination with crustal materials as magmas rise and are emplaced within epizonal depths.

Pb isotopes

Tilton and Barreiro (1980) demonstrated that a component of lead from the Precambrian basement rocks occurs in the Plio-Pleistocene volcanics of south Peru. Studies by De Paolo and Wasserburg (1977) and James (1982a) using Nd, and Sr and O isotopes, respectively, have also detected a Precambrian component in the Quaternary lavas. Considering the wide areal extent of these basement rocks in the southern part of the Western Cordillera and in the entire Coastal Cordillera, we envisioned that perhaps they also contributed materials to magmas from which some of the older volcanics and plutonic rocks in the area formed.

We present here Pb isotope results on carefully picked and prepared feldspars from the Coastal Batholith in the areas of Arequipa Department and the Huaura river valley just north of Lima, the northern sector of the San Nicolás Batholith between San Juan and Pisco, and from the Cordillera Blanca Batholith and the stocks that share its trend. These results are graphically summarized in Figure 3.

The $^{207}Pb/^{204}Pb$ versus $^{206}Pb/^{204}Pb$ correlation diagram shows that all sample fields lie well above the ORL. This indicates that a depleted mantle source such as those observed in many oceanic settings and some continental areas is not dominant here. Out of the 31 samples analysed from the Coastal Batholith 27 fall in two distinct groups: those which lie slightly to the left (group I) and those (group II) which plot to the right of, but also overlap with, the highly homogeneous source identified in the central and southern Chilean Quaternary rocks and ores by Tilton (1979), Tilton and Barreiro (1980) and Barreiro and Stern (1982). The remaining four samples are the extremes of group II, and they are treated separately below.

Group I samples are all from the Arequipa segment where Precambrian basement exposures are very abundant. Those in group II are from the Lima segment where no old basement exposures have been observed. Because most of the batholithic rocks have ages of 80 ± 20 Ma (see Table 1), the mantle reservoir isotope ratios are adjusted to 80 million years ago by moving the field 80 million years along the Pb evolution curves in Figure 3. Although the exact U/Pb and Th/Pb ratios in the reservoir are unknown, the adjustment is small and should not produce a very significant uncertainty. The isotopic data shows that the group II samples have higher $^{206}Pb/^{204}Pb$ ratios than the presumed reservoir. This relationship appears to be compositionally controlled. Gabbros of the Patap super-unit are close to the central and southern Chilean field at 80 Ma, but diorites, tonalites and granodiorites of the Paccho and Santa Rosa super-units have slightly lower $^{207}Pb/^{204}Pb$ and considerably higher $^{206}Pb/^{204}Pb$ ratios. Analyses of the more evolved super-units of the centred complexes which are granitic to monzogranitic in composition (open circles in Figure 3), give extremely radiogenic ratios defining a linear regression array with a slope of 0.1179 having a correlation coefficient of 0.9972, which corresponds to a Pb isochron age of 1.92 Ga. We believe that the Pb isotope trend displayed by these silica-rich super-units is

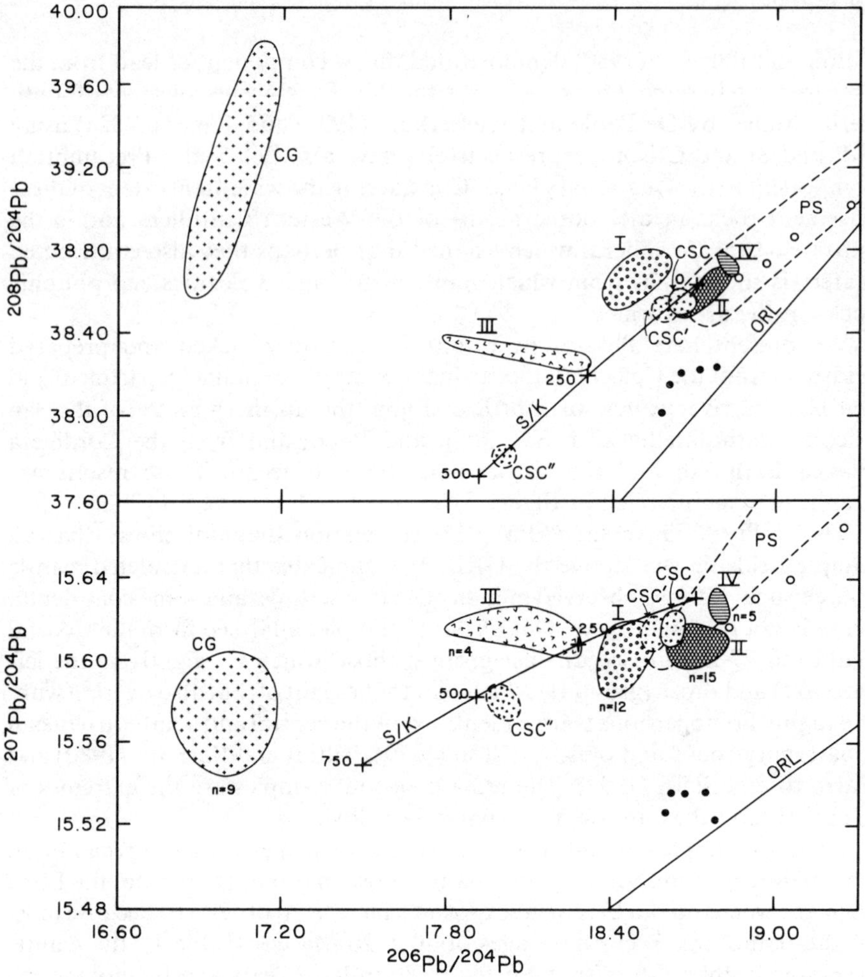

Figure 3. **Pb isotope compositions of the Arequipa segment (group I) and Lima segment (group II and open circles) of the Coastal Batholith, the San Nicolás Batholith (group III) and the Cordillera Blanca Batholith and associated stocks (group IV). These are evaluated in the framework of the Nazca plate basalts (closed circles), Pacific Ocean sediments (PS), time adjusted central and southern Chile homogeneous source (CSC, CSC′ and CSC″), and the Charcani gneisses (CG). The Pb evolution curve of Stacey and Kramers (1975) (S/K) is shown for comparison**

indicative of high level assimilation of crustal materials such as those found in the Jurassic to Upper Cretaceous marginal basin of Peru. The Pb isochron age seems to have no real geologic meaning although it could represent sedimentary debris derived from crustal materials with an age of 1.92 Ga.

We feel that the Group I data require a more complex explanation involving a three-component source consisting of the mantle reservoir, the upper crustal

Pb seen in the Group II samples, plus a lower crustal component typified by the Charcani gneiss data. Such a source adequately explains the isotope fields for Group I leads in both parts of Figure 3.

The San Nicolás Batholith data require an explanation similar to that for the Group I leads. The presumed homogeneous mantle reservoir must be adjusted to an age of 400 Ma in this case. Again there is no obvious two-component model which will explain the isotope relations, using the mantle reservoir as one of the components. An adequate model assumes mixing between the reservoir lead and lower crustal Pb, similar to that in the Charcani gneiss, followed by addition of upper crustal lead, presumably at higher levels. The lower crustal source requires a somewhat higher $^{207}Pb/^{204}Pb$ ratio than any actually observed in the Charcani gneiss field. This could be due to the limited sampling done thus far.

Finally, the reconnaissance data we have gathered from the Cordillera Blanca Batholith and the Carhuish and Surco stocks are presented as group IV in Figure 3. Present sampling is too limited to allow complete assessment of the data. It is clear, however, that no detectable Precambrian basement Pb is present in these rocks. Considering the mineralogy of the sample suite (hornblende granodiorite and single- and two-mica leucogranodiorites) it is conceivable that the cluster may represent a mixture of varying amounts of mantle-derived material and some relatively young sediments.

Conclusions

1. It is apparent that there is a spatial association between the occurrence of a Precambrian basement component of Pb in the Phanerozoic intrusives and exposures of Precambrian granulites and gneisses. The old Pb component has been detected in all samples from the northern sector of the San Nicolás Batholith and in the Tiabaya, Incahuasi and Linga super-units of the Arequipa segment of the Coastal Batholith. In contrast, super-units in the Lima segment where Precambrian exposures are totally lacking exhibit no detectable old crustal Pb component.
2. The dominant Pb component in all super-units appears to have come from a (time-adjusted) isotopically homogeneous reservoir very similar to that identified by Tilton (1979); Tilton and Barreiro (1980) and Barreiro and Stern (1982) for the Quaternary ones, and volcanic and plutonic rocks in central and southern Chile. If this reservoir represents a mantle source, then the sub-Andean mantle in Peru does not have a depleted character. Comparable Pb isotopic results from the Columbia River basalts (Church, 1983) and Yellowstone Plateau volcanics (Doe *et al.,* 1982), give credence to the occurrence of such enriched mantle sources in numerous areas.
3. Assuming the same degree of fractional crystallization of Pb-rich phases in all pre-contaminated magmas, mixing models suggest that there is more Precambrian crustal component in the Plio-Pleistocene volcanics of southern Peru (Tilton and Barreiro, 1980) than there is in the Arequipa segment plutons of the same area.

Geochemical Characteristics of the Late Mesozoic Andean Volcanics

M. BOILY, C. BROOKS and D. E. JAMES

Abstract

We have evaluated new trace element, Nd and Sr isotope analyses for two late Mesozoic volcanic series located in south Peru: the Jurassic Chocolate Formation and late Cretaceous Toquepala Group. The trace element results confirm the close affinity of the Chocolate extrusives with other calc-alkaline volcanics erupted in mature volcanic arcs or active continental margins. The lavas are divided in four groups; within each group the basaltic andesites and andesites can be generated by fractional crystallization from a basaltic parent enriched in LIL elements. Although Sr isotope data may show some dispersion due to alteration processes, Nd isotopes clearly indicate a depleted source for the entire suite ($\varepsilon Nd_i = +2.7$ to $+7.9$). The Toquepala rhyolites and dacites display great variation in their trace element ratios, but in general have high LIL element concentrations. Whilst the Nd isotopes indicate a depleted source ($\varepsilon Nd_i = +3.0$ to $+6.5$) their origin remains unclear. Derivation by crystal fractionation from an andesitic parent or by partial melting of underlying Precambrian crust is not supported by the combined trace element and isotopic data. However, the complete volcanic suite must have undergone a different geochemical evolution to that of the Cenozoic Arequipa and Barosso Volcanics which assimilated Precambrian crustal material during ascent in the upper crust.

Introduction

Recent geochemical studies concerning subduction-related volcanism in the Andean orogenic belt have focused primarily on Cenozoic volcanic rocks. Extensive sets of major, trace element and isotopic analyses have been used to decipher the origin of the calc-alkaline andesite-rhyolite suite exposed in the CVZ (see James 1982a; Deruelle, 1982; Thorpe and Francis, 1979). In contrast very few geochemical studies are reported for the Mesozoic volcanics which are thought to represent the earliest volcanic events associated with the subduction of the Pacific seafloor beneath the western edge of the South American plate (see James *et al.*, 1974, 1975; Dostal *et al.*, 1977b).

In this summary paper we present new trace element and isotopic analyses for two late Mesozoic volcanic sequences localized in south Peru; the Chocolate Formation and the Toquepala Group. The aim of these current investigations is to add a new geochemical and temporal perspective to the overall geological evolution of the andesite-rhyolite magmatism of the central Andes.

Geological framework

The volcanic rocks of the Chocolate Formation crop out along the south Peruvian coast (Figure 1). Both marine and subaerial in character, they are associated in place with marine clastic sediments and limestones, suggesting eruption near sea level in a coastal area. The volcanics, mainly basaltic andesitic to basaltic in composition (Figure 2), are thought to represent the initial stage of development of an Andean island arc. New and previous geochronological work with Sr isotopes indicate Jurassic ages between 185 to 150 Ma for the volcanic suite (see James *et al.*, 1974, 1975).

The Toquepala Group is late Cretaceous in age (70 Ma, James *et al.*, 1974) and consists of volcanic flows, breccias and agglomerates of andesitic to rhyolitic composition. In close spatial association with older Cretaceous volcanics (100 Ma; Boily, unpublished data), they are exposed on the Andean flanks just west of the zone of Cenozoic volcanism and overlie Precambrian sialic crust.

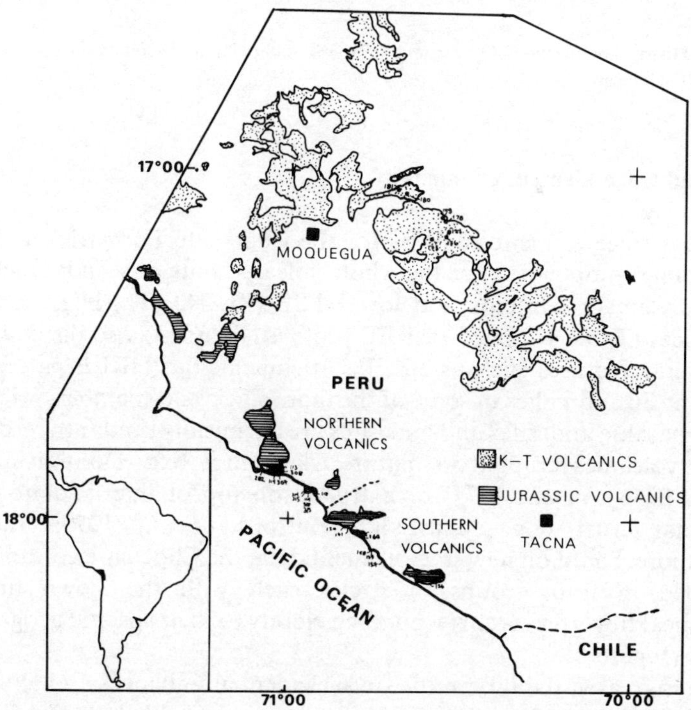

Figure 1 Geologic and sample locality map showing exposed volcanic rocks of the Chocolate Formation (Jurassic) and Toquepala Group (Cretaceous), modified from James *et al.* (1974)

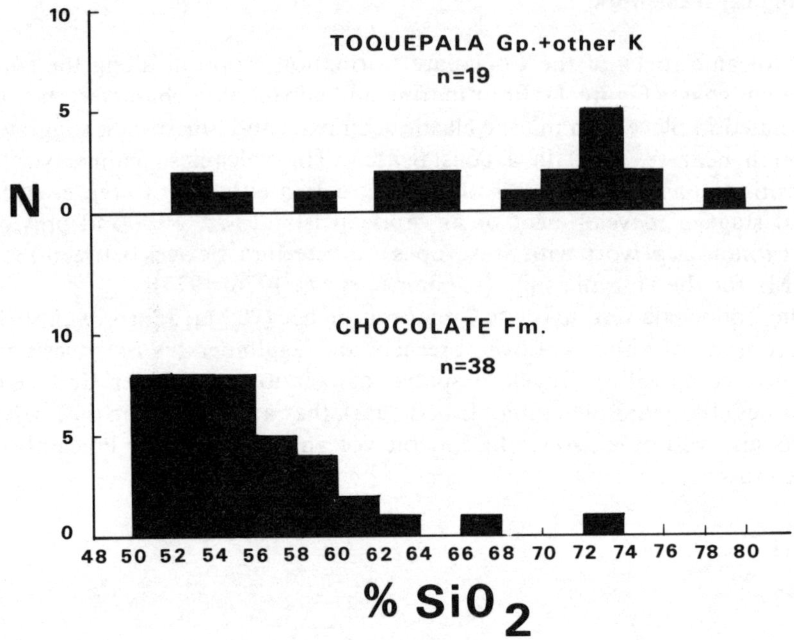

Figure 2 Histogram of the SiO_2 content of the Toquepala, mid-Cretaceous volcanics, and Chocolate volcanics

Major and trace element chemistry

Major and trace element analyses for the Chocolate Formation indicate a calc-alkaline composition for the whole volcanic suite as seen by high Al_2O_3 and LIL element contents and low HFSE (Zr, Nb, Ta, Hf), Cr and Ni abundances (Table 1). Selected REE profiles (Figure 3) also display typical calc-alkaline features such as LREE enrichment, flat HREE patterns and significant Eu anomalies in some of the more silicic end-members. Hence the Jurassic basaltic andesites and andesites are chemically similar to recent calc-alkaline volcanics erupted in mature island arcs (e.g. Dominica, Lesser Antilles; Brown *et al.*, 1977) or extruded on top of intermediate or thin continental crust (e.g. Java-Bali; Whitford *et al.*, 1979; Table 1). Furthermore, based on new trace element data, the Chocolate volcanics were subdivided in four groups or cycles each with their own fine-scale geochemical fingerprints corresponding closely to their specific geographical location (Figure 3).

There is great variability in the trace element abundances and abundance ratios of the Toquepala and mid-Cretaceous rocks (Table 1); but as expected, the dacites and rhyolites possess an overall highly incompatible trace element content. Their REE profiles (Figure 3d) display LREE enrichment and flat

Table 1 **Summary of the major and trace element data for the Chocolate, Toquepala and other mid-Cretaceous volcanics**

	Jurassic volcanics (Chocolate Formation)			Cretaceous volcanics (Toquepala Group and older mid-Cretaceous rocks)		Modern arc volcanics	
	Basalts n = 8	Basaltic andesites n = 16	Andesites n = 11	Dacites and rhyolites n = 19		Java-Bali	Dominica
SiO_2 (%)	51.01	54.02	58.54	69.86		59.6	57.2
Al_2O_3	18.77	17.80	18.25	13.99		18.1	17.8
TiO_2	0.84	0.88	0.72	0.45		0.74	0.77
Ba (ppm)	495	479	424	727	(164–1627)	420	210
Rb	56	67	72	101	(36–128)	62	36
Sr	431	385	332	209	(39–536)	360	292
Zr	100	130	137	145	(73–264)	151	89
Nb	5	6	8	11	(9–14)	9.4	5.0
Y	23	28	25	27	(19–37)	28	27
Ni	14	18	17	—		8	35
Cr	109	92	87	—		18	122
Th	2.0	4.5	7.93T	11.6	(1.7–38.6)	—	—
Hf	2.72	2.73	3.33T	3.84	(2.02–4.64)	—	—
Ta	0.13	0.31	0.37T	0.67	(0.32–1.37)	—	—
Th/Hf	0.7	1.64	2.4	3.02		—	—
$La/Sm_{e.f.}$	1.69	1.91	2.17	3.36		3.37	—
Zr/Nb	20	22	18	13		16	18
Zr/Y	4	5	6	5		5	3
Ti/Zr	50	41	32	25		29	52
Rb/Sr	0.13	0.17	0.22	0.48		0.17	0.12
H_2O (%)	1.47	1.70	1.55	—		—	—
CO_2	0.24	0.28	0.22	—		—	—

Major elements from James *et al.*, 1974 and unpublished data. Rb, Sr, Y, Ba, Zr, Nb, Cr and Ni were determined by XRF at Université de Montréal using standard techniques (Schroeder *et al.*, 1980). REE, Ta, Hf and Th were analysed by INAA at Université de Montréal and Ecole Polytechnique following the procedure of Bergerioux *et al.* (1979). Data from recent volcanics are from Whitford *et al.* (1979) and Brown *et al.* (1977). T—n = 7.

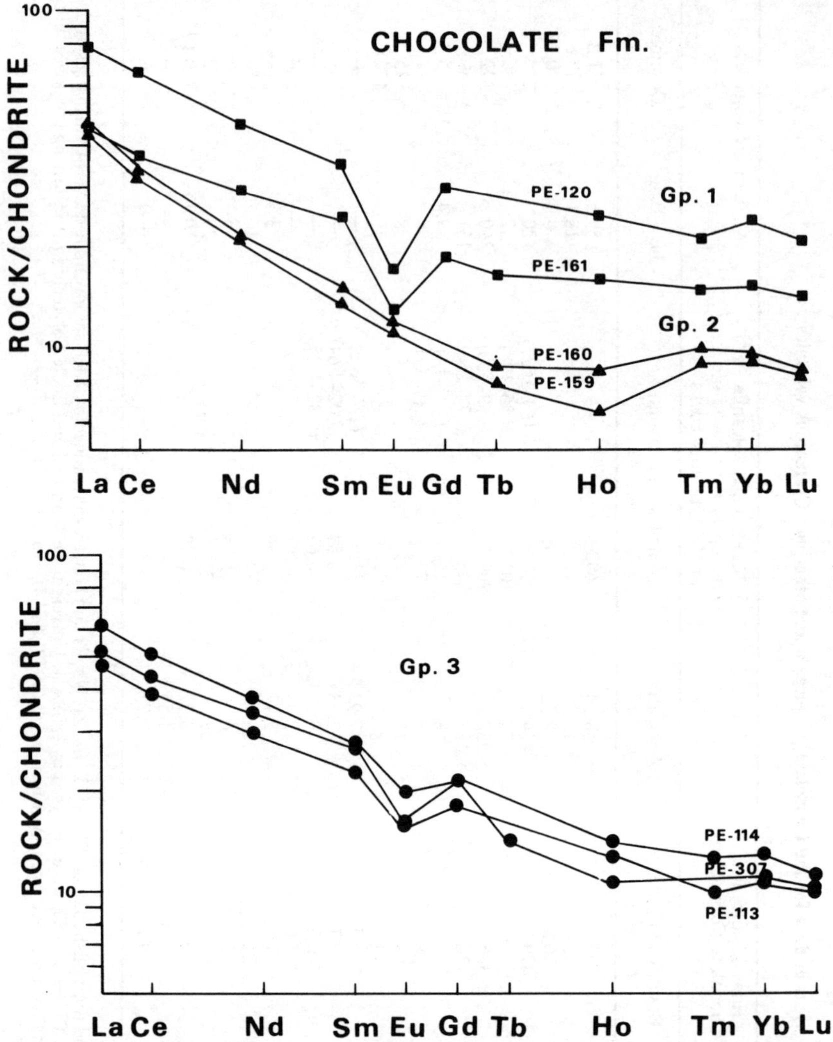

Figure 3 Selected REE profiles for the Chocolate Formation and Toquepala Group. Data are. normalized to the chondritic values of Taylor and Gorton (1977)

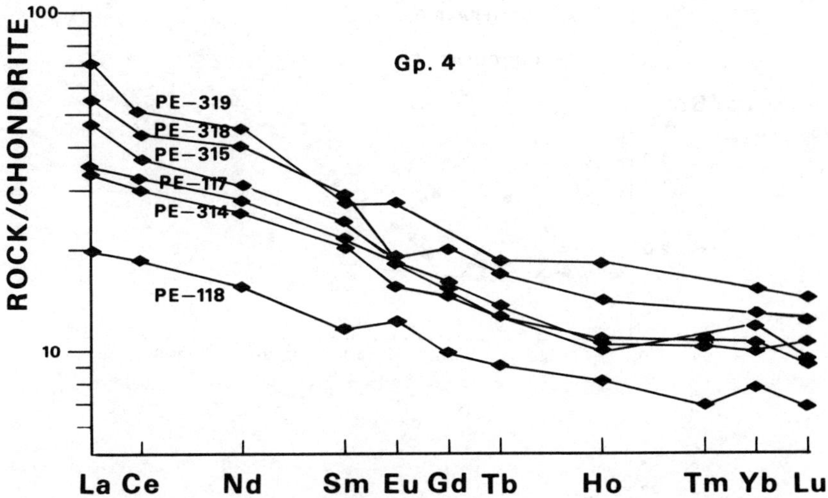

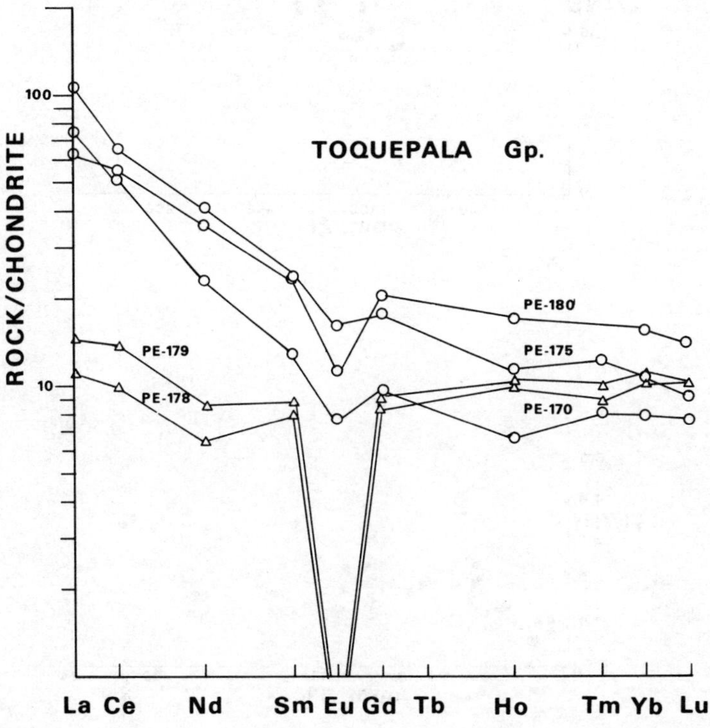

Figure 3 (contd.)

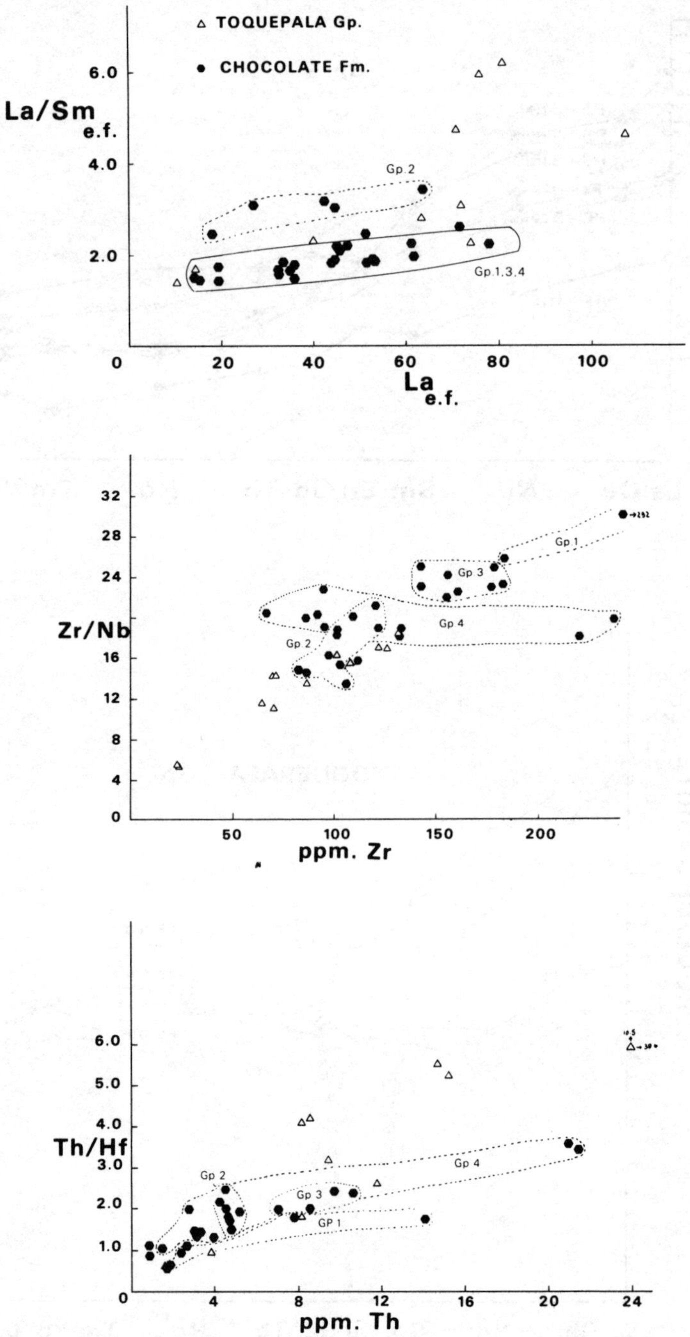

Figure 4 Binary diagrams showing selected incompatible trace element ratios plotted against one incompatible element ($La_{e.f.}$, Zr and Th) used as differentiation index

HREE patterns while some rhyolites have very low REE abundances (~10× chondrite normalized) and huge negative Eu anomalies.

Selected incompatible trace element ratios are reported in binary plots similar to the Joron and Treuil (1977) diagrams (Figure 4). Here $La_{e.f.}$; Zr and Th are plotted on the abcissa and used as differentiation indices. While the trace element ratios of the Chocolate volcanics increase only slightly during differentiation (because $D_{La} < D_{Sm}$, $D_{Th} < D_{Hf}$, $D_{Zr} < D_{Nb}$ for basaltic and andesitic rocks; D—bulk distribution coefficient), the results are more difficult to interpret for the Toquepala rocks. For example, Figure 4b shows a good linear correlation between Zr/Nb and Zr while Figure 4a and 4c exemplify the greater dispersion found in most of the data. These results may actually reflect the variable incompatibility of certain trace elements in silicic magmatic system in response to complex magmatic processes (see below). Hence, care is needed when trying to interpret the Toquepala data in terms of partial melting or crystallization trends since our studies have also shown no systematic variations between trace element abundance ratios and major element content.

Taking into account the data base and the unique differentiation trend of each Chocolate subgroup or cycles, we propose that the Chocolate basaltic andesites and andesites are related by a process of fractional crystallization. By this model, variable amounts of plag. + cpx + ti magnetite ± opx are separated from a slightly LILE enriched basaltic parent which had undergone previous fractionation of olivine ± cpx (low Ni and Cr).

Petrogenesis of the Toquepala volcanics, as judged by trace element abundances, remains unclear. The two most obvious possibilities—formation of the siliceous volcanics by small amounts of partial melting or direct derivation by lower pressure crystal fractionation from an andesitic parent—are ruled out by isotopic data in the first instance (see below) and by trace element variations in the second. We speculate that complex magmatic processes, possibly involving liquid state differentiation in a magma chamber and trace element transport by fluids and volatiles may play an important role in the formation of some of the rhyolites (see Hildreth, 1981).

Isotopic data

This paper brings together isotope data previously obtained at the Department of Terrestrial Magnetism and new Sr and Nd isotope analyses from Université de Montréal. While the new Sr isotope data essentially confirms the results reported earlier by James *et al.*(1974, 1975) for the two volcanic sequences, we stress that the analysed samples, though carefully chosen for their relative freshness (i.e. low CO_2, H_2O), may have suffered significant Sr and other alkalis mobility through seawater alteration, metamorphism and in some cases silicification. Hence care is necessary in interpreting the individual age or initial isotopic ratios of some of the samples. Despite this, the Toquepala volcanics have been shown to display a convincing linear array on a $^{87}Sr/^{86}Sr$ versus Rb/Sr plot (James *et al.*, 1974). However, when the samples are regrouped by their geographical location, ages of 63 and 74 Ma are obtained giving initial ratios of 0.7048 ± 2 (Group A) and 0.7042 ± 2 (Group B) at 2 σ confidence level. The preferred age adopted for the Chocolate volcanics is 185 Ma, giving initial $^{87}Sr/^{86}Sr$ around 0.7038, (James *et al.*, 1974, 1975; this study). However, regression analyses made on several subgroups of samples yield ages ranging from 140 to 205 Ma. This suggests that some volcanics may be in fact younger or that alteration and metamorphism strongly affected some samples and considerably changed their respective Rb/Sr and $^{87}Sr/^{86}Sr$ values. We believe the first hypothesis to be true at least for the southern volcanic lavas (see Figure 2) as they form distinct geographical and geochemical groups apart from their northern counterparts. A mean age of 150 Ma (giving an initial $^{87}Sr/^{86}Sr$ value around 0.7046) was thus adopted for these volcanics.

As REE are less sensitive to alteration processes, we have analysed 13 samples for their Nd isotope ratios and present the results on an isochron plot (Figure 5). In this diagram, a tight clustering of the Toquepala samples is apparent if we exclude two discordant samples. These two, which belong to Group A, yield an age of 85 Ma (65 Ma with the Rb/Sr techniques) with an initial $^{143}Nd/^{144}Nd$ ratio of 0.51273 while the other samples (Group B), when corrected back for a 74 Ma age, possess a mean initial $^{143}Nd/^{144}Nd$ ratio of 0.51286. Thus the isotopic differences found in both groups with Sr isotopes are corroborated by the Nd data. Also evident is the net division of the

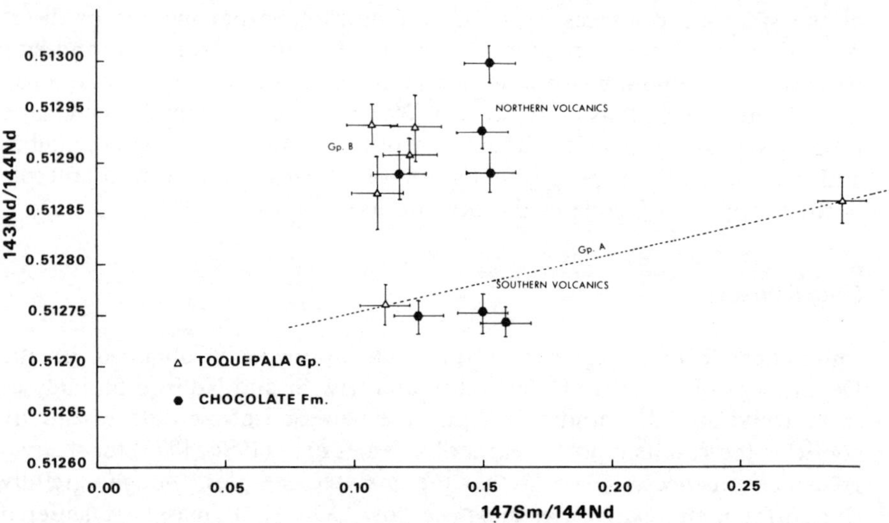

Figure 5 $^{143}Nd/^{144}Nd$ versus $^{147}Sm/^{144}Nd$ for the late Mesozoic volcanics of south Peru. Symbols as in Figure 3. Nd isotope ratios are normalized against $^{143}Nd/^{144}Nd = 0.7219$; errors quoted on $^{143}Nd/^{144}Nd$ are 2 standard errors on the mean. Sm and Nd were analysed by INAA. Errors on $^{147}Sm/^{144}Nd$ are 10% or less

Chocolate volcanics into two clusters. It is noticeable that the lowest uncorrected Nd isotope values (0.51274 to 0.51275) occur solely in the southern volcanics, while the other group located further north on the Peruvian sea coast displays higher values (0.51230 to 0.51289) and shows a slight dispersion on the isochron plot (Figure 5). No meaningful ages are obtained either by the Sm–Nd and Rb–Sr method when these 4 samples are taken together, and we think it highly improbable that this is related to alteration processes because these samples are amongst the freshest rocks sampled in the field. We therefore believe that it represents some small scale source heterogeneity or even source contamination (by sea water and subducted sediments?); this last process now being widely postulated in the magmagenesis of recent calc-alkaline volcanic rocks (see Hawkesworth *et al.*, 1979a). Moreover, if ages of 185 Ma and 150 Ma are assumed for the northern and southern volcanics, then initial $^{143}Nd/^{144}Nd$ ratios ranging from 0.512705 to 0.512812 are obtained for the first group, while a mean of 0.51261 is obtained for the latter. This implies at least two distinct isotopic sources, an assumption which seems also supported by the Sr isotopes.

When reported on a time corrected ϵSr versus ϵNd diagram (Figure 6a) all the Nd data fall in the 'depleted' quadrant, while the Sr data trend towards the 'enriched' portion. The scatter observed may in part reflect the combined effect of age uncertainties, alteration and possible source or crustal contamination, although the Nd isotope data clearly imply a depleted source for the parental magma of both series. Futhermore, note that the time corrected ϵ values are similar to those of recent calc-alkaline volcanics erupted in mature island arcs or in magmatic arcs underlain by thin or intermediate continental crust (e.g. Grenada, Ecuador; Figure 6b).

Possible depleted sources can be found in the upper mantle (subcontinental or oceanic), the subducted oceanic plate or parts of the lower crust. More specifically, mantle melts coming from the wedge overlying the descending slab, and previously enriched in LIL elements by fluxing or melting of the upper subducted plate, would possess the required isotopic and trace element composition of the Chocolate volcanics source (Saunders *et al.*, 1980). This may also be true for the Toquepala volcanics if these siliceous rocks are derived mainly by crystal fractionation from an andesitic parent originating in the upper mantle. However, if crustal material is involved in their magmatic evolution, it must be fairly unradiogenic in character or relatively young in age. Hence, the underlying Precambrian sialic crust is immediately ruled out as a source due to its high $^{87}Sr/^{86}Sr$ (0.730–0.740) and low $^{143}Nd/^{144}Nd$ (0.5115) ratios (James, 1982a). Alternatively, melting of a 'calc-alkaline' source (whether basaltic or peridotitic) preserved in the lower Andean crust or in the subcontinental mantle since Jurassic times (130–180 Ma) would produce a parental magma with the right isotopic values at 70 Ma (i.e. $^{87}Sr/^{86}Sr = 0.7042$; $^{143}Nd/^{144}Nd = 0.51286$) provided it kept initial Rb/Sr and Sm/Nd ratios of 0.20 and 0.25 with $^{87}Sr/^{86}Sr$ and $^{143}Nd/^{144}Nd$ ratios of 0.7038 and 0.5128, values which are in fact thought to represent the source material of the northern Chocolate volcanics.

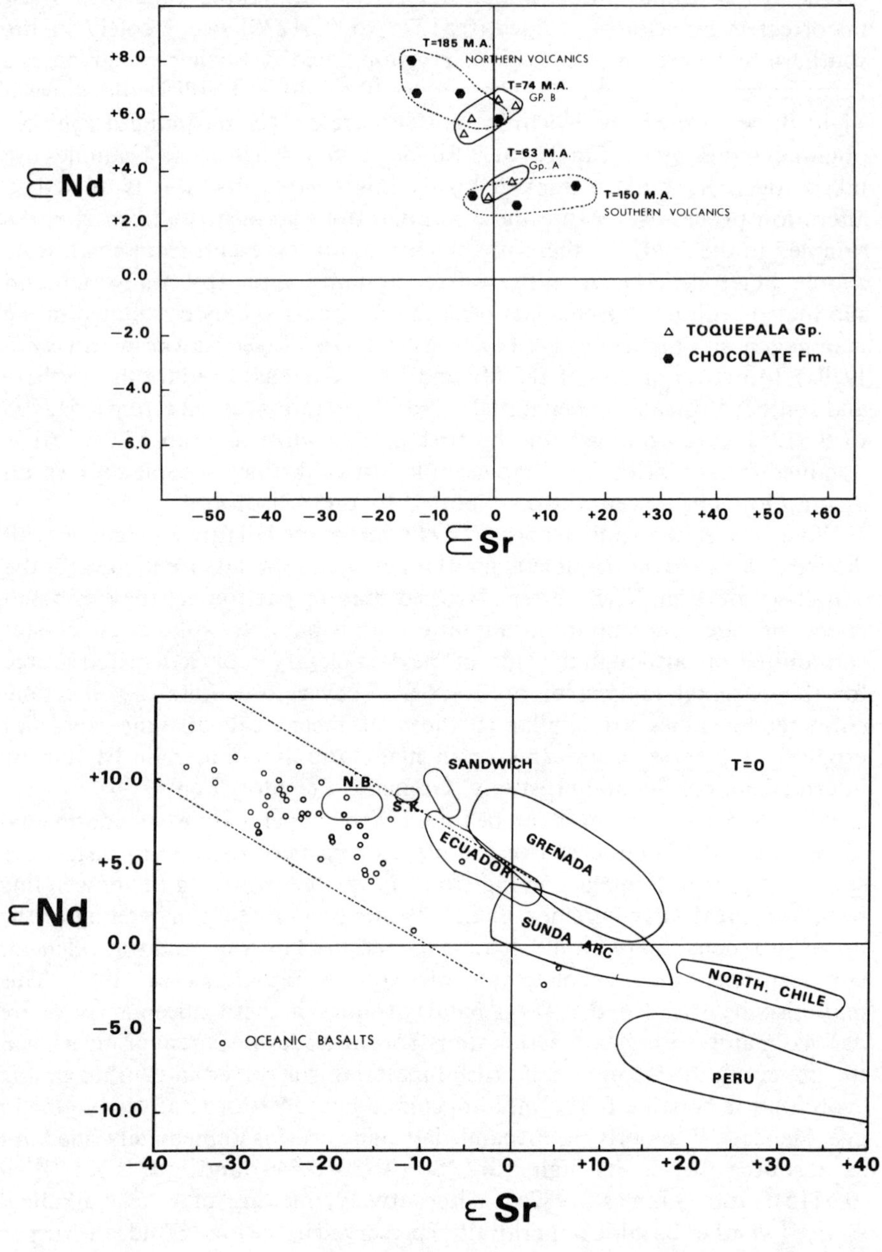

Figure 6 (a) Time corrected εNd versus εSr diagram for the Chocolate and Toquepala volcanics. Symbols as in Figure 3. (b) Present day εNd versus εSr plot for selected orogenic volcanic rocks. Data sources are: N.B.—New-Britain, De Paolo and Johnson (1979); Sandwich—Sandwich Islands, Hawkesworth *et al.* (1977); S.K.—St. Kitts, Lesser Antilles, Hawkesworth and Powell (1980); Grenada—Grenada, Lesser Antilles, Hawkesworth *et al.* (1979a); Ecuador and Northern Chile, Hawkesworth *et al.* (1979b); Sunda arc, Whitford *et al.* (1981); Peru, James (1982a)

When looking at Figures 6a and b, one must be impressed by the large variations found in the isotopic composition of the Peruvian volcanic rocks from Mesozoic through Recent times. Even more puzzling, is the close spatial association of the Late Cretaceous volcanics with the Plio-Pleistocene to Recent lavas (e.g. Barroso and Arequipa volcanics). While the isotopic and trace element variations found in the latter groups have been ascribed to a fractional crystallization-assimilation process, the contaminant being the underlying Precambrian Charcani gneiss (Tilton and Barreiro, 1980; James, 1982a), it is clear that such a process must be ruled out in the magmatic evolution of the Toquepala volcanics. On the other hand, the Chocolate volcanics probably formed part of a Jurassic magmatic arc lying on the western continental margin near the Pacific coast. Although they are emplaced within the Precambrian Palaeozoic Arequipa Massif (Shackleton *et al.*, 1979), their geochemical signature suggests minimal contamination by such an 'enriched' material. We suggest that a rapid rate of ascent through a thinner Mesozoic crust (at that time), coupled with short residence periods in the upper crust during differentiation, may possibly explain the absence of extensive crustal contamination in both volcanic series. One problem with this model is that the highly differentiated nature of the Toquepala rocks implies a longer stay in the upper crust than was experienced by the Chocolate basaltic andesites and andesites.

Concluding remarks

Major and trace element and isotopic analyses have confirmed the geochemical affinity of the Chocolate and Toquepala volcanics with other recent calc-alkaline volcanics erupted at destructive margins. Nd isotope data indicate a depleted source for the volcanics from 180 to 70 Ma, but contrary to their Cenozoic counterparts, they show little or no contamination by Precambrian sialic material.

It seems that rapid changes in the isotopic composition of the central Andean magmas occurred only through Cenozoic times. For example, prior to crustal contamination the parental magma(s) of the Barroso and Arequipa volcanics were clearly isotopically enriched (e.g. $\varepsilon Sr = +7$ to $+34$, $\varepsilon Nd = -2$ to -6; James, 1982a) while the Toquepala volcanics had a depleted signature at just 70 Ma.

Possible causes for these abrupt changes may be the development of partial melting sites at deeper level in the lower crust or in the upper mantle as crustal thicknesses increased during the late Cenozoic times, thus permitting the sampling of an enriched reservoir. While it is difficult to conceive such a reservoir in a 'normal' upper mantle, old enriched and heterogeneous material could be preserved in the lower Andean crust or in the subcontinental mantle, to be later melted, mixed or used to contaminate a more depleted magma. Alternatively increases in elevation accompanying crustal thickening could also enhance crustal contamination as magmas would need to undergo more extensive crystal fractionation before attaining densities low enough to permit eruption from greatly elevated volcanic cones. In any case, further geochemical and geological studies are needed to document the significance of this 'transition gap' between 70 Ma to present.

The Magmatic Evolution of the Cordillera Oriental, Southeastern Peru

D. J. KONTAK, A. H. CLARK and E. FARRAR

Abstract

Magmatism within the Central Andean Inner Arc of the *Cordillera Oriental* of southeastern Perú is characterized both by its episodicity (*ca.* 350, 270–250, 240–230, 180, 90–70 and 26–4 Ma) and its variable chemical/petrological signature. In both respects, marked contrasts are evident with the quasi-continuous and petrochemically unified subduction-related magmatism of Triassic-Quaternary age in the *Cordillera Occidental.* The geochemical and petrographic features of the volcanic and plutonic rocks reveal highly variable crustal and mantle contributions within, and between, episodes of magmatism. For example, the 270–250 Ma Mitu Group volcanics (Sr_i = 0.704), and the 180 Ma Macusani syenite complex represent mantle-derived melts which experienced negligible crustal modification. In contrast, the 26–20 Ma (K–Ar, biotite) peraluminous (A/CNK $>$ 1.2) cordierite-sillimanite-bearing granitoids (Sr_i = 0.718), and 17–4 Ma (K–Ar, biotite) peraluminous (A/CNK $>$ 1.2) andalusite-sillimanite-muscovite ± cordierite-bearing rhyolitic ash-flow tuffs (Sr_i = 0.722) of the Macusani area represents melts of crustal derivation. The remaining suites, however, are chemically and petrographically more ambiguous, and in some cases variable proportions of mantle and crustal components were probably involved. The Permo-Triassic Coasa granitoid batholith, for example, comprises suites of both mantle and crustal derivation which may be distinguished on the basis of their mineralogy and whole-rock and mineral chemistry.

The relationship between the periods of Inner Arc magmatism and regional Andean tectonics (e.g. subduction, deformation) remains obscure. However, magmatism was associated with compressional, extensional and 'neutral' tectonic regimes. We attribute the continued juxtapositioning of multifarious igneous rocks in this relatively small area (8800 km^2) to the existence of a zone of structural weakness, probably of Precambrian origin. The structure was variably reactivated during the Phanerozoic in response to regional stress systems and acted as a zone along which magmas of varying origin were channelled, and in some instances (i.e. the crustal-derived suites) may have been partially responsible for generating some of these magmas.

Introduction

The main focus of research into the magmatic evolution of the Central Andes has been on those igneous suites which form the 'Main Arc System' (Clark *et al.*, 1983; Figure 1). In contrast, comparatively little attention (e.g. Carlier *et al.*, 1982) has been paid to the distinctive rocks which constitute the areally restricted 'Inner Arc', cospatial with the *Cordillera Oriental* (eastern

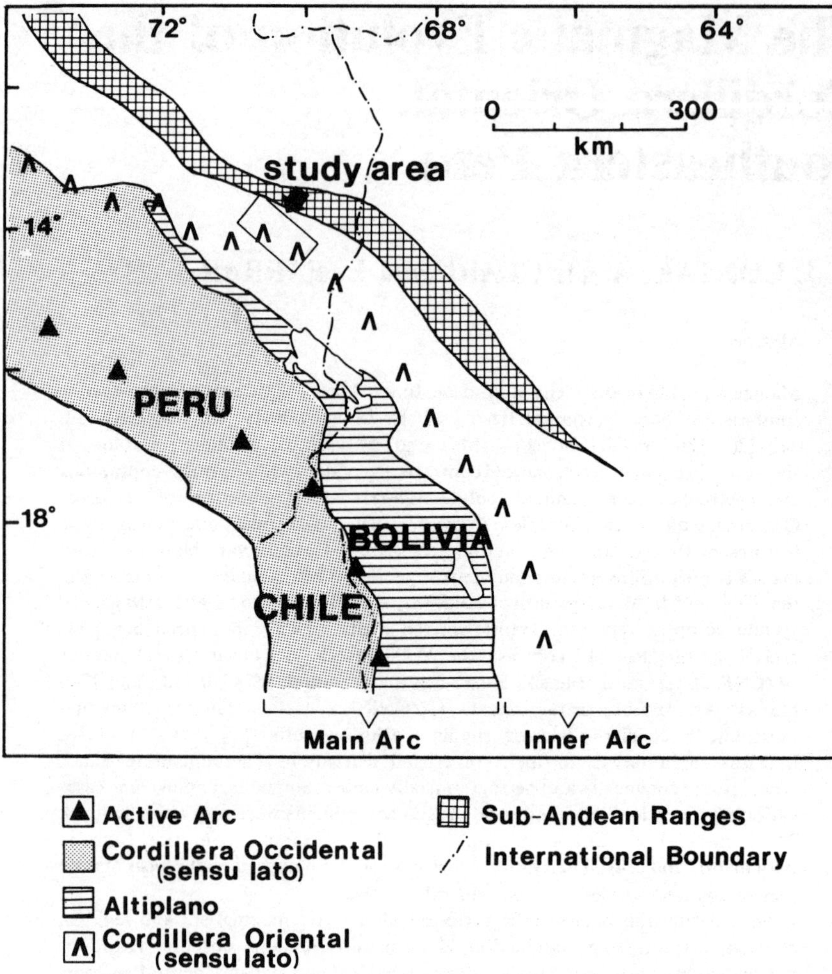

Figure 1 Location of the study-area, the Cordillera de Carabaya region, in the *Cordillera Oriental* of southeast Peru. The characteristics of the 'Main Arc' and the 'Inner Arc' systems are discussed in the text.

cordillera) of Peru and Bolivia. Whereas the Main Arc magmatism was, overall, quasicontinuous (Clark *et al.*, 1976; Aguirre, 1983), mantle-dominated (this volume), and subduction zone-related (Stauder, 1973; Barazangi and Isacks, 1976; James, 1971b), Inner Arc magmatism was instead episodic and, periodically, crustal source-dominated (Palma, 1981; Pitcher, 1983), and can only be tenuously related to subduction.

The Cordillera de Carabaya region of southeastern Peru (Figures 1 and 2), displays several superimposed igneous suites which well illustrate the characteristics of the Inner Arc. This paper summarizes our field,

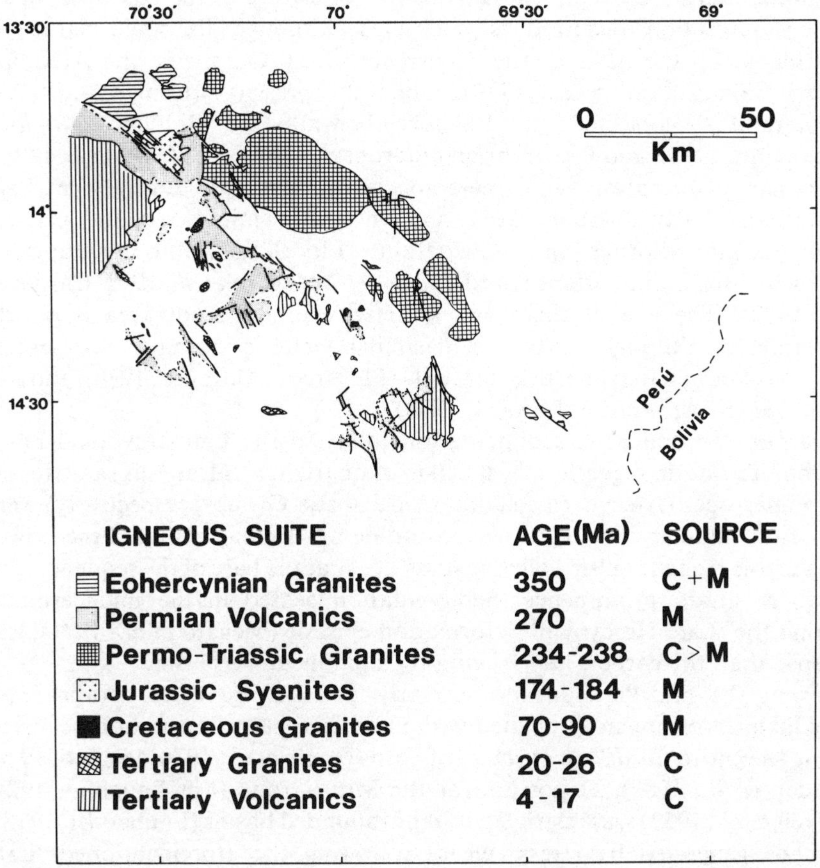

Figure 2 Geological (after Laubacher, 1978) and geochronological (U–Pb zircon; K–Ar biotite; Rb–Sr whole-rock) outline of the igneous suites of the Cordillera de Carabaya area. C and M represent the crust and mantle, respectively, and refer to the source region assigned to that particular magmatic episode as discussed in the text

petrological, geochronological, isotopic and geochemical studies of these suites. The new data suggest that markedly different processes controlled Inner Arc magmatism at different times. Magma compositions display wide variations in response to the tectonic regime, whether compressional, extensional or 'neutral', to the relative importance of crustal and mantle contributions, and to changes in source regions in the crust and mantle.

Regional geology, tectonic setting and geochronological data

The exposed geological history of southeastern Peru (Figure 2) was initiated in the Early Palaeozoic (Laubacher, 1978; Newell, 1949), when eugeosynclinal-type sedimentation was dominant and generated over 10 km of pelitic and psammitic strata by the Late Devonian. The nature of the basement upon which deposition occurred is undefined, although its sialic nature is postulated on the basis of the occurrence of 2.0 Ga gneiss, the Arequipa Massif (Shackleton *et al.*, 1979), along the present littoral, *ca.* 1.0 Ga metagranite intersected in drill-core beneath the Bolivian *Altiplano* (Lehmann, 1978), and Precambrian outcrops west of Cuzco (Marocco, 1978). Regional deformation and metamorphism occurred during the Late Devonian—Early Carboniferous interval, with initial igneous activity (diorites, granodiorites, and monzogranites) localized within metamorphic and structural culminations (Bard *et al.*, 1974; Mégard *et al.*, 1971; Carlier *et al.*, 1982). The age of this igneous activity in the study-area is poorly constrained, although intrusions of similar tectonic setting in east-central Peru have been dated at 330 ± 10 Ma (U–Pb, zircon; Marocco, 1978), and 346 ± 10 Ma (K–Ar, biotite; Mégard, 1978).

Permo-Carboniferous sedimentation followed the Late Devonian-Early Carboniferous deformation: 3–40000 m of quartzites, calcareous sandstones, and limestones (Ambo, Tarma and Copacabana Groups, respectively) were deposited in an ensialic basin whose outline was similar to that of the Lower Palaeozoic trough. Felsic volcanic units occur at the base of the sequence, but nowhere attain prominence. Sedimentation ceased in the mid-Permian, during the 'Late Hercynian' deformation episode (Mégard *et al.*, 1971), less intense than the earlier, and lacking regional metamorphism.

From the mid-Permian to the Early Triassic, extensive alkaline and subalkaline volcanism associated with molasse-type sedimentation occurred along the entire *Cordillera Oriental* of Peru (Noble *et al.*, 1978; Mégard, 1978; Carlier *et al.*, 1982). Deposition of the Mitu Group (MacLaughlin, 1924; Newell *et al.*, 1953) was restricted to fault-bounded basins (Laubacher, 1978), perhaps generated by stress release following the aforementioned 'Late Hercynian' deformation episode, although a relationship to larger-scale tectonic events such as the break up of Gondwanaland cannot be excluded (cf. Noble *et al.*, 1978). There is, however, little preserved geological evidence for eastward subduction beneath this 6–16°S segment of the Central Andes during the interval 300–250 Ma, and it is unlikely that the Mitu Group represents part of a back-arc setting. Our Rb–Sr whole-rock data define an age of 250–270 Ma for this volcanism, in broad agreement with two whole-rock K–Ar dates of 280 Ma and 245 Ma (McBride *et al.*, 1983) for similar basaltic volcanics in northwestern Bolivia, and a maximum Leonardian (i.e. 265 Ma) age assigned by Laubacher (1978) and Newell *et al.* (1953) on palaeontological grounds.

Large (200–1200 km^2) Permo-Triassic granitoid intrusions were emplaced immediately following the Mitu volcanism and sedimentation in southeastern Peru, forming part of an extensive belt of granitoid rocks of similar age extending from northern Peru to western Bolivia (Lancelot *et al.*, 1978; Capdevila *et al.*, 1977; McBride *et al.*, 1983). The age of initial intrusion of the southeastern Peruvian batholiths is given by U–Pb zircon dates of 234–238 Ma (Laubacher, 1978; Lancelot *et al.*, 1978). However, a notable discordance between these dates and both Rb-Sr whole-rock (205 Ma) and whole-rock-biotite (205–211 Ma), and K–Ar biotite (215–180 Ma) ages is evident. These data, in conjunction with contact metamorphic assemblages, crystallization history, and Al–Si order/disorder relations in alkali feldspar (Kontak *et al.*, in press), suggest that an initial period of deep intrusion (at 6–10 km), was followed at *ca.* 200 Ma by rapid uplift and exhumation. The tectonic regime during this period was 'cryptic' (Clark *et al.*, 1982 and in press), with little evidence for either strong extension or compression. However, considering the inferred crustal nature of these granitoids, this may imply (Clark and Robertson, 1978) that some mechanism of essentially linear, transcurrent dislocation may have been active deep within the crust (cf. Strong and Hanmer, 1981). Important in this regard is the restriction of the intrusions to the faulted margins (in particular the eastern) of the basin in which the Mitu Group was earlier deposited and where there may have been a high residual geothermal gradient.

The occurrence in the Macusani area (13°49′S, 70°30′W) of a large (150 km^2) peralkaline syenite complex (Douglas, 1920; Francis, 1956) of mid-Jurassic age (K–Ar biotite) suggests a return to an extensional tectonic regime (Bailey, 1974, 1978), albeit of more local extent than that prevailing during the Permian (see Figure 2). This magmatic episode was contemporaneous with subduction-related magmatism in the Main Arc to the west (James, 1971b; James *et al.*, 1974; McBride, 1977). Spatially associated with this intrusive complex are large volumes of peralkaline 'rhomb porphyry' and nepheline- and alkali feldspar-phyric and aphanitic volcanics. These rocks have been considered as part of the Mitu Group (Laubacher, 1978), but the following relationships suggest they may, in fact, be essentially coeval and cogenetic with the syenite complex: (a) the volcanics form a carapace overlying the syenite; (b) the volcanics show a gradual transition downwards into the intrusive syenite with no clear-cut boundary; (c) the suites are very similar in chemistry with respect to major elements (Figures 3 and 4; Table 1), trace elements and REE abundances and patterns (Table 1); and (d) Rb–Sr whole-rock dating defines a 200–180 Ma age for the volcanics which overlaps the 174–184 Ma (K–Ar biotite) age of the syenite. We therefore consider the peralkaline volcanic suite to be related to the Jurassic syenite complex. In stratigraphic terms, this would mean that these volcanics lie outside of the mid-Permian to Early Triassic age range assigned to the Mitu Group on palaeontological grounds (Newell *et al.*, 1953; Mégard, 1978). However, these age constraints are uncertain in southeastern Peru in the absence of detailed stratigraphic data, and we herein tentatively include the peralkaline volcanics with the Mitu Group.

Intrusive activity during the mid-Cretaceous was also areally restricted, generating a few small stocks north of Crucero (14°15′S, 70°00′W). The age of this event is inferred from two whole-rock K–Ar ages (87 Ma and 69 Ma) determined for fine-grained, hypabyssal granodiorites, and overlaps with that reported for andesitic and basaltic flows and dykes in contiguous northern Bolivia by McBride *et al.* (1983). The tectonic regime during this

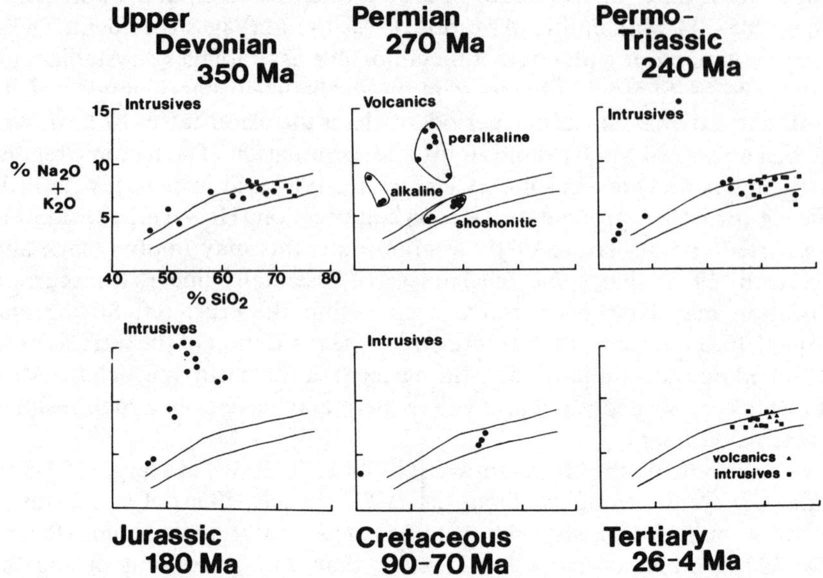

Figure 3 Alkali ($Na_2O + K_2O$) versus SiO_2 diagrams for the various igneous suites discussed in the text. The boundaries between the alkali series, high-Al basalt (calc-alkaline) series and tholeiitic series are from Kuno (1969). In this, and subsequent diagrams, we use the following symbols: ●—suites of mantle derivation; ▲, ■—suites of crustal derivation. All chemical data were obtained from XRF analyses, but FeO was determined using wet chemical techniques. Data from Carlier *et al.* (1982) for the Late Devonian (N = 6) Permo-Triassic (N = 4) and Jurassic (N = 5) intrusives are also included on this and subsequent diagrams

period is again uncertain, although this magmatic episode may be related to Main Arc broadening (Clark and McNutt, 1982).

Extensive magmatism resumed in this area during the Late Oligocene, continuing, essentially uninterrupted, into the Pliocene. This event, characterized by the strongly peraluminous chemistry of the rock suites, is divided into plutonic (26–20 Ma; K–Ar, biotite) and volcanic (17–4 Ma; K–Ar biotite) episodes. Although the crust-derived nature of these suites implies a tectonic regime similar to that active during the Permo-Triassic plutonism, the magmatism was coeval with, and probably related to, the sensibly instantaneous broadening of the Main Arc at 25 Ma (Clark *et al.*, 1976; Grant *et al.*, 1979; Clark and McNutt, 1982; McBride *et al.*, 1983). This magmatic episode was thus broadly associated with subduction-related tectonism, but we hesitate to postulate a direct link between it and magma generation related to '... low-angle subduction', as suggested by Noble *et al.* (1982). The presence of mafic intrusions of similar age (Audebaud *et al.*, 1979) near Ayaviri (15°00′S, 70°34′W), immediately west of the study-area in the *Altiplano* does, however, suggest that mantle-derived melts may have contributed thermally towards the generation of the Mio-Pliocene peraluminous magmatism.

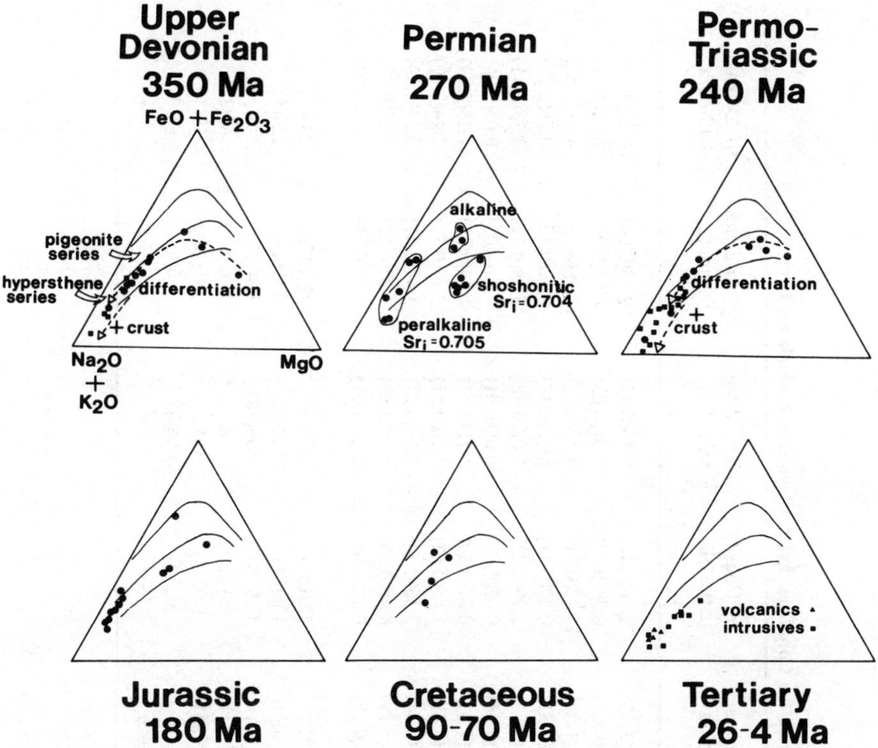

Figure 4 AFM plots for the igneous suites discussed in the text. Boundaries for the pigeonite (tholeiitic series) and hypersthene (calc-alkaline or high-Al basalt series) are from Kuno (1969). For the Upper Devonian and the Permo-Triassic suites we distinguish the chemical trends attributed to crystal fractionation processes in mantle-derived melts ('differentiation') and trends influenced by crustal processes (this includes contamination in the case of the Upper Devonian suite)

Petrographic and geochemical constraints on magma sources

On the basis of geochemical and petrographic characteristics, we propose to classify the igneous rocks of the study-area into mantle-, crustal- and mixed (mantle and crust) types. There is evidence to suggest that several distinct mantle and crustal source regions were involved, each characterized by geochemical, mineralogical and isotopic signatures. We now summarize the characteristics of each episode of magmatism, in chronological order; reference should be made to Figures 3, 4 and 5.

LATE DEVONIAN (~350 Ma)

Intrusive rocks of this age occur near San Gabán and Limacpampa at, respectively, the northwestern and southeastern limits of the study-area. The

Table 1 **Chemical analyses of igneous rocks from the Cordillera de Carabaya region, southeast Peru**

	(a) Mitu Gp. Alkaline basalts	(b) Mitu Gp. shoshonite suite	(c) Mitu Gp. peralkaline suite	(d) Jurassic syenites	(e) Coasa batholith	(f) Limbani batholith	(g) Aricoma batholith	(h) Ayapata batholith
SiO_2	45.21(0.37)	59.22(0.43)	54.52(1.70)	53.26(1.85)	72.16(1.44)	69.53(0.22)	69.53(0.94)	66.28(3.73)
TiO_2	3.12(0.06)	0.67(0.06)	0.83(0.44)	1.18(0.85)	0.33(0.07)	0.05(0.03)	0.54(0.09)	0.71(0.31)
Al_2O_3	17.35(0.62)	16.46(0.23)	20.05(1.77)	19.53(1.59)	14.02(0.51)	15.13(0.27)	14.98(0.33)	14.98(1.32)
Fe_2O_3	11.47(0.91)	2.83(1.15)	3.95(3.08)	3.02(0.31)	0.30(0.11)	0.22(0.13)	0.32(0.12)	0.78(0.21)
FeO	2.12(0.64)	2.39(0.90)	1.36(1.31)	2.05(0.87)	1.99(0.33)	3.02(0.18)	3.08(0.51)	3.78(2.41)
MnO	0.14(0.06)	0.07(0.01)	0.13(0.05)	0.17(0.04)	0.05(0.01)	0.06(0.01)	0.07(0.01)	0.05(0.04)
MgO	4.73(0.40)	4.99(0.41)	1.20(0.37)	1.57(1.21)	0.81(0.39)	1.12(0.21)	1.14(0.14)	1.13(0.29)
CaO	6.27(1.07)	4.55(0.30)	2.87(1.11)	3.05(1.81)	1.45(0.43)	2.20(0.51)	2.13(0.32)	2.44(0.60)
Na_2O	3.02(0.28)	2.92(0.34)	4.95(2.05)	7.61(2.34)	3.56(0.37)	3.14(0.27)	3.44(0.51)	3.52(0.49)
K_2O	3.45(0.30)	3.59(0.27)	6.75(2.72)	5.27(1.39)	4.48(0.25)	4.15(0.63)	4.20(0.63)	4.96(0.87)
P_2O_5	0.46(0.42)	0.26(0.04)	0.30(0.20)	0.28(0.12)	0.13(0.05)	0.21(0.01)	0.17(0.05)	0.18(0.07)
Rb	190(20)	222(51)	143(80)	179(59)	281(69)	196(9)	217(37)	231(17)
Sr	785(50)	342(38)	313(185)	446(104)	106(41)	148(5)	141(18)	224(55)
Ba	326(20)	814(99)	423(230)	526(96)	432(101)	596(38)	604(82)	448(55)
Nb	54(1)	12(4)	260(105)	287(111)	21(4)	18(1)	18(4)	43(15)
Zr	296(3)	160(5)	830(327)	605(335)	145(15)	165(2)	196(44)	389(150)
V	235(3)	95(4)	25(14)	30(16)	36(10)	56(4)	60(9)	68(37)
Zn	160(1)	83(7)	71(45)	78(40)	—	—	—	—

	(a) Mitu Gp. Alkaline basalts	(b) Mitu Gp. shoshonite suite	(c) Mitu Gp. peralkaline suite	(d) Jurassic syenites	(e) Coasa batholith	(f) Limbani batholith	(g) Aricoma batholith	(h) Ayapata batholith
Sr_i	0.7042	0.7044	0.705	—	0.708–0.709	—	—	—
Σ REE (ppm)	377	164	179	110	93	141	147	321
$(La/Yb)_N$	20.8	10.8	8.8	3.45	2.60	7.50	8.20	14.88
A/CNK	—	—	—	—	1.05	1.11	1.07	0.98
N	2	7	7	8	8	4	5	5

Chemical analyses (mean with one standard deviation) using XRF technique. The headings refer to the following suites discussed in the text: (a) plagioclase-phyric alkali basalts, Mitu Group; (b) orthopyroxene-phyric shoshonites, Mitu Group; (c) peralkaline volcanics, Mitu Group; (d) Macusani peralkaline syenite complex; (e) Coasa batholith; (f) Limbani batholith; (g) Aricoma batholith; (h) Aricoma region, northwest part of the Coasa Batholith (analyses only include granodiorites and monzogranites). Data from Carlier *et al.* (1982) are included in the averages of the Coasa (N = 2), Limbani (N = 1), Aricoma (N = 1), and Macusani syenite (N = 5) intrusives.

N—number of analyses for the major and trace elements only; REE—La, Ce, Pr, Nd, Sm, Eu, Gd, Dy, Er, Yb; $(La/Yb)_N$—REE data are normalized to the Leedey chondrite (Masuda *et al.*, 1973); A/CNK—molecular ratio ($Al_2O_3/K_2O + Na_2O + CaO$).

plutons are surrounded by highly metamorphosed and multiply-deformed Lower Palaeozoic sediments. The granitoid rocks include both syn- and post-kinematic bodies in the San Gabán area, but are of entirely synkinematic nature around Limacpampa.

The granitoid rocks of the San Gabán area comprise a suite ranging from pyroxene-hornblende-biotite diorite, through hornblende-biotite granodiorite, to biotite monzogranite and biotite-muscovite syenogranite. This association displays continuous chemical trends on variation diagrams and suggests that the rocks represent a comagmatic suite generated by processes of crystal fractionation involving the major silicate phases (pyroxene, hornblende, biotite and feldspars). The rocks are subalkaline (see Figure 3), correspond to the alkali-calcic classification of Peacock (1931), define an Fe-enrichment trend on an AFM diagram (see Figure 4), and show increasingly peraluminous compositions with progressively higher silica contents (Figure 5). Although peraluminous compositions may be generated by diverse mechanisms including fractional crystallization, crustal anatexis or assimilation, and vapour phase transfer (Halliday *et al.*, 1981), we believe that fractional crystallization involving amphibole (Cawthorn *et al.*, 1976) was probably responsible for the observed trend. However, the presence of clearly xenocrystic garnet (Alm_8 Py_{12} Sp_8) with 'reverse' zoning (i.e. Mn-rich rims), muscovite, andalusite, sillimanite, zircon and, possibly, biotite (Laubacher (1978) also reports staurolite and cordierite) indicates that some crustal contamination was involved. This is supported by the observation that only the more differentiated members of the suite (i.e. the granodiorites and monzogranites) contain crustal xenoliths, suggesting that the more peraluminous compositions of these rocks may be in part due to assimilation (see Figures 4 and 5). The following features indicate, however, that crustal input to the magmas may not have been extensive: (a) the xenocrystic material is volumetrically minor (i.e. $<0.1\%$); (b) the xenocrystic garnets are generally euhedral to subhedral, implying that only a small amount of corrosion has occurred; (c) biotites from microxenoliths in the granitoid rocks are compositionally distinct from magmatic biotites in the same sample, showing that chemical equilibrium with the melt was not attained; and (d) the good overall trends in variation diagrams suggest that insufficient contamination occurred to obscure the effect of crystal fractionation processes.

We therefore interpret the San Gabán suite as the product of an initial mantle-derived melt, possibly of tholeiitic affinity, and propose that moderate (2–3 kbar) pressure crystal fractionation processes involving amphibole crystallization was *mainly* responsible for the peraluminous compositions of the later differentiates. Crustal contamination occurred, but not to an extent sufficient to modify crystal fractionation processes.

In contrast to the San Gabán area rocks, the Limacpampa granitoids are biotite-muscovite monzogranites and muscovite syenogranites, with no associated more basic rocks at the presently exposed structural levels. They are characterized by very low MgO, Ca and Fe_2O_3/FeO, and high K_2O, Rb, Sn and A/CNK (Figure 5). These are features of S-type granites (Chappell and

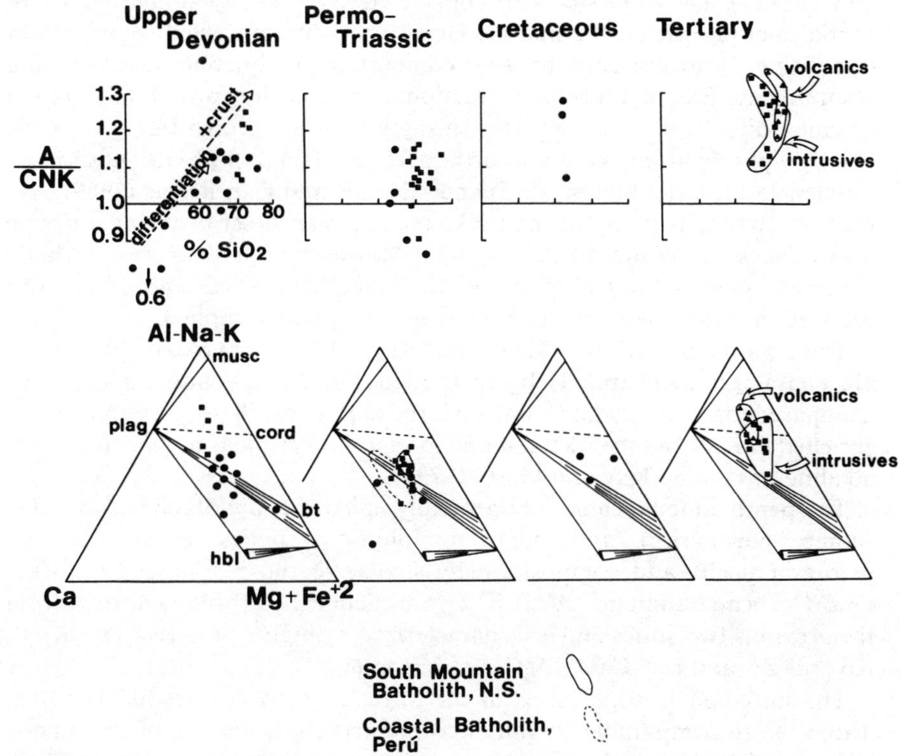

Figure 5 A/CNK (molecular ratio ($Al_2O_3/Na_2O + K_2O + CaO$) versus SiO_2, and ACF diagrams (after White and Chappell, 1977) for the Upper Devonian, Permo-Triassic, Cretaceous and Tertiary suites. The I and S fields were defined using data for the Coastal Batholith of Peru (Barreiro, 1982c, Johan *et al.*, 1980), and South Mountain batholith of Nova Scotia (Clarke and Halliday, 1980), respectively, and are compared to the Permo-Triassic granitoids.

Note the strong correlation between peraluminous index (A/CNK) and silica contents of the Upper Devonian rocks. We attribute this trend to a combination of amphibole fractionation (Cawthorn *et al.*, 1976), and crustal contamination by metasedimentary material of an originally mantle-derived melt (see text for further discussion)

White, 1974) and we therefore interpret these rocks as being derived from a sedimentary protolith in the continental crust.

The Late Devonian magmatism thus involved both mantle- and crust-derived melts. The San Gabán and Limacpampa suites are associated with envelopes of highly deformed and metamorphosed metasediments, and are themselves variably deformed, indicating that compressional tectonism accompanied both mantle and crustal magmatism.

PERMIAN (250–270 Ma)

The Mitu Group volcanic rocks may be subdivided on mineralogical and chemical criteria into alkaline, peralkaline and shoshonitic suites (see Figure 3 and Table 1). The three suites are considered to represent variable degrees of partial melting within the mantle. However, their contrasted compositions (see Table 1) do not conform to a common source therein, and variation diagrams, REE data, and crystal fractionation modelling prohibit derivation of one suite from another. The shoshonitic suite, consisting of olivine (Fo_{70-84})-phyric absarokites and orthopyroxene (En_{70-92})-phyric shoshonites (nomenclatures of Iddings (1895) and Nicholls and Carmichael (1969)), are chemically similar to high-K calc-alkaline andesites described from orogenic areas (Jakes and White, 1972; Gill, 1981). However, such rocks are not wholly confined to such settings (Joplin, 1968) and therefore the existence of an active arc system further west at this time is not necessarily implied.

The alkaline basalts are plagioclase (An_{40-80}) ± clinopyroxene-phyric and are characterized (Table 1) by their high TiO_2, Sr, V and Zn contents. Compared to other alkaline basalts (Carmichael *et al.*, 1974), they are notably enriched in K_2O, but their REE abundances are very similar to those found for alkaline basalts by Kay and Gast (1973).

The peralkaline volcanics are generally aphanitic, but alkali feldspar (i.e. rhomb porphyries) and nepheline phyric varieties occur and are petrographically and compositionally similar to those described from the Oslo Graben (Neumann, 1978). They are chemically (Table 1) distinct from the previous two suites and are characterized by high Al_2O_3, ($Na_2O + K_2O$), Nb and Zr, and low CaO, MgO and V contents.

The initial Sr isotope ratios of the Mitu Group volcanics (0.704–0.705: Table 1) are comparable to values considered characteristic of the mantle (Faure, 1977). Although the ratios are perhaps slightly high for unmodified, pristine mantle, they are in agreement with Sr_i values reported for other alkaline suites (Powell and Bell, 1974), and are consistent with the generally high-LIL character of all the suites. However, the Sr_i values do not support a crustal source, or even significant amounts of crustal contamination, as might be inferred from the presence of embayed quartz megacrysts in the shoshonitic rocks, and the high K_2O contents of the alkaline basalts.

We interpret the geochemical variation among the volcanics assigned to the Mitu Group to reflect several distinct source regions, all in the mantle. Such diversity of magma types within such a small area is not unique, comparable situations having been described from Hawaii (Carmichael *et al.*, 1974; Stille *et al.*, 1983) and the Oslo Graben (Neumann, 1978). The occurrence (Noble *et al.*, 1978) of comenditic-pantelleritic volcanics in the east-central Peru segment of the Mitu Group province further emphasizes the wide range of magma sources which characterized this igneous episode.

PERMO-TRIASSIC (240–230 Ma)

The Permo-Triassic batholiths are characteristically medium- to coarse-grained, K-feldspar-megacrystic, rapakivi-textured, biotite granodiorites and monzogranites, associated with volumetrically minor amounts of syenogranites and cordierite-muscovite monzogranites. They are notably barren of metasedimentary xenoliths, except along the southwestern margin of the Coasa batholith, and structures within the adjacent metasedimentary country rocks suggest forceful emplacement rather than passive intrusion accompanied by stoping. Of petrogenetic significance are: (a) the persistent presence of doleritic dykes, similar in composition to the Permian alkali basalt suite; (b) the presence of corroded mafic xenoliths along the southwestern margin of the Coasa batholith; and (c) the I-type character of the northern portion of the Coasa batholith (i.e. in the Ayapata area), where hornblende-biotite (± pyroxene) granodiorite and monzogranite are the dominant lithologies.

In chemical terms (Table 1: e, f, g, h), the granitoid rocks comprise two petrological groups, one interpreted to be of crustal origin and the other to represent mantle-derived melts. The first suite (Coasa, Limbani, Aricoma) is subalkaline (see Figure 3) and peraluminous (see Figure 5), and is characterized by a restricted range of high silica contents (Figures 3 and 5); chemical variation diagrams define continuous trends which can be modelled by plagioclase- and biotite crystal fractionation. Comparison (Figure 5) of the chemical data to those of the S-type, peraluminous, South Mountain Batholith (Clarke and Halliday, 1980; Longstaffe *et al.*, 1980) on the White and Chappell (1977) ACF plot shows that the Peruvian granitoid rocks conform to the S-type field. In contrast, the second petrological suite (Ayapata) straddles the alkaline-subalkaline boundary (see Figure 3), shows a continuous trend on an AFM plot (see Figure 4), is metaluminous to peraluminous (see Figure 5), and is relatively enriched in TiO_2, Sr, Nb and ($Na_2O + K_2O$). On variation diagrams this suite defines distinct trends when compared to the others, indicating a different crystal fractionation-melt evolution. This difference in composition and crystallization history is reflected in contrasting REE abundances and patterns (see Table 1); notable is the high REE abundance in the Ayapata suite, indicative of it's alkaline affinity (cf. Taylor *et al.*, 1981). The Ayapata granitoids, in contrast to the crustal suite, correspond to the field defined by the I-type Coastal Batholith of Peru (Pitcher, 1978, 1983) on the White and Chappell (1977) ACF plot (see Figure 5).

The associated, but contrasted, 'crustal' and 'mantle' suites are also distinguished by their mineral compositions. For example, biotites in the former are enriched in Al_2O_3, Mg and Rb, and depleted in ($FeO + Fe_2O_3$) and TiO_2 relative to those in the latter suites. The chemical features of the biotites are consistent with the whole-rock compositions and modal mineralogies of the suites, and are therefore considered to reflect the melt compositions from which they crystallized. However, use of biotite compositions to characterize the granitoids according to Ishihara's (1977, 1981) scheme assigns all of these rocks to his ilmenite, or reduced, series. The bipartite division of ilmenite and magnetite series has been shown by Takahashi *et al.* (1980) do not invariably correspond to the S- and I-type division of Chappell and White (1974). Therefore, although use of the biotites to classify the granitoid suites appears to agree with the whole-rock chemistry for the 'crustal' suite, there is a discrepancy with the mantle suite.

The isotopic data (see Table 1) are consistent with a crustal source for the southern portion of the Coasa batholith. Considering the geochemical similarity of the Limbani and Aricoma batholiths (see Table 1), we tentatively assign these suites to a similar source region, within the lower crust and probably of metasedimentary nature. A dolerite dyke cutting the Coasa batholith along its southwestern margin has a Sr_i value of 0.7060–0.7065. Thus, mantle-derived melts were apparently generated at the same time as the crustal granitoid suites.

We interpret the granitoid rocks of the Ayapata area to represent, in contrast to the above rocks, an I-type suite in the sense of Chappell and White (1974). This would imply a source region in the lower crust or upper mantle. Our data at present do not permit us to discriminate between these two models, but the source region of this suite was clearly quite different from that of the larger part of the Coasa batholith.

JURASSIC (184–174 Ma)

An episode of peralkaline magmatism is represented by the mid-Jurassic Macusani syenite complex (Francis, 1956; Laubacher, 1978), which is associated with peralkaline volcanics of the Mitu Group. This intrusion consists predominantly of medium- to coarse-grained, pyroxene-amphibole-biotite syenite and nepheline syenite, cross-cut by minor leucocratic, feldspathic pegmatoids. Other peralkaline intrusions in the study-area include several small plugs and dykes of nepheline syenite in the Antauta district (14°14′S, 70°22′W). Although these have not been dated they are compositionally similar to the Macusani intrusion. The peralkaline syenite is characterized by high (Na_2O + K_2O), Al_2O_3, Nb, Zr and Zn contents (see Figure 4 and Table 1) and, as noted above, is very similar in composition to the peralkaline volcanics assigned to the Mitu group (Table 1).

The genesis of peralkaline rocks has been extensively discussed (e.g. Sorensen, 1974; Bailey, 1978). A mantle origin is clearly indicated, although minor crustal contamination may have been involved in some cases (e.g. Taylor *et al.*, 1980). If, however, the syenite is indeed comagmatic with the peralkaline volcanics, then the Sr data for the volcanics (i.e. $Sr_i = 0.705$) indicate that this latter process was of little petrogenetic importance.

CRETACEOUS (90–70 Ma)

Cretaceous magmatism is represented in the study-area by three small (i.e. <0.5 km^2), hypabyssal intrusions and several dolerite dykes located north of Crucero (14°15′S, 70°00′W). Although volumetrically insignificant, the occurrence of these rocks indicates that magmatism of this age, although areally restricted, did penetrate eastwards into the Inner Arc, forming part of a quasi-continuous zone of igneous activity which extended over this portion of the Central Andes (Clark *et al.,* 1983).

The dolerite dykes are similar in composition to those of the Permo-Triassic and to some alkaline volcanics of the Mitu Group, with high TiO_2, CaO, MgO, V and Ni contents. In contrast, the stocks are granodioritic in composition and subalkaline (see Figure 3), and their high K_2O, Rb and Ba contents suggest affinities with the shoshonitic clan (Joplin, 1968; Jakes and White, 1972). Although no other igneous rocks of similar age have been reported in the *Cordillera Oriental* of Peru, andesitic to basaltic volcanic rocks and dykes of Late Cretaceous age (McBride *et al.*, 1983; Evernden *et al.*, 1977) have been reported in northern Bolivia. These volcanics and feeder dykes form part of the Cretaceous Puca Group (Russo and Rodrigo, 1965) deposited in a back-arc trough referred to as the Titicaca Basin. The Cretaceous intrusives may therefore represent mantle-derived melts related to an extensional regime.

NEOGENE (26–24 Ma)

The most recent episode of magmatism in the area comprises two distinct suites: an older (26–20 Ma) group of medium-grained, biotite-cordierite-sillimanite-K-feldspar-megacrystic monzogranites; and a younger (17–14 Ma) succession of biotite-andalusite-sillimanite-muscovite-cordierite-bearing rhyolitic ash-flow tuffs. The markedly peraluminous mineralogies of both suites, suggestive of a crustal origin, are reflected by their high and essentially constant SiO_2 contents (66–72 wt. %), A/CNK values generally exceeding 1.20 (see Figure 5), low CaO ($<$ 1.5 wt. %) and high average Sr_i (0.718 for the granites and 0.722 for the volcanics), all features of S-type granitoid rocks (Chappell and White, 1974). Although the two suites may comprise a coherent magmatic episode, their ages have yet to be shown to overlap, and they are chemically and mineralogically distinct. This difference is clear in Figure 5, in which the more aluminous character of the volcanic rocks is emphasized.

Comparison of the two suites with the I- and S-type fields delineated on the ACF diagram (Figure 5) indicates that, although they are both strongly peraluminous, corresponding to the field of S-type magmas, they markedly differ in bulk rock composition. We interpret this variation to be a reflection of different source regions within the crust. The chemical data for these suites, and the older, Permo-Triassic crust-derived suite, are compared in Figure 6. Considering the relatively high and little-varying silica contents of each suite (see Figure 5), which we would interpret to be a reflection of equilibrium melting in the respective source regions, the differences in bulk rock and mineral compositions must, therefore, reflect different melt geochemistries. We infer that, in relative terms, the Permo-Triassic granitoid suite originated from a protolith enriched in Fe, the Tertiary granitoids from one enriched in Mg and the Tertiary volcanics from one enriched in Al. This diversity in crustal source regions is supported by the variable Sr_i values for the three suites (Figure 6c).

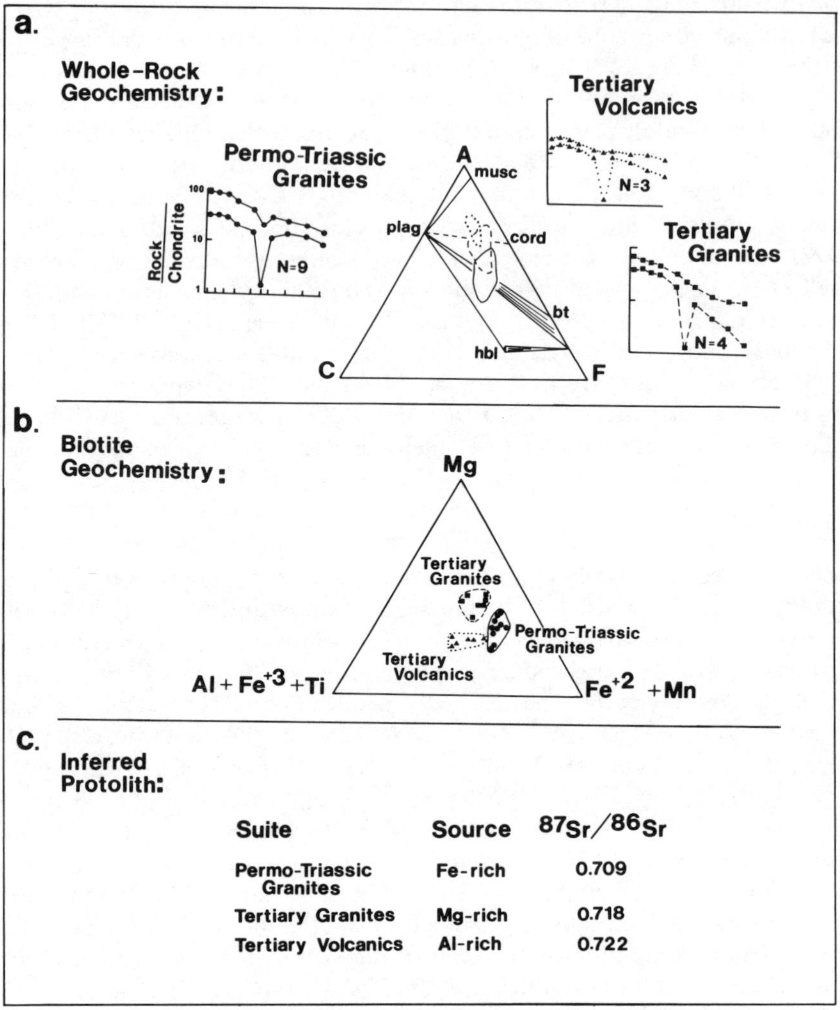

Figure 6 Whole-rock and mineral (biotite) chemistry for the Permo-Triassic and Tertiary granitoid, and Tertiary volcanic rocks. The REE data are normalized to the Leedey chondrite (Masuda *et al.*, 1973). The biotite data are from XRF and wet chemical analyses of mineral separates

Conclusions

The Cordillera de Carabaya region, southeastern Peru, displays a remarkably diverse history of episodic, plutonic and volcanic igneous activity (at *ca.* 350, 250, 240, 180, 90–70, and 26–24 Ma), in which mutually 'incompatible' igneous rocks have been juxtaposed in a relatively small area (8800 km^2). Represented are rocks of apparently unmodified, and slightly modified, mantle derivation (i.e. the Upper Devonian (San Gabán), Permian, Jurassic and Cretaceous suites), and rocks of entirely crustal origin (the Upper Devonian (Limacpampa), Permo-Triassic (excluding those in the Ayapata area) and Tertiary suites). The emplacement of these magmas occurred in association with a variety of tectonic regimes, ranging from compressional to extensional, as well as 'cryptic'.

It remains speculative why this region should have been the locus of such multifarious magmatism, which displays analogies to those of provinces as contrasted as the Oslo Graben and the granitic batholith terrains of the European Hercynian orogen and southeast Australia, but also at times merged with that of the Central Andean Main Arc. However, we suggest that the area may have acted as a 'boundary layer' between the essentially stable craton to the east, and the relatively unstable orogenic belt to the west. This tectonic domain probably originated during the Precambrian and has since acted as a zone of structural weakness responding in a variety of ways (i.e. extension, compression, shear) to the regional stresses active during Phanerozoic time.

ACKNOWLEDGEMENTS

The authors gratefully acknowledge technical assistance from Dr D. A. Archibald (K–Ar dating), and F. Dunphy and J. R. Foster (chemical analysis). The Rb–Sr data and REE analyses were kindly provided by Dr H. Baadsgaard (University of Alberta) and Dr D. F. Strong (Memorial University), respectively, to whom we are grateful. Field studies in Peru were supported by operating grants to A. H. Clark from the Natural Sciences and Engineering Research Council of Canada, and laboratory work was funded by NSERC grants to E. Farrar and A. H. Clark. Field work was carried out with the logistical support of Minsur. S. A. I. Wolfson, J. Jeffrey and L. Harris were responsible for figure drafting and manuscript typing.

Geochemical Characteristics of Magmatism in the Southern Andes (45–46° S)

D. S. BARTHOLOMEW and J. TARNEY

Abstract

Along the continental margin of southern Chile (45–46°S), Mesozoic to Recent magmas of the Patagonian Batholith were emplaced through a metasedimentary basement which had developed as an accretionary sedimentary wedge during the late Palaeozoic-early Mesozoic. Despite the rather radiogenic character of this basement, initial $^{87}Sr/^{86}Sr$ ratios of the plutons are generally low (0.7038–0.7047), suggesting only limited involvement of basement in their petrogenesis. A zonation of magma types is observed, but there is no unidirectional migration with time in the locus of magmatism, a consequence of the Andean margin being subject to alternate periods of compression and extension. The main Mesozoic zones of plutonism are typically calc-alkaline. Alkaline or tholeiitic magmas were emplaced in the back-arc regions, including a bimodal volcanic suite erupted in a back-arc basin which formed in the late Jurassic-early Cretaceous. On the oceanward side of the calc-alkaline arc, an intra-arc basin developed during the mid-Tertiary into which basalts with transitional calc-alkaline-tholeiitic characteristics were erupted. Tertiary tonalite-trondhjemite plutons emplaced within this basin have tholeiitic trace element characteristics and flat rare-earth patterns. Closure of the basin resulted in major uplift and exposure of deep levels of the main calc-alkaline arc, made up of minor gabbros, diorites and tonalites (often gneissified), some of the latter having REE patterns with HREE depletion indicative of hornblende fractionation.

Introduction

Problems of magma genesis at convergent plate margins centre not only on whether the source lies in the subducted ocean crust or the overlying mantle wedge, but also on whether the range of observed compositions is governed by fractional crystallization or by partial melting and whether sediment subduction and element transfer within the subduction zone exert important controls on the chemistry of the magmas. The abundance of granitoid magmas at continental margins such as the Andes, compared with intraoceanic arcs, also begs the question of the extent of crustal involvement. Many of these problems remain intractable and can only be solved by comparative studies of different arc settings employing a range of geochemical and isotopic data. It is, of course, commonly assumed that convergent plate margins are dominated by compressional tectonics, although this is not the case in intraoceanic arc systems where extensional marginal basins are common. Here we consider the relationship between magma chemistry and extensional and compressional tectonics in a section of the Andean margin in southern Chile and try to assess the importance of crustal involvement in the petrogenesis of the granitoid plutons. In contrast to central and northern Chile, where the Mesozoic and Tertiary magmatic arcs are essentially superimposed upon the earlier Palaeozoic arc, the Jurassic to Recent calc-alkaline arc in southern Chile is located well to the west of the Palaeozoic magmatic arc (Figure 1), and was emplaced within a metasedimentary basement which had formed as a broad accretionary sedimentary wedge in the late Palaeozoic-early Mesozoic (Forsythe, 1982; Bartholomew and Tarney, in press). This basement at latitude 45–46°S is composed of highly deformed metasediments which, prior to their burial and deformation, consisted largely of greywackes and shales. These two rock types have suffered strong tectonic intermixing along foliation planes which verge toward the continent at low angles, with the result that the basement is relatively uniform in character over the whole area and it is relatively easy to establish its chemical and isotopic composition and thus assess its potential as a contaminant in magma genesis.

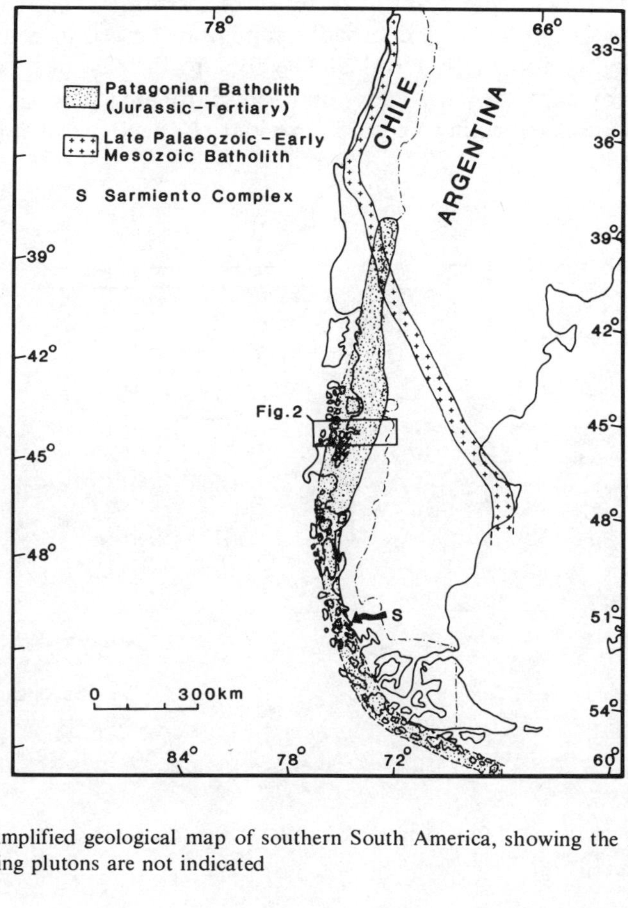

Figure 1 Simplified geological map of southern South America, showing the main plutonic belts. Flanking plutons are not indicated

Most of the Mesozoic and later magmatic activity between 45–46°S has taken place within the region covered by Figure 2. Plutons on the western margin of the Patagonian batholith are emplaced into the basement metasediments; on the eastern side (Zone F, Figure 2) only volcanics and sediments of Jurassic and younger age are exposed.

Figure 2 illustrates that there is a zonation of magma types in linear belts parallel to the continental margin. However field and geochronological evidence indicates that there has been no unidirectional migration with time of the locus of magmatism, either from west to east or vice versa. Geochemical studies also demonstrate that there is no clear trend of increasing K_2O with distance from the trench: the plutons of intermediate composition with the highest K_2O/SiO_2 ratios in fact are the oldest, middle- to late-Jurassic, tonalities and granodiorites forming a belt (Zone B, Figure 2) on the trenchward side of the main arc.

The central zone of the batholith (Zone D, Figure 2) has suffered major uplift and erosion as a result of a strong Miocene phase of compression, and has exposed deeper sections of the main magmatic arc. This zone, some 50 km wide, is a complex of gabbros, diorites, quartz diorites and tonalites intruded by numerous amphibolite dykes in all stages of disruption. The rocks in this zone are variably foliated and broadly similar in appearance to those in Archaean high-grade gneiss terrains. The zone to the east of this (Zone E, Figure 2) is by contrast much more uniform, unfoliated potassic granite of mid-Cretaceous age forming a belt 25 km wide. The back-arc region (Zone F,

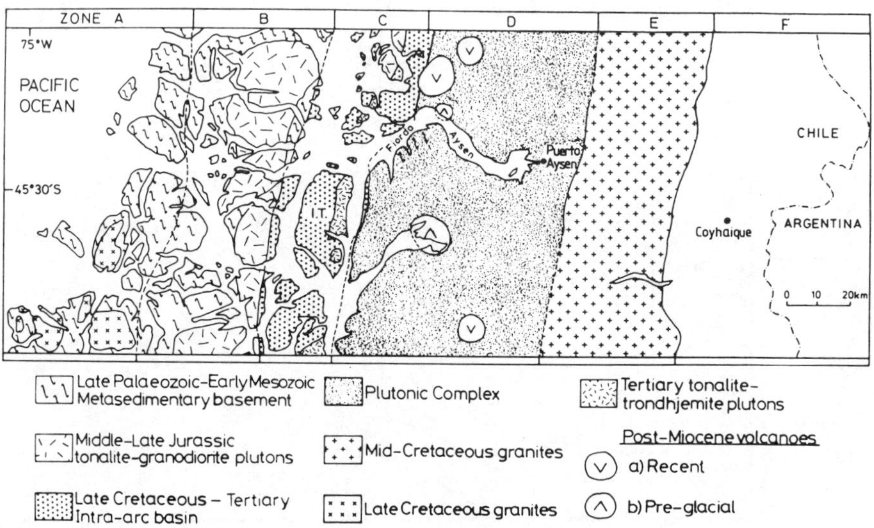

Figure 2 Geological map of Chile between 45 and 46°S. Zone F—Jurassic to Recent volcanics and sediments; I.T.—Isla Traiguen

Figure 2) is characterized by minor intrusions of alkaline or 'within-plate' character, which also cut the potassic granite (Zone E). Zone C (Figure 2), lying immediately west of the plutonic complex, was the site of extensive tholeiitic volcanism in the late Cretaceous or early Tertiary and represents an intra-arc basin which was active during this period. Recent volcanic activity is concentrated in, or slightly to the east, of this zone.

In this region of the Andean margin there are not only high-level calc-alkaline plutons exposed, but also deeper levels of the magmatic arc and volcanics, and plutons emplaced under an extensional regime. It appears that the deep levels of the plutonic complex may have been uplifted as a consequence of partial closure of the intra-arc basin during the Miocene. There is also evidence of earlier back-arc extension, equivalent in age to the 'rocas verdes' marginal basin in southernmost Chile (Dalziel *et al.*, 1974) which opened in the late Jurassic but which closed again in the mid-Cretaceous. These alternating phases of extension and compression may be responsible for the changing locus of magmatic activity in the region. They also appear to correlate with differences in magma chemistry. We will deal with magmatism associated with back-arc or intra-arc extension first before considering the petrogenesis of the major calc-alkaline super-units.

Back-arc magmatism

Lavas ranging from quartz tholeiites through olivine tholeiites and alkali basalts to basanites have been extruded in the back-arc region (Zone F, Figure 2) intermittently from late Cretaceous to Recent times (Baker *et al.,* 1981). However, prior to this a back-arc marine basin existed in the late Jurassic-early Cretaceous in which extensive bimodal mafic-silicic volcanism occurred; the Tobifera silicic volcanics here are broadly equivalent to those associated with the 'rocas verdes' marginal basin further south at Sarmiento (Saunders *et al.*, 1979). The basin was terminated during mid-Cretaceous deformation that was followed closely by the emplacement of the major high-level belt of potassic granites of Zone E (Figure 2).

The mafic magmas associated with this back-arc basin are silica-saturated and many have a clear 'within-plate' geochemical signature. On a multi-element diagram, with abundances normalized to typical MORB values (cf. Pearce, 1982), many samples show enrichment not only in the large-ion-lithophile (LIL) elements and light rare earths, but also in the high field strength (HFS) elements, including Nb (Figure 3a). The more fractionated rocks have high concentrations of Nb, Ce, Zr and Y. In other samples relative enrichment in the LIL elements and the development of negative Nb anomalies suggests a transition towards arc magma types. These features are common in marginal basin basalts (Tarney *et al.*, 1981), and in continental tholeiites (Weaver and Tarney, 1983), and can be interpreted in terms of addition of a subduction component to the mantle source regions of the basalts (Tarney *et al.*, 1981; Saunders and Tarney, in press).

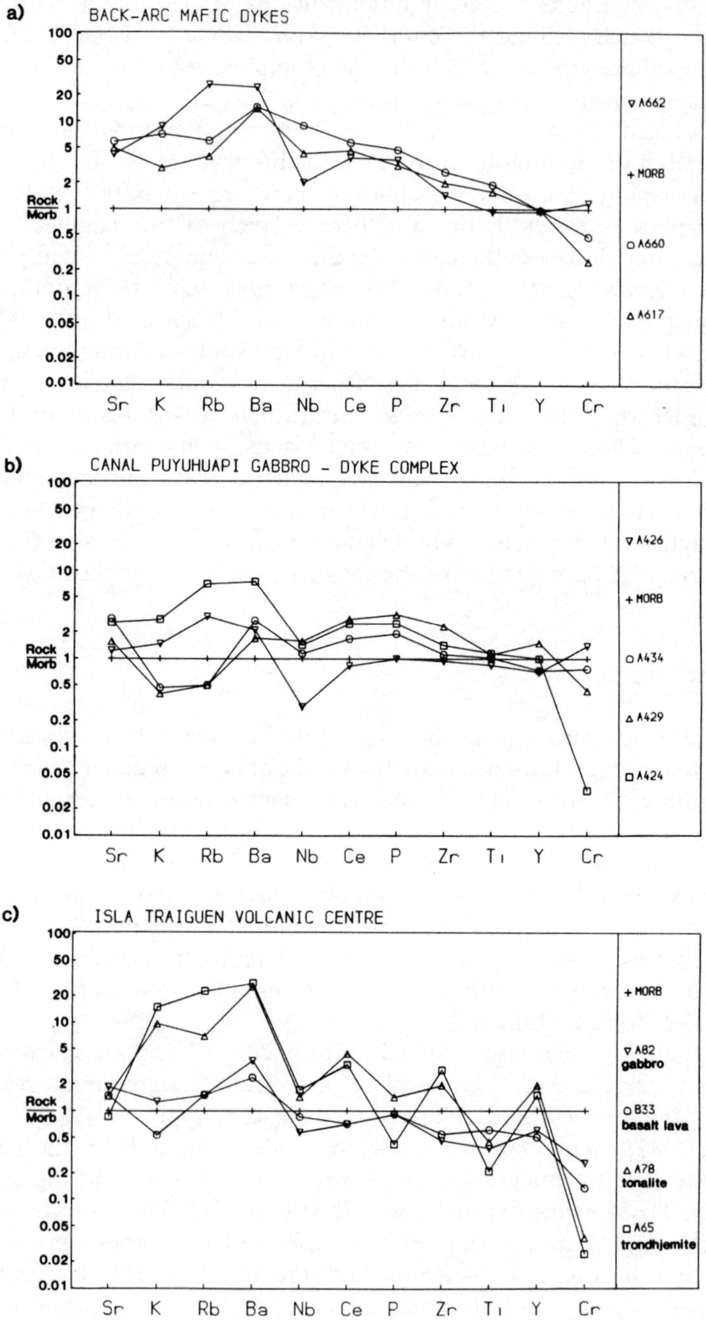

Figure 3 Trace element patterns of igneous rocks between 45 and 46°S, normalized against the mid-ocean ridge basalt of Pearce (1982)

Tertiary intra-arc basin

Much of this basin (Zone C, Figure 2) still lies below sea level. However, thrust slices of gabbros and mafic volcanics are exposed along the eastern margin of the basin where it is in tectonic contact with the plutonic complex of Zone D. In Canal Puyuhuapi (44°55′S) there is a complex of medium grained gabbros intruded by numerous basic dykes, which locally constitute up to 40% of the outcrop. The mafic rocks show a wide range of Fe/Mg ratios and Zr contents with only minor variation in silica, and thus are tholeiitic rather than calc-alkaline. Most samples from Canal Puyuhuapi show slight enrichment in Ce relative to Y compared with MORB (Figure 3b). Although some of these rocks have suffered depletion in selected LIL elements (K, Rb, Ba), this may be a consequence of hydrothermal fluid circulation; the remainder are enriched in LIL elements relative to MORB and display 'negative' Nb anomalies, suggestive of a subduction-related magma component. These features are

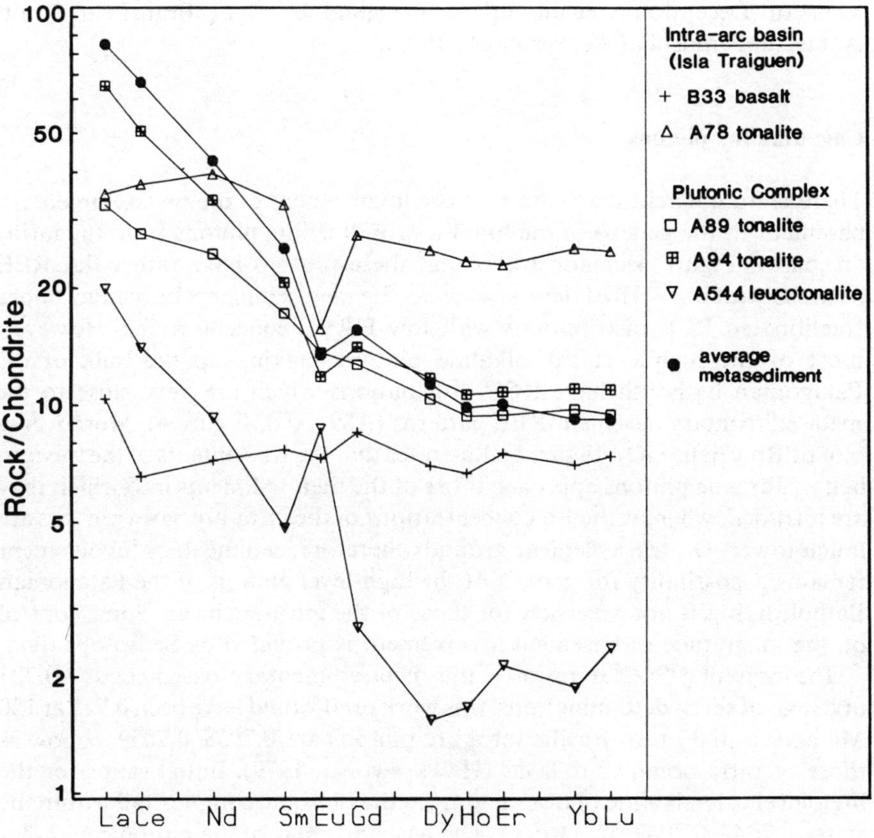

Figure 4 Chondrite-normalized rare earth patterns of samples from the intra-arc basin (Zone C) and plutonic complex (Zone D)

quite similar to those of the mafic rocks of the Sarmiento marginal basin complex in southernmost Chile (Saunders *et al.*, 1979). Some of the lavas in the marginal thrust slices are, however, richer in high field strength elements such as Nb and P, and may have been erupted during the initial stages of basin formation.

Within this intra-arc basin there are a number of younger volcanic centres which are intruded by plutons with a chemistry quite distinct from that of the main calc-alkaline batholith. On Isla Traiguen, in the centre of the basin (see Figure 2), the lavas are intruded by two high-level tonalite-trondhjemite plutons. The mafic lavas are low-K tholeiites with essentially flat REE patterns (B33, Figure 4). The silicic plutons also have relatively flat REE patterns with large negative Eu anomalies (A78, Figure 4), indicative of plagioclase fractionation. Many LIL and HFS elements show comparable degrees of enrichment over their concentrations in the associated mafic lavas (see Figure 3c) and there is a trend toward high Fe/Mg ratios. Interestingly, the chemistry of these plutons is very similar to that of the plagiogranites in the Sarmiento marginal basin complex (Saunders *et al.*, 1979) and to the silicic lavas of Deception volcano in the marginal basin of Bransfield Strait, Antarctic Peninsula (Weaver *et al.*, 1979).

Calc-alkaline plutons

There is no inherent reason to propose involvement of the metasedimentary basement in the genesis of the tonalite-trondhjemite plutons from the intra-arc basin. Figure 4 demonstrates that these plutons have rather flat REE patterns with high HREE levels, whereas the metasediments have much more fractionated REE distributions with low HREE concentrations. However, most of the high-level calc-alkaline plutons making up the bulk of the Patagonian batholith have REE distributions which are very close to the metasedimentary basement REE patterns (A89, A94, Figure 4). Moreover, a plot of Rb versus SiO_2 (Figure 5) illustrates that the Rb contents of the western belt of Jurassic plutons approach those of the metasediments into which they are intruded, whereas the Rb concentrations of the intra-arc basin plutons are much lower. On trace element grounds therefore, sedimentary involvement remains a possibility for genesis of the high-level plutons of the Patagonian Batholith, but is not necessary for those of the intra-arc basin. Some control on the magnitude of basement involvement is provided by Sr isotope data.

The present $^{87}Sr/^{86}Sr$ ratio of the metasedimentary basement is ~0.721 (average of seven determinations, this work) and would have been 0.715 at 150 Ma ago. Initial ratios for the intra-arc plutons are 0.7038–0.7039, as low as those of intra-oceanic arc lavas (Hawkesworth, 1979). Initial ratios for the high-level calc-alkaline plutons and from the plutonic complex fall within the range 0.7042–0.7049 (this work). The basic magmas of the plutonic complex have fairly high Sr contents (~500 ppm) compared with the metasedimentary basement (~120 ppm) and would be able to assimilate a significant fraction of

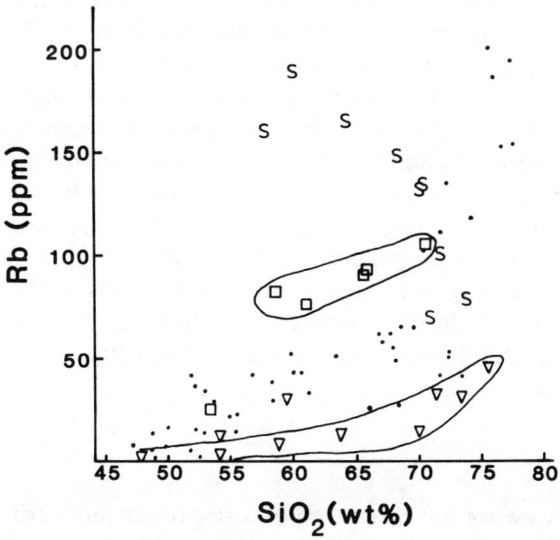

Figure 5 Plot of Rb versus SiO_2 for plutonic rocks and metasediments between 45 and 46°S. Squares – zone B granitoids; triangles – zone C plutonic rocks; dots – plutonic rocks from zones A, D and E (Figure 2); S – basement metasediments

basement with only moderate increase in $^{87}Sr/^{86}Sr$ ratio. However, even with this favourable circumstance, and assuming a mantle source with $^{87}Sr/^{86}Sr$ ratios as low as 0.7038 (the intra-arc basin value) for the primary mafic magma, it is clear that it would be difficult to involve assimilation of more than 10–30% of basement without increasing the $^{87}Sr/^{86}Sr$ ratios above the observed values. This discounts any radiogenic Sr component from the subducted slab. In reality it may be much less: the mantle sources of some ocean islands and in particular some subcontinental sources may have $^{87}Sr/^{86}Sr$ ratios well in excess of 0.705 (McCulloch *et al.*, 1983). The evidence for significant basement involvement in the genesis of the granitoids of the Patagonian Batholith is therefore not compelling.

Rare-earth patterns of the type displayed by A89 and A94 (see Figure 4) are fairly typical of high-level plutons of the Andean belt (Atherton *et al.*, 1979). Rather different REE patterns however are exhibited by some of the tonalities within the deep-level plutonic complex, of which A544 (Figure 4) is an example. This tonalite has a concave-upwards REE pattern with very low HREE levels and a marked positive Eu anomaly. Such patterns are common in many tonalitic high-grade gneiss terrains of Archaean age (Weaver and Tarney, 1981) and occur in the Proterozoic gabbro-diorite-tonalite-trondhjemite suite of southwestern Finland (Arth *et al.*, 1978). There is general agreement that such distinctive patterns reflect the dominant role played by hornblende, either during partial melting or fractional crystallization (Arth and Barker, 1976; Weaver and Tarney, 1980). This further implies that the complex may have evolved under moderately high water pressures in order to maintain hornblende as the dominant liquidus or residual phase. Hornblende is certainly the dominant mafic mineral phase throughout the plutonic complex of Zone D although there are no obvious cumulates. It is worth reiterating that parts of the complex are strongly foliated and superficially resemble Precambrian gneiss terrains. The complex intermixture of dioritic and tonalitic rock types within this zone is in marked contrast to the rather uniform and large scale nature of the high-level plutons of zones A, B and E. This suggests that the rocks have remained ductile, deep beneath the arc, for long periods. Harris *et al.* (1970) have stressed that hydrous granitoid magmas cannot rise to high levels in the crust because reducing $p(H_2O)$ will rapidly induce crystallization. Such rock types may occupy much of the deeper zones of the Andean Batholith, but are rarely exposed by erosion.

Conclusions

In this brief review we have focussed on some of the more critical features of the Patagonian Batholith. As in most other continental margin calc-alkaline complexes there is a range of composition between gabbro and potassic granite, with tonalite being the dominant rock type. However, the geochemical character of the emplaced or erupted magmas varies according to whether the Andean margin was under compression or extension, and this also influences the locus of magmatic activity. During extension, marginal basins form and the associated magmas are dominantly basaltic, with rather low $^{87}Sr/^{86}Sr$ ratios and with these and any associated silicic differentiates having rather flat REE patterns. During the normal compressional situation, high-level calc-alkaline plutons were emplaced, some, such as the potassic granites of Zone E, being relatively uniform large-scale bodies. Basin closure led to uplift and eventual exposure of the deeper, central parts of the magmatic arc, a complex intermixture of tonalites and diorites, variably foliated. These rocks appear to have evolved under relatively high $p(H_2O)$; by contrast the high-level plutons are devoid of pegmatites and associated hydrothermal features and may have evolved under relatively low $p(H_2O)$.

The basement into which these magmas were emplaced was an accretionary metasedimentary wedge of continental provenance, with a rather radiogenic isotopic signature. Sr isotope data for the plutons indicates however that, despite the convenience of this basement as a potential source for the granitoids, only very limited assimilation or mobilization of this basement could have taken place. The problem remains how large masses of uniform potassic granite, 25 km wide and perhaps several hundred kilometres in lateral extent, could have been generated.

ACKNOWLEDGEMENTS

This work has been supported by the Natural Environment Research Council. We should like to thank Dr M. Suarez for logistic support in southern Chile and Dr A.N. Halliday for making available laboratory facilities for the isotope determinations. Full details of the isotope and geochemical results will be published elsewhere.

References

Agar, R.A. (1978). The Peruvian Coastal Batholith: Its monzanitic rocks and their related mineralisation, *PhD Thesis, University of Liverpool* (unpubl.).

Agar, R.A. (1979). La mineralización de cobre y la Super-Unidad Linga del Batolito de la Costa, *Bol. Soc. Geol. Peru,* **62**, 45–62.

Agar, R.A. (1981). Copper mineralization and magmatic hydrothermal brines in the Rio Pisco section of the Peruvian Coastal Batholith, *Econ. Geol.*, **78**, 677–693.

Aguirre, L. (1983). Granitoids in Chile. In: *Circum-Pacific Plutonic Terranes* (Ed. J.A. Roddick), pp. 293–316, Geol. Soc. Am. Mem. 159.

Aguirre, L., Levi, B. and Offler, R. (1978). Unconformities as mineralogical breaks in the burial metamorphism of the Andes, *Contr. Mineral. Petrol.,* **66**, 361–366.

Anderson, A.T., Clayton, R.N. and Mayeda, T.K. (1971). Oxygen isotope thermometry of mafic igneous rocks, *J. Geol.,* **79**, 715–729.

Arculus, R.J. and Johnson, R.W. (1981). Island-arc magma sources: a geochemical assessment of the roles of slab-derived components and crustal contamination, *J. Geochem.,* **15**, 109–133.

Arculus, R.J. and Wills, K.J.A. (1980). The petrology of plutonic blocks and inclusions from the lesser Antilles Island Arc, *J. Petrol.,* **21**, 743–799.

Arth, J.G. and Barker, F. (1976). Rare-earth partitioning between hornblende and dacitic liquid and implications for the genesis of trondhjemitic-tonalitic magmas, *Geology*, **4**, 534–536.

Arth, J.G., Barker, F., Peterman, Z.E. and Friedman, I. (1978). Geochemistry of the gabbro-diorite-tonalite-trondhjemite suite of southwest Finland and its implications for the origin of tonalitic and trondhjemitic magmas, *J. Petrol.,* **19**, 289–316.

Atherton, M.P. (1981). Horizontal and vertical zoning in the Peruvian Coastal Batholith, *J. Geol. Soc.*, **138**, 343–349.

Atherton, M.P. and Brenchley, P.J. (1972). A preliminary study of the structure, stratigraphy and metamorphism of some contact rocks of the western Andes, near the Quebrada Venado Muerto, Peru. *J. Geol.,* **8**, 161–178.

Atherton, M.P., McCourt, W.J., Sanderson, L.M. and Taylor, W.P. (1979). The geochemical character of the segmented Peruvian Coastal Batholith and associated volcanics. In: *Origin of Granite Batholiths: Geochemical Evidence* (Eds. M.P. Atherton and J. Tarney), pp. 45–64; Shiva Publishing, Nantwich.

Atherton, M.P., Pitcher, W.S. and Warden, V. (1983). The Mesozoic marginal basin of central Peru, *Nature*, **305**, 303–306.

Audebaud, E., Bernard, D., Vatin-Perignon, N. and Vivier, G. (1979). Quelques âges K/Ar sur des roches ignées cénozoiques du sud péruvien. Consequences géodynamiques. *Réunion Annual, Science de la Terre.*

Auer, V. (1974). The isorhytnmicity subsequent to the Feugo-Patagonian and Fennoscandian ocean level transgressions and regressions of the latest glaciation, *Annales Academia Scientiarum Fennicae A III,* **115**, 1–188.

Bailey, D.K. (1974). Continental rifting and alkaline magmatism. In: *The Alkaline Rocks,* pp. 148–159; J. Wiley and Sons, New York.

Bailey, D.K. (1978). Continental rifting and mantle degassing. In: *Petrology and Geochemistry of Continental Rifts* (Eds. E.R. Neumann and I.B. Ramberg), Reidel, Holland.

Baker, M.C.W. (1981). The nature and distribution of Upper Cenozoic ignimbrite centres in the central Andes. *J. Volc. Geoth. Res.,* **11**, 293–315.

Baker, P.E., Rea, W.J., Skarmeta, J., Caminos, R. and Rex, D.C. (1981). Igneous history of the Andean Cordillera and Patagonian Plateau around latitude 46°S, *Phil. Trans. R. Soc. Lond.*, **A303**, 105–149.

Barazangi, M. and Isacks, B.L. (1976). Spatial distribution of earthquakes and subduction of the Nazca plate beneath South America, *Geology*, **4**, 686–692.

Barazangi, M. and Isacks, B.L. (1979). Subduction of the Nazca plate beneath Peru: evidence from spatial distribution of earthquakes. *Geophys. J. R. Astr. Soc.*, **57**, 537–555.

Bard, J.P., Botello, R., Martinez, C. and Subieta, T. (1974). Relations entre tectonique, métamorphisme et mise en place d'un granite eohercynian à deux micas dans la Cordillère Real de Bolivie (Massif de Zongo-Yani). Cahiers ORSTOM Sér Géol., **6**, 3–18.

Barreiro, B. (1982a) Lead isotope data from Late Cenozoic volcanic rocks of the Andes, *Proc. 5th Int. Conf. Geochron., Cosmochron., Isotope Geol.*, **16**.

Barreiro, B. (1982b). Lead isotopic evidence for crustal contamination of Andean magmas, *Trans. Am. Geophys. Union*, **63**, 456.

Barreiro, B. (1982c). Lead isotope evidence for crust-mantle interaction during magmagenesis in the South Sandwich Island Arc and in the Andes of South America, *PhD Thesis, University of California* (unpubl.).

Barreiro, B. (1983). Lead isotopic compositions of South Sandwich Island volcanic rocks and their bearing on magmagenesis in intra-oceanic island arcs, *Geochim. et Cosmochim. Acta*, **47**, 817–822.

Barreiro, B. and Clark, A.H. Lead isotopic evidence for evolutionary changes in magma-crust interaction, central Andes, southern Peru, *Earth Planet. Sci. Lett.*, in press.

Barreiro, B. and Stern, C.R. (1982). Pb isotopic composition of recent calc-alkaline volcanic rocks from the southernmost Andes, *Trans. Am. Geophys. Union*, **63**, 1148.

Bartholomew, D.S. and Tarney, J. Crustal extension in the Southern Andes (45-46°S). In: *Volcanic Processes in Marginal Basins* (Eds. B.P. Kokelaar, M.F. Howells and R.A. Roach); Geol. Soc. Lond. Spec. Publ., in press.

Bateman, P.C. and Dodge, F.C.W. (1970). Variations of major chemical constituents across the central Sierra Nevada batholith, *Bull. Geol. Soc. Am.*, **81**, 409–420.

Bellido, E. (1969). Sinopsis de la geología del Peru, *Ser. Geol. Min. Perú Bol.*, **22**.

Bellido, E. and De Montreuil (1972). Aspectos generales de la metalogenia del Peru, *Serv. Geol. Min. Peru, Geologia Economica*, **1**.

Bergerioux, C., Kennedy, G. and Zikovsky, L. (1979). Use of the semi-absolute method in neutron activation analyses, *J. Radioanal. Chem.*, **50**, 229–234.

Blake, S. (1981). Eruptions from zoned magma chambers, *J. Geol. Soc. Lond.*, **138**, 281–287.

Briqueu, L. and Lancelot, J.R. (1979). Rb–Sr systematics and crustal contamination models for calc-alkaline igneous rocks, *Earth Planet. Sci. Lett.*, **43**, 385–396.

Brooks, C., James, D.E. and Hart, S.R. (1976). Ancient lithosphere: its role in young continental volcanism, *Science*, **193**, 1086–1094.

Brown, G.M., Holland, J.G., Sigurdsson, H., Tomblin, J.F. and Arculus, P.J. (1977). Geochemistry of the Lesser Antilles volcanic arc, *Geochim. et Cosmochim. Acta*, **41**, 785–801.

Brown, L., Klein, J., Middleton, R., Sacks, I.S., and Tera, F. (1982). ^{10}Be in island-arc volcanoes and implications for subduction, *Nature*, **299**, 718–720.

Bussell, M.A. (1976). Fracture control of high level plutonic contacts in the Coastal Batholith of Peru, *Proc. Geol. Ass.*, **87**, 237–46.

Bussell, M.A. (1983). Timing of tectonic and magmatic events in the Central Andes of Peru, *J. Geol. Soc. Lond.*, **140**, 279–286.

Bussell, M.A. and McCourt, W.S. (1977). The Iglesia Irca intrusion and the role of gas brecciation in the emplacement of the Coastal Batholith of Peru, *Geol. Mag.*, **114**, 375–87.

Bussell, M.A., Pitcher, W.S. and Wilson, P.A. (1976). Ring complexes of the Peruvian Coastal Batholith: a long standing subvolcanic regime, *Can. J. Earth Sci.*, **13**, 1020–30.

Caldas, J. (1979). Complejo metamorfico de Illescas, *Bol. Soc. Geol. Peru.*, **63**, 217–235.

Capdevilla, R., Mégard, F., Paredes, J. and Vidal, Ph. (1977). Le batholite de San Ramon (Cordillère orientale du Pérou central). Un granite tardihercynian mis en place à la limite Permien-Trias. Données geologiques et radiométriques, *Geol. Rundsch.*, **66**, 434–46.

Cardozo, M. and Vidal, C. (1981). Metalogenesis y magmatisma en el flanco occidental de los Andes centrales: revision, *Bol. Soc. Geol. Peru*, **68**, 13–24.

Carlier, G., Gerardin, G., Laubacher, G., Marocco, R. and Mégard, F. (1982). Present knowledge of the magmatic evolution of the Eastern Cordillera of Peru, *Earth Science Rev.*, **18**, 253–283.

Carmichael, I.S.E., Turner, F.J. and Verhoogen, J. (1974). *Igneous Petrology;* McGraw-Hill, New York.

Carr, M.J. (1983). Symmetrical and segmented variation of physical and geochemical characteristics of the Central American volcanic front, *J. Volc. Geoth. Res.*

Carr, M.J., Rose, W.I. and Stoiber, R.E. (1982). Volcanism in Central America. In: *Andesites: Orogenic Andesites and Related Rocks* (Ed. R.S. Thorpe); John Wiley and Sons, New York.

Carr, M.J., Stoiber, R.E. and Drake, C.L. (1974). The segmented nature of some continental margins. In: *The Geology of Continental Margins* (Eds. C.A. Burk and C.L. Drake), pp. 105–116; Springer-Verlag, New York.

Case, J.E., Duran, L.G., Lopez, R. and Moore, W.R. (1971). Tectonic investigations in western Colombia and eastern Panama, *Geol. Soc. Am. Bull.*, **82**, 2685–2712.

Cawthorn, R.G., Strong, D.F. and Brown, P.A. (1976). Origin of corundum normative intrusive and extrusive magmas, *Nature*, **259**, 102–104.

Chappell, B.W. and White, A.J.R. (1974). Two contrasting granite types, *Pacific Geology*, **8**, 173–174.

Charrier, R. (1981). Geologic der chilenischen Hanptkordillere zwischen 34° und 34° 30′ süollischer Breite und ibre tektonisher magmatische und paläegeographische Entwicklung, *Berliner Geowissenschattliche Abhandlwingen*, **A36**.

Charrier, R., Linares, E., Niemeyer, H. and Skarmeta, J. (1979). K–Ar ages of basalt flows of the Meseta Buenos Aires in southern Chile and their relation to the southeast Pacific triple junction, *Geology*, **7**, 436–439.

Chow, T.J. and Patterson, C.C. (1962). The occurrence and significance of lead isotopes in pelagic sediments, *Geochim et. Cosmochim. Acta*, **26**, 263–308.

Church, S.E. (1983). Lead isotope relationships for the Columbia River Basalts and their bearing on petrogenesis of volcanic rocks in the Pacific northwest, USA, *Earth Planet Sci. Lett.*, in press.

Church, S.E. and Tatsumoto, M. (1975). Lead isotope relations in oceanic ridge basalts from the Juan de Fuca-Gorda Ridge area, NW Pacific Ocean, *Contr. Mineral. Petrol.*, **53**, 253–279.

Clark, A.H. and Farrar, E. (1973). The Bolivian tin province: notes on the available geochronological data, *Econ. Geol.*, **68**, 102–116.

Clark, A.H. and McNutt, R.H. (1982). Interrelated arc-broadening, topographic uplift and crustal contamination of magmas in two transects of the Mesozoic-Cenozoic central Andes, *Abs. 5th Int. Conf. Geochron. Cosmochron. Isotope Geol.*

Clark, A.H. and Robertson, R.C.R. (1978). The evolution and origin of the northern, plutonic, sub-provinces of the Bolivian tin belt: (Abstr.) International Symposium on Geology of Tin Deposits, *Geol. Soc. Malaysia*, **4(2)**, 42–43.

Clark, A.H., Farrar, E., Caelles, J.C., Haynes, S.J., Lortie, R.B., McBride, S.L., Quirt, G.S., Robertson, R.C.R. and Zentilli, M. (1976). Longitudinal variations in the metallogenetic evolution of the Central Andes: A Progress Report. In: *Metallogeny and Plate Tectonics* (Ed. D.F. Strong) pp. 23–58; Geol. Ass. Canada Spec. Pap. No. 14.

Clark, A.H., Farrar, E. and McNutt, R.H. (1983). Evolution of the Central Andean Magmatic Arcs (Abstr.), *Trans. Am. Geophys. Union*, **64**, 326.

Clark, A.H., Kontak, D.J. and Farrar, E. (1982). A comparative study of the metallogenetic and geochronological relationships in the northern part of the Central Andean Tin Belt, SE Perú and NW Bolivia. In: *International Association for the Genesis of Ore Deposits, VI Symposium, Tbilisi, USSR*, pp. 64–65.

Clarke, D.B. and Halliday, A.N. (1980). Strontium isotope geology of the South Mountain batholith, Nova Scotia, *Geochim. et Cosmochim. Acta*, **44**, 1045–1058.

Clayton, R.N., Goldsmith, J.R., Karel, K.J., Mayeda, T.K. and Newton, R.C. (1975). Limits on the effect of pressure on isotopic fractionation, *Geochim. et Cosmochim. Acta*, **39**, 1197–1201.

Cobbing, E.J. (1976). The geosynclinal pair at the continental margin of Peru, *Tectonophysics*, **36**, 157–185.

Cobbing, E.J. (1978). The Andean geosyncline in Peru and its distinction from Alpine geosynclines, *J. Geol. Soc. Lond.* **135**, 207–218.

Cobbing, E.J. The Central Andes: Peru and Bolivia, in press.

Cobbing, E.J. and Pitcher, W.S. (1972a). Plate tectonics and the Peruvian Andes, *Nature*, **240**, 51–53.

Cobbing, E.J. and Pitcher, W.S. (1972b). The Coastal Batholith of central Peru, *J. Geol. Soc. Lond.*, **128**, 421–460.

Cobbing, E.J. and Pitcher, W.S. (1983). Andean plutonism in Peru and its relationship to volcanism and metallogenesis at a segmented plate edge. In: *Circum-Pacific Plutonic Terranes* (Ed. J.A. Roddick), pp. 277–291; Mem. Geol. Soc. Am. 159.

Cobbing, E.J., Ozard, J.M. and Snelling, N.J. (1977a). The reconnaissance geochronology of the crystalline basement rocks of the Coastal Cordillera of Southern Peru, *Bull. Geol. Soc. Am.*, **88**, 241–246.

Cobbing, E.J., Pitcher, W.S. and Taylor, W.P. (1977b). Segments and super-units in the Coastal Batholith of Peru, *J. Geol.*, **85**, 625–631.

Cobbing, E.J., Pitcher, W.S., Wilson, J.J., Baldock, J.W., Taylor, W.P., McCourt, W.J. and Snelling, N.J. (1981). The geology of the Western Cordillera of northern Peru, Overseas Mem. Inst. Geol. Sci. No. 5.

Cohen, R.S. and O'Nions, R.K. (1982). Identification of recycled continental material in the mantle from Sr, Nd and Pb isotope investigation, *Earth Planet. Sci. Lett.*, **61**, 73–84.

Cohen, R.S., Evensen, N.M., Hamilton, P.J. and O'Nions, R.K. (1980). U–Pb, Sm–Nd and Rb–Sr systematics of mid-ocean ridge basalt glasses, *Nature*, **283**, 149–153.

Couch, R., Whitsett, R., Huehn, B. and Briceno-Guarupe, L. (1981). Structures of the continental margin of Peru and Chile, *Mem. Geol. Soc. Am.*, **154**, 703–726.

Coulon, C. and Thorpe, R.S. (1981). Role of continental crust in petrogenesis of orogenic volcanic associations, *Tectonophysics*, **77**, 79–93.

Cumming, G.L. (1976). Lead isotope ratios in DSDP Leg 37 basalts, *Earth Planet. Sci. Lett.*, **31**, 179–183.

Cummings, D. and Schiller, G.I. (1971). Isopach map of the Earth's crust, *Earth Sci. Rev.*, **7**, 97–125.

Dalmayrac, B., Lancelot, J.R. and Leyreloup, A. (1977). Evidence of 2 b.y. granulites in the late Precambrian metamorphic basement rocks along the southern Peruvian coast, *Science*, **198**, 49–51.

Dalmayrac, B., Laubacher, G. and Marocco, K. (1980). Géologie des Andes Péruviennes, *Trav. Doc. ORSTOM*, **122**.

Dalziel, I.W.D. (1981). Back-arc extension in the southern Andes: a review and critical reappraisal, *Phil. Trans. R. Soc. Lond.*, **A300**, 319–335.

Dalziel, I.W.D., de Wit, M.J. and Palmer, K.F. (1974). Fossil marginal basin in the southern Andes, *Nature*, **250**, 291–294.

Dasch, E.J. (1981). Lead isotopic composition of metalliferous sediments from the Nazca plate. In: *Nazca Plate: Crustal Formation and Andean Convergence* (Eds. L.D. Kulm *et al.*), pp. 199–209; Mem. Geol. Soc. Am. 154.

De Paolo, D.J. (1981). Trace element and isotopic effects of combined wallrock assimilation and fractional crystallization, *Earth Planet. Sci. Lett.*, **53**, 189–202.

De Paolo, D.J. and Johnson, R.W. (1979). Magma genesis in the New Britain island-arc: Constraints from Nd and Sr isotopes and trace element patterns, *Contr. Mineral. Petrol.*, **70**, 367–379.

De Paolo, D.J. and Wasserburg, G. (1977). The sources of island arcs as indicated by Nd and Sr isotopic studies, *Geophys. Res. Lett.*, **4**, 465–468.

Deruelle, B. (1977). Subduction et zonation du volcanisme Plio-Quaternaire des Andes du Sud (18–55°S), *Rev. Geographic Phys. Geol. Dyn.*, **19**, 179–188.

Deruelle, B. (1978). Calc-alkaline and Shoshonitic lavas from five Andean volcanoes (between latitudes 21° 45′ and 24° 30′S) and the distribution of the plio-quaternary volcanism of the South-Central and Southern Andes, *J. Volc. Geoth. Res.*, **3**, 281–298.

Deruelle, B. (1982). Petrology of the Plio-Quaternary volcanism of the south-central and meridional Andes, *J. Volc. Geoth. Res.*, **14**, 77–124.

Deruelle, B. and Deruelle, J. (1974). Géologie des volcans quaternaires des Nevados de Chillán (Chili), *Bull. Volc.*, **38**, 425–444.

Deruelle, B., Harmon, R.S. and Moorbath, S. (1983). Combined Sr–O isotope relationships and petrogenesis of Andean volcanics of South America, *Nature*, **302**, 814–816.

Dixon, T.H. and Batiza, R. (1979). Petrology and chemistry of recent lavas in the northern Marianas: Implications for the origin of island arc basalts, *Contr. Mineral. Petrol.*, **70**, 167–181.

Doe, B.R. (1967). The bearing of lead isotopes in the source of granitic magma, *J. Petrol.*, **8**, 51–83.

Doe, B.R. and Delevaux, M.H. (1973). Variations in lead-isotopic composition in Mesozoic granitic rocks of California: a preliminary investigation, *Bull. Geol. Soc. Am.*, **84**, 3513–3526.

Doe, B.R. and Zartman, R.E. (1979). Plumbotectonics: the Phanerozoic. In: *Geochemistry of Hydrothermal Ore Deposits* (Ed. H.L. Barnes), pp. 22–70.

Doe, B.R., Leeman, W.P., Christiansen, R.L. and Hedge C.E. (1982). Lead and strontium isotopes and related trace elements as genetic tracers in the Upper Cenozoic rhyolite-basalt association of the Yellowstone Plateau volcanic field, *J. Geophys. Res.*, **87**, 4785–4806.

Doe, B.R., Tilling, R.I., Hedge, C.E. and Klepper, M.R. (1968). Lead and strontium isotope studies of the Boulder Batholith, southwestern Montana, *Econ. Geol.*, **63**, 884–906.

Domenick, M.A., Kistler, R.W., Dodge, F.C.W. and Tatsumoto, M. (1983). Nd and Sr isotopic study of crustal and mantle inclusions from the Sierra Nevada and implications for batholith petrogenesis, *Bull. Geol. Soc. Am.*, **94**, 713–719.

Dosso, L., Vidal, P., Cantagrel, J.M., Lameyre, J., Marot, A. and Zimine, S. (1979). Kerguelen: Continental fragment or oceanic island?: Petrology and isotope geochemistry evidence, *Earth Planet. Sci. Lett.*, **43**, 46–60.

Dostal, J., Dupuy, C. and Lefèvre, C. (1977a). Rare earth elements distribution in Plio-Quaternary volcanic rocks from Southern Peru, *Lithos*, **10**, 173–183.

Dostal, J., Zentilli, M., Caelles, J.C. and Clark, A.H. (1977b). Geochemistry and origin of volcanic rocks of Andes (26–28°S), *Contr. Mineral. Petrol.*, **63**, 113–128.

Douglas, J.A. (1920). Geological section through the Andes of Peru and Bolivia: II. From the Port of Mollendo to the Inambari River, *Q. J. Geol. Soc. Lond.*, **89**, 308–347.

Drake, R.E. (1976). Chronology of Cenozoic igneous and tectonic events in the central Chilean Andes—Latitudes 35° 30′ to 36°S, *J. Volc. Geoth. Res.*, **1**, 285–295.

Drake, R. (1981). Sr and Pb isotopic data bearing on the origin of late Cenozoic igneous rocks and associated ore deposits from the central Chilean Andes (Lat. 33° and 36°S), *Geol. Soc. Am.*, Abstracts with Programs, **13**, 56.

Drake, R. (1983). Geochronological relationships between magmatism and tectonics in the Andes of Central Chile, *Trans. Am. Geophys. Union*, **64**, 327.

Dupré, B. and Allègre, C.J. (1980). Pb–Sr–Nd isotopic correlation and the chemistry of the North Atlantic mantle, *Nature*, **286**, 17–22.

Dupré, B. and Allègre, C.J. (1983). Lead and strontium isotope variations in Indian Ocean basalts and mixing phenomena, *Nature*, **303**, 142–146.

Dupré, B., Lambret, B. and Allègre, C.J. (1982). Isotopic variations within a single oceanic island, the Terceira case, *Nature*, **299**, 620–622.

Dupuy, C., Dostal, J., Marcelot, G., Bougault, H., Joron, J.L. and Treuil, M. (1982). Geochemistry of basalts from central and southern New Hebrides arc: implication for their source rock composition, *Earth Planet. Sci. Lett.*, **60**, 207–225.

Egeler, C.G. and De Booy, T. (1956). Geology and petrology of part of the southern Cordillera Blanca, Peru, *Verk. K. Ned. Geol. Mijnb. Genoots. Geol. Ser.*, **17**, 11–86.

Eichelberger, J.C. (1980). Vesiculation of mafic magma during replenishment of silicic magma reservoirs, *Nature*, **288**, 446–450.

Estrada, F. (1978). Edades K–Ar de los principales eventos geológicos de Cerro Verde, *Bol. INCITEMI, Peru*, 1–15.

Evernden, J.F., Kriz, S.J. and Cherroni, C. (1977). Potassium-argon ages of some Bolivian rocks, *Econ. Geol.*, **72**, 1042–1061.

Ewart, A. (1976). Mineralogy and chemistry of modern orogenic lavas. Some statistics and implications, *Earth Planet. Sci. Lett.*, **31**, 417–432.

Ewart, A. (1982). The mineralogy and petrology of tertiary-recent orogenic volcanic rocks: with special reference to the andesitic-basaltic compositional range. In: *Andesites* (Ed. R.S. Thorpe), pp. 25–95; John Wiley and Sons, New York.

Ewart, A., Brothers, R.N. and Mateen, A. (1977). An outline of the geology and geochemistry, and the possible petrogenetic evolution of the volcanic rocks of the Tonga-Kermadec-New Zealand island arc, *J. Volc. Geoth. Res.*, **2**, 205–250.

Farmer, G.L. and De Paolo, D.J. (1983). Origin of Mesozoic and Tertiary granite in the Western United States and implications for Pre-Mesozoic crustal structure: 1 Nd and Sr isotopic studies in the Geocline of the northern Great Basin, *Geophys. Res.*, **88**, 3379–3401.

Faure, G. (1977). *Principles of Isotope Geology;* J. Wiley and Sons, New York.

Forsyth, D.W. (1975). Fault plane solutions and tectonics of the South Atlantic and Scotia Sea, *J. Geophys. Res.*, **80**, 1492–1543.

Forsythe, R.D. (1982). The Late Palaeozoic to Early Mesozoic evolution of southern South America: a plate tectonic interpretation, *J. Geol. Soc. Lond.*, **139**, 671–682.

Francis, G.A. (1956). La geologia de la zona entre Macusani y Ollachea, Departamento de Puno. Instituto Nacional de Investigaciones y Fomento Minero., *Boletín,* **21**, 5–10.

Francis, P.W., Moorbath, S. and Thorpe, R.S. (1977). Strontium isotope data in andesites from Ecuador and north Chile, *Earth Planet. Sci. Lett.*, **37**, 197–202.

Francis, P.W., Roobol, M.J., Coward, M.P., Cobbold, P.R. and Walker, G.P.L. (1974). The San Pedro and San Pablo volcanoes of northern Chile and their hot avalanche deposits, *Geol. Rundsch.*, **63**, 357–388.

Francis, P.W., Thorpe, R.S., Moorbath, S., Kretschmar, G.A. and Hammill, M. (1980). Strontium isotope evidence for crustal contamination of calc-alkaline volcanic rocks from Cerro Galan, northwest Argentina, *Earth Planet. Sci. Lett.*, **48**, 257–267.

Frey, F.A., Bryan, W.B. and Thompson, G. (1974). Atlantic Ocean floor: Geochemistry and petrology of basalts from Legs 2 and 3 of the Deep Sea Drilling Project, *J. Geophys. Res.*, **79**, 5507–5527,

Frey, F.A., Gerlach, D.C., Hickey, R.L., Lopez-Escobar, L. and Munizaga, F. Petrogenesis of the Laguna del Maule volcanic complex, Chile (36°), *Contr. Mineral. Petrol.*, in press.

Fricker, P. and Weibel, M. (1960). Zur Kenntnis der Eruptivgesteine in der Cordillera Vicabamba, Peru, *Schweiz. Miner. Petrog. Mitt.*, **40**, 359–382.

Friedman, I. and O'Neil, J.R. (1977). Compilation of stable isotope fractionation factors of geochemical interest. In: *Data in Geochemistry* (Ed. M. Fleischer), US Geological Survey Professional Paper, 440-KK.

Friedman, I., Lipman, P., Obradovich, J.D., Gleason, J.D. and Christiansen, R.L. (1974). Meteoric water in magmas, *Science,* **184**, 1069–1072.

Frutos, J. (1981). Andean tectonics as a consequence of sea-floor spreading, *Tectonophysics,* **72**, 21–32.

Fuenzalida, R. (1974). The Magellan fault zone, In: *Andean and Antarctic Volcanology Problems* (Ed. O. Gonzales-Ferran), pp. 373–391; Stabilimento Tipografico Francesco Giannini and Figli, Napoli.

Fultz, L.A. (1979). Sr isotopic determinations of Guatemalan basalts, *MSc Thesis, Michigan Technological University* (unpubl.).

Garcia, W. (1968). Geología de los cuadrángulos de Mollendo y la Joya, *Bol. Ser. Geol. Min. Peru,* **19**.

Garlick, G.D. (1966). Oxygen isotope fractionation in igneous rocks, *Earth Planet. Sci. Lett.*, **1**, 361–368.

Gerlach, D., Hickey, R., Frey, F., Moreno, H., Munizaga, F. and Lopez, L. (1982). Petrologic and geochemical variations in volcanic centers of central South Chile (33–42°S), *Trans. Am. Geophys. Union,* **63**, 1148.

Gerlach, D.C., Frey, F.A., Hickey, R., Moreno-Roa, H. and Hildreth, W. (1983). Geochemistry of Puyehue Volcano and Cordon Caulle, Southern Andes (40.5°S), *Trans. Am. Geophys. Union,* **64**, 326.

Gill, J.V. (1974). Role of underthrust oceanic crust in the genesis of a Fijian calc-alkaline suite, *Contr. Mineral. Petrol.*, **43**, 29–45.

Gill, J.V. (1981). *Orogenic Andesites and Plate Tectonics;* Springer-Verlag, Berlin.

Godoy, E., Dobbs, M., Stern, C. (1981). El volcán Hudson, primeros datos químicos e isotópicos en coladas interglaciales, *Comunicaciones Dep. Geol. Univ. Chile*, **32**, 1–9.

González-Diaz, E.F. (1961). La petrografia del Cerro Tupungato y otras rocas efusivas de la region, *Revista de la Asociacion Geologica Argentina*, **16**, 205–234.

González-Ferrán, O. (1972). *Volcanos Activos de Chile;* Instituto Geografico Militar de Chile, Santiago.

Gorton, M.P. (1977). The geochemistry and origin of quaternary volcanism in the New Hebrides, *Geochim. et Cosmochim. Acta*, **41**, 1257–1270.

Grant, J.N., Halls, C., Avila, W. and Snelling, N.J. (1977). Edades Potasio-Argon de las rocas igneas y la mineralización de parte de la Cordillera Oriental, Bolivia, *Bol. Ser. Geol. Bolivia, Ser. A.*, **1**, 33–60.

Grant, J.N., Halls, C., Salinas, W.A. and Snelling, N.J. (1979). K–Ar ages of igneous rocks and mineralization in part of the Bolivian tin belt, *Econ. Geol.*, **74**, 838–851.

Green, T.H. (1969). High-pressure experimental studies on the origin of anorthosite, *Can. J. Earth Sci.*, **6**, 427–440.

Green, T.H. (1980). Island-arc and continent-building magmatism—a review of petrogenetic models based on experimental petrology and geochemistry, *Tectonophysics*, **63**, 367–385.

Green, T.H. and Ringwood, A.E. (1968). Genesis of the calc-alkaline igneous rock suite, *Contr. Mineral. Petrol.*, **18**, 105–162.

Gregory, R.T. and Taylor, H.P. (1981). An oxygen isotope profile in a section of cretaceous oceanic crust, Samail ophiolite, Oman: Evidence for $\delta^{18}O$ buffering of the oceans by deep ($>$ 5 km) seawater-hydrothermal circulation at mid-ocean ridges, *J. Geophys. Res.*, **86**, 2737–2755.

Grunder, A.L. (1983). The Calabozos volcanic system; a major Quaternary silici center in the central Chilean Andes, *Trans. Am. Geophys. Union*, **64**, 326.

Guest, J.E. (1968). Banded pumice in a Chilean ignimbrite, *Geol. Mag.*, **105**, 177–184.

Guest, J.E. (1969). Upper Tertiary ignimbrites in the Andean Cordillera, Antologasta province, north Chile, *Geol. Soc. Ann. Bull.*, **80**, 337–362.

Gustafson, L. (1979). Porphyry copper deposits and calc-alkaline volcanism. In: *The Earth: Its Origin, Structure and Evolution* (Ed. M.W. McElhinny), pp. 427–468; Academic Press, London.

Gustafson, L.B. and Hunt, J.P. (1975). The porphyry copper deposit at El Salvador, Chile, *Econ. Geol.*, **70**, 857–912.

Halliday, A.N., Stephens, W.E. and Harmon, R.S. (1981). Isotopic and chemical constraints on the development of peraluminous Caledonian and Acadian granites, *Can. Mineral.*, **19**, 205–216.

Halpern, M. (1973). Regional geochronology of Chile south of 50°S latitude, *Bull. Geol. Soc. Am.*, **84**, 2407–2422.

Halpern, M. (1979). Strontium isotope composition of rocks from the Disputada copper mine, Chile, *Econ. Geol.*, **74**, 129–130.

Hanus, V. and Vanek, J. (1976). Intermediate aseismicity of the Andean subduction zone and recent andesitic volcanism, *J. Geophys.*, **42**, 219–223.

Hanus, V. and Vanek, J. (1978). Morphology of the Andean Wadati-Benioff zone, andesitic volcanism, and tectonic features of the Nazca plate, *Tectonophysics*, **44**, 65–77.

Harmon, R.S., Barreiro, B., Moorbath, S., McHugh, J., Thorpe, R.S., Francis, P.W., Hoefs, J. and Deruelle, B. Regional O-, Sr-, Pb-isotope relationships in Late Cenozoic calc-alkaline lavas in the Andean Cordillera, *J. Geol. Soc. Lond.*, in press.

Harmon, R.S., Moorbath, S. and McHugh, J.M. (1983). O-, Sr- and Pb-isotope relationships in Recent Andean volcanics, *Trans. Am. Geophys. Union*, **64**, 325.

Harmon, R.S., Thorpe, R.S. and Francis, P.W. (1981). Petrogenesis of Andean andesites from combined O–Sr relationships, *Nature*, **290**, 396–399.

Harris, C., Bell, J.D. and Atkins, F.B. (1982). Isotopic composition of lead and strontium in lavas and coarse-grained blocks from Ascension Island, South Atlantic, *Earth Planet. Sci. Lett.*, **60**, 79–85.

Harris, P.G., Kennedy, W.Q. and Scarfe, C.M. (1970). Volcanism versus plutonism—the effect of chemical composition. In: *Mechanism of Igneous Intrusion* (Eds. G. Newall and N. Rast), pp. 187–200; Geological Journal of Liverpool Special Issue No. 2.

Hart, S.R. (1976). LIL-Element geochemistry, Leg 34, basalts, *Init. Rep. DSDP* (Yeats, R.S., Hart, G.R., *et al.*, eds.) Washington. US Gov. Printing Off. **34**, 283–288.

Hart, S.R., Erlank, A.J. and Kable, E.J.D. (1974). Sea floor basalt alteration: some chemical and isotopic effects, *Contr. Mineral. Petrol.*, **44**, 219–230.

Hawkesworth, C.J. (1979). $^{143}Nd/^{144}Nd$, $^{87}Sr/^{86}Sr$ and trace element characteristics of magmas along destructive plate margins. In: *Origin of Granite Batholiths: Geochemical Evidence* (Eds. M.P. Atherton and J. Tarney), pp. 76–89; Shiva Publishing, Nantwich.

Hawkesworth, C.J. (1982). Isotopic characteristics of magmas erupted along destructive plate margins. In: *Orogenic Andesites* (Ed. R.S. Thorpe); J. Wiley and Sons, New York.

Hawkesworth, C.J. and Powell, M. (1980). Magma series in the Lesser Antilles island arc, *Earth Planet. Sci. Lett.*, **51**, 297–308.

Hawkesworth, C.J., Hammill, M., Gledhill, A.R., van Calsteren, P. and Rogers, G. (1982). Isotope and trace element evidence for late-stage intra-crustal melting in the High Andes, *Earth Planet. Sci. Lett.*, **58**, 240–254.

Hawkesworth, C.J., O'Nions, R.K. and Arculus, R.J. (1979a). Nd and Sr isotope geochemistry of island arc volcanics, Grenada, Lesser Antilles, *Earth Planet. Sci. Lett.*, **45**, 237–248.

Hawkesworth, C.J., Norry, M.J., Roddick, J.C., Baker, P.E., Francis, P.W. and Thorpe, R.S. (1979b). $^{143}Nd/^{144}Nd$, $^{87}Sr/^{86}Sr$ and incompatible trace element variations in calc-alkaline andesitic and plateau lavas from South America, *Earth Planet. Sci. Lett.*, **42**, 45–57.

Hawkesworth, C.J., O'Nions, R.K., Pankhurst, R.J., Hamilton, P.J. and Evensen, N.M. (1977). A geochemical study of island-arc and back-arc tholeiites from the Scotia Sea, *Earth Planet. Sci. Lett.*, **36**, 253–262.

Hayes, D.E. (1966). A geophysical investigation of the Peru-Chile trench, *Marine Geology*, **4**, 309–351.

Herron, E.M. (1981). Chile margin near 38°S: Evidence for a genetic relationship between continental and marine geologic features or a case of curious coincidences, *Mem. Geol. Soc. Am.*, **154**, 755–760.

Herron, E.M., Bruhn, R., Winslow, M. and Chuaqui, L. (1977). Post-Miocene tectonics of the margin of southern Chile. In: *Island Arcs, Back-Arc Basins, and Deep Sea Trenches* (Eds. M. Talwani and W.C. Pitman III), pp. 273–284; American Geophysical Union, Washington.

Herron, E.M., Cande, S.C. and Hall, B.R. (1981). An active spreading center collides with a subduction zone: A geophysical survey of the Chile Margin triple junction, *Mem. Geol. Soc. Am.*, **154**, 683–702.

Hickey, R.L., Frey, F.A., Gerlach, D. and López-Escobar, L. (1983). Isotopic and trace element data bearing on the origin of the volcanic rocks from central south Chile, *Trans. Am. Geophys. Union*, **64**, 326.

Hickey, R.L., Frey, F.A., López-Escobar, L. and Munizaga, F. (1982). Nd and Sr isotopic data bearing on the origin of Andean volcanic from S. central Chile, *Geol. Soc. Am. Ann. Meetings*, **14**, 514.

Hildreth, W. (1981). Gradients in silicic magma chambers: implications for lithospheric magmatism, *J. Geophys. Res.*, **86**, 10153–10192.

Hildreth, W. and Drake, R.E. (1983). 1932 eruption of Quizapu, central Chilean Andes, *Geol. Soc. Am.*, Abstracts with Programs, **15**, 390.

Hildreth, W., Drake, R.W. and Sharp, W.D. (1981). Voluminous Late Pleistocene ash-flow and caldera complex in the Andes of central Chile, *Geol. Soc. Am.*, Abstracts with programs, **13**, 61.

Hildreth, W., Grunder, A.L. and Drake, R.E. (1983). The Loma Seca Tuff and the Calabozos Caldera: a major ash-flow and Caldera complex in the Southern Andes of Central Chile, *Bull. Geol. Soc. Am.*, in press.

Hoefs, J., Faure, G. and Elliot, D.H. (1980). Correlation of $\delta^{18}O$ and initial $^{87}Sr/^{86}Sr$ ratios in Kirkpatrick Basalt on Mt. Falla, Transantarctic Mountains, *Contr. Mineral. Petrol.*, **75,** 199–203.

Hofmann, A.W. and White, W.M. (1983). Ba, Rb and Cs in the earth's mantle, *Z. Naturforsch*, **38a,** 256–266.

Hudson, C. (1979). Zoneamiento de la metalogenia andina del Perú, *Bol. Soc. Geol. Peru,* **60,** 61–72.

Huppert, H.E. and Sparks, R.S. (1980). The fluid dynamics of a basaltic magma chamber replenished by influx of hot, dense ultrabasic magma, *Contr. Mineral. Petrol.*, **75,** 279–289.

Huppert, H.E., Sparks, R.S.J. and Turner, J.S. (1982). Effects of volatiles on mixing in calc-alkaline magma systems, *Nature,* **297**, 554–557.

Hussong, D.M. and Wipperman, L.K. (1981). Vertical movement and tectonic erosion of the continental wall of the Chile-Peru trench near 11°30′S latitude, *Mem. Geol. Soc. Am.*, **154,** 509–524.

Iddings, J.P. (1895). Absarokite-shoshonite-banakite series, *J. Geol.,* **3**, 935–959.

Irvine, T.N. and Baragar, W.R.A. (1971). A guide to the chemical classification of the common Volcanic rocks, *Can. J. Earth Sci.*, **8**, 523–548.

Ishihara, S. (1977). The magnetite-series and ilmenite-series granite rocks, *Mining Geology,* **27**, 273–305.

Ishihara, S. (1981). The granitoid series and mineralization. In: *Economic Geology Seventy-fifth Anniversary volume* (Ed. B.J. Skinner), pp. 458–484.

Jakes, P. and Smith, I.E. (1970). High potassium calc-alkaline rocks from Cape Nelson, eastern Papua, *Contr. Mineral. Petrol.*, **28**, 259–271.

Jakes, P. and White, A.J.R. (1972). Major and trace element abundances in volcanic rocks of orogenic areas, *Bull. Geol. Soc. Am.*, **83**, 29–40.

James, D.E. (1971a). Andean crustal and upper mantle structure, *J. Geophys. Res.*, **76**, 3246–3271.

James, D.E. (1971b). Plate tectonic model for the evolution of the central Andes, *Bull. Geol. Soc. Am.*, **82**, 3325–3346.

James, D.E. (1981a). The combined use of oxygen and radiogenic isotopes as indicators of crustal contamination, *Ann. Rev. Earth Planet. Sci.*, **9**, 311–344.

James, D.E. (1981b). Role of subducted continental material in the genesis of calc-alkaline volcanics of the central Andes, *Mem. Geol. Soc. Am.*, **154**, 769–790.

James, D.E. (1982a). A combined O, Sr, Nd and Pb isotopic and trace element study of crustal contamination in central Andean lavas: I. Local geochemical variations, *Earth Planet. Sci. Lett.*, **57**, 47–62.

James, D.E. (1982b). Isotopic evidence for the petrogenesis of modern andesites of the northern most Andes, *Yb. Carnegie Inst. Wash.*, **81**, 485–489.

James, D.E. (1983). Crustal contamination in northern Andean volcanics, *Yb. Carnegie Inst. Wash.*, **82**, in press.

James, D.E. An O-, Sr-, Nd- and Pb-isotopic study of andesite magmagenesis in the northern Andes, *J. Geol. Soc. Lond.*, in press.

James, D.E., Brooks, C. and Cuyubamba, A. (1974). Strontium isotopic composition and K, Rb, Sr. Geochemistry of Mesozoic volcanic rocks of the Central Andes, *Yb. Carnegie Inst. Wash.*, **73**, 970–983.

James, D.E., Brooks, C. and Cuyubamba, A. (1975). Early evolution of the Central Andean Volcanic Arc, *Yb. Carnegie Inst. Wash.*, **74**, 247–250.

Jenks, W.F. (1951). Triassic to Tertiary stratigraphy near Cerro do Pasco, Peru, *Bull. Geol. Soc. Am.*, **62**, 203–220.

Jensen, L.S. (1976). A new cation plot for classifying subalkalic volcanic rocks. Ontario Division of Mines, Misc. Pap., **66,** 1–22.

Johan, Z., Le Bel, L. and McMillan, W.J. (1980). Evolution géologique et pétrologique des complexes granitoides fertiles. In: *Minéralisations Lièes aux Granitoïds* (Ed. Z. Johan), pp. 21–70; Mémoires du Bureau de Recherches géologiques et minières, no. 99.

Johnson, R.W. (1981). Island arc basalts. In: *Basaltic Volcanism on the Terrestrial Planets* (Basaltic Volcanism Study Project, pp. 193–213, Pergamon Press, New York.

Johnson, R.W., Jaques, A.L., Hickey, R.L., McKee, C.O. and Chappell, B.W. (1984). Manam Island, Papua, New Guinea: petrology and geochemistry of a low TiO_2 basaltic island arc volcano, Submitted to *J. Petrol.*

Jones, P.R. (1981). Crustal structures of the Peru continental margin and adjacent Nazca plate, 9°S latitude, *Mem. Geol. Soc. Am.*, **154**, 423–443.

Joplin, G.A. (1968). The shoshonitic association: a review. *J. Geol. Soc. Australia*, **15**, 275–294.

Jordan, T.E., Isacks, B.L., Allmendinger, R.W., Brewer, J.A., Ramos, V.A. and Ando, C.J. (1983). Andean tectonics related to geometry of subducted Nazca plate, *Bull. Geol. Soc. Am.*, **94**, 341–361.

Joron, J.L. and Treuil, M. (1977). Utilisation des propriétés des éléments fortement hygromagmatophiles pour l'ètude de la composition chimique et de l'hétérogéneité du manteau, *Bull. Soc. Geol. France*, **6**, 1197–1205.

Katsui, Y. and Vergara, M. (1966). Antecedentes petrológicos del Cerro San Cristóbal, Santiago, Chile, *Rev. Minerales*, **92**, 9–25.

Kay, R. (1977). Geochemical constraints on the origin of Aleutian magmas. In: *Island Arcs, Deep Sea Trenches and Back-Arc Basins*, (Eds. M. Talwani and W.C. Pittman III), pp. 229–242; Am. Geophys. Union, Washington.

Kay, R.W. (1978). Aleutian magnesian andesites: melts from subducted Pacific ocean crust, *J. Volc. Geoth. Res.*, **4**, 117–132.

Kay, R.W. (1980). Volcanic arc magmas: implications of a melting-mixing model for element recycling in the crust-upper mantle, *J. Geol.*, **88**, 497–522.

Kay, R.W. and Gast, P.W. (1973). The rare-earth content and origin of alkali-rich basalts, *J. Geol.*, **81**, 653–682.

Kay, R.W., Sun, S.S. and Lee-Hu C.N. (1978). Pb and Sr isotopes in volcanic rocks from the Aleutian Islands and Pribilof Islands, Alaska, *Geochim. et Cosmochim. Acta.*, **42**, 263–273.

Klerkx, J., Deutsch, S., Pichler, H. and Zeil, W. (1977). Strontium isotopic composition and element data bearing on the origin of Cenozoic volcanic rocks of the Central and Southern Andes, *J. Volc. Geoth. Res.*, **2**, 49–71.

Knox, G.J. (1971). The structure and emplacement of the Rio Fortaleza centred acid ring complex, Ancash, Peru, *J. Geol. Soc. Lond.*, **130**, 295–308.

Kontak, D.J., Clark, A.H. and Farrar, E. Order/disorder relationships of alkali feldspars in granitoid rocks: implications for the distribution of Sn-Cu-Pb-Zn-Ag deposits, SE Perú, *Bulletin de Minéralogie*, in press.

Kuno, H. (1950). Petrology of Hakone volcano and adjacent areas, Japan, *Bull. Geol. Soc. Am.*, **61**, 967–1020.

Kuno, H. (1966). Lateral variation of basalt magma type across continental margin and island arcs, *Bull. Volc.*, **29**, 195–222.

Kuno, H. (1968a). Origin of andesite and its bearing on the island arc structure, *Bull. Volc.*, **32**, 141–176.

Kuno, H. (1968b). Differentiation of basalt magmas. In: *Basalts* (Eds. Hess and Poldervaart), pp. 623–688; Interscience, New York.

Kuno, H. (1969). Andesite in time and space. Oregon Department of Geological and Mineralogical Industries, *Bulletin*, **65**, 13–20.

Kyser, T.K., O'Neil, J.R. and Carmichael, I.S.E. (1981). Oxygen isotope thermometry of basic lavas and mantle modules, *Contr. Mineral. Petrol.* **77**, 11–23.

Kyser, T.K., O'Neil, J.R. and Carmichael, I.S.E. (1982). Genetic relations among basic lavas and ultramafic nodules: Evidence from oxygen isotope compositions, *Contr. Mineral. Petrol.*, **81**, 88–102.

Lancelot, J.R., Laubacher, G., Marocco, R. and Renaud, U. (1978). U/Pb radio-chronology of two granitic plutons from the Eastern Cordillera (Peru): Extent of Permian magmatic activity and consequences, *Geol. Rundsch.*, **67**, 236–243.

Langmuir, G., Goldstein, S., Zindler, A., Weaver, H., Staudigel, H. and Carr, M.J. (1981). Nd and Sr isotopic composition of Central American volcanoes, *Trans. Am. Geophys. Union*, **62**, 440.

Larsen, E.S. (1948). Batholith and associated rocks of Corona, Elsinore and San Luis Rey quadrangles, Southern California, *Mem. Geol. Soc.*, **A29**.

Larsson, W. (1941). Petrology of interglacial volcanics from the Andes of northern Patagonia, *Bull. Geol. Upsala*, **28**, 191–405.

Laubacher, G. (1978). Estudio Geológico de la Region Norte Del Lago Titicaca, Boletin No. 5, *Instituto de Geologia y Mineria, Peru.*

Laughlin, A.W., Damon, P.E. and Watson, B.N. (1968). Potassium-argon ages from Toquepala and Michiquillay, Peru, *Econ. Geol.*, **63**, 166–168.

Le Bel, L.M., Cocherie, A., Baubron, J.C., Fouillac, A.M. and Hawkesworth, C.J. A high-K, mantle-derived, plutonic suite from Linga, near Arequipa, Peru, in press.

Lefèvre, C. (1973). Les caractères magmatiques du volcanisme plioquaternaire des Andes dans le Sud de Perou, *Contr. Mineral. Petrol.*, **41**, 259–272.

Lehmann, B. (1978). A Precambrian sample from the Altiplano, Bolivia, *Geol. Rundsch.*, **67**, 270–278.

Levi, B. and Aguirre, L. (1981). Ensialic spreading-subsidence in the Mesozoic and Palaeogene Andes of Central Peru, *J. Geol. Soc. Lond.*, **138**, 75–81.

Lewis, J.F. (1973). Petrology of the ejected plutonic blocks of the Soufriere volcano, St. Vincent, West Indies, *J. Petrol.*, **14**, 81–112.

Lomnitz, C. (1962). On Andean structure, *J. Geophys. Res.*, **76**, 351–363.

Longstaffe, F.J., Clark, A.H., McNutt, R.H. and Zentilli, M. (1983). Oxygen isotope compositions of Central Andean plutonic and volcanic rocks, latitudes 26–29° south, *Earth Planet. Sci. Lett.*, **64**, 9–18.

Longstaffe, F.J., Smith, T.E. and Muehlenbachs, K. (1980). Oxygen isotope evidence for the genesis of Upper Paleozoic granitoids from southwestern Nova Scotia, *Can. J. Earth Sciences*, **17**, 132–141.

López-Escobar, L. (1978). Consideraciones petrogenéticas sobre las granodioritas terciarias de los Andes del Sur entre los 33–34°S (Chile Central), *Comunicaciones Dep. Geol. Univ. Chile*, **22**, 1–31.

López-Escobar, L. (1982). Características geoquímicas de rocas ígneas asociadas con pórfidos cupríferos chilenos, *Rev. Geol. Chile*, **17**, 3–19.

López-Escobar, L. and Frey, F.A. (1976). Rocas volcánicas cuaternarias de Chile central-sur (33–41°S): Modelos petrogenéticos sugeridos por las Tierras Raras, *Actas I Congreso Geológico Chileno*, **2**, F223–F255.

López-Escobar, L., Frey, F.A. and Vergara, M. (1976). Andesites from central-south Chile: Trace element abundances and petrogenesis. In: *IAVCEI Proc. Symp. on Andean and Antarctic Volcanology Problems* (Ed. O. González F.), pp. 725–761; Santiago-Chile.

López-Escobar, L., Frey, F.A. and Vergara, M. (1977). Andesites and high-alumina basalts from the central-south Chile High Andes: Geochemical evidence bearing on their petrogénesis, *Contr. Mineral. Petrol.*, **63**, 199–228.

López-Escobar, L., Moreno, H., Tagiri, M., Notsu, K. and Onuma, N. Geochemistry of lavas from San José volcano, Southern Andes 33° 45′S, central Chile, in prep.

López-Escobar, L., Vergara, M. and Frey, F.A. (1981). Petrology and geochemistry of lavas from Antuco Volcano, a basaltic volcano of the southern Andes (37° 25′S), *J. Volc. Geoth. Res.*, **11**, 329–352.

Lowrie, A. and Hey, R. (1981). Geological and geophysical variations along the western margin of Chile near lat. 33° to 36°S and their relation to Nazca plate subduction, *Mem. Geol. Soc. Am.*, **154**, 741–754.

MacDonald, G.A. (1972). *Volcanoes;* Prentice-Hall, Inc., New Jersey.

MacLaughlin, D.H. (1924). Geology and physiography of the Peruvian Cordillera. Departments of Junín and Lima, *Bull. Geol. Soc. Am.*, **35**, 591–680.

Magaritz, M., Whitford, D.J. and James, D.E. (1978). Oxygen isotopes and the origin of high $^{87}Sr/^{86}Sr$ andesites, *Earth Planet. Sci. Lett.*, **40**, 220–230.

Mahoney, J., Macdougall, J.D., Lugmair, G.W., Murali, A.V., Sankar Das, M. and Gopalan, K. (1982). Origin of the Deccan Trap flows at Mahabaleshwar inferred from Nd and Sr isotopic and chemical evidence, *Earth Planet. Sci. Lett.*, **60**, 47–60.

Marocco, R. (1971). Etude Geologique de la chaîne andine au niveau de la deflexion d'Abancy (Perou), *Cahiers ORSTOM, sér. géol.*, **3**, 45–58.

Marocco, R. (1978). Un segment est-ouest de la chaîne des Andes péruviennes: la deflexion d'Abancay, *Travailles ORSTOM,* **94**.

Martínez, M. (1970). Geología del Basamento Paleozoico en las montañas Amotapes y posible orígen del petróleo en las rocas paleozoicas del Noroeste del Perú, *Bol. 1er. Cong. Latinoamericano de Geologia, Tomo II.*

Mason, G. (1982). The mineralogy and textural history of the Coastal Batholith of Peru, *PhD Thesis, University of Liverpool* (unpubl.).

Masuda, A., Nakamura, N. and Tanaka, T. (1973). Fine structures of mutually normalized rare-earth patterns of chondrites, *Geochim. et Cosmochim. Acta,* **37**, 239–248.

Matsuhisa, Y. (1979). Oxygen isotopic compositions of volcanic rocks from the east Japan island arcs and their bearing on petrogenesis, *J. Volc. Geoth. Res.*, **5**, 271–296.

Matsuhisa, Y., Goldsmith, J. and Clayton, R.N. (1979). Oxygen isotopic fractionation in the system quartz-albite-anorthite-water, *Geochim. et Cosmochim. Acta,* **43**, 1131–1140.

Matsuhisa, Y., Matsubaya, O. and Sakai, H. (1973). Oxygen isotope variations in magmatic differentiation processes of the volcanic rocks in Japan, *Contr. Mineral. Petrol.*, **39**, 277–288.

McBride, S.L. (1977). A K–Ar Study of the Cordillera Real, Bolivia, and its Regional Setting, *PhD Thesis, Queen's University, Kingston, Ontario* (unpubl.).

McBride, S.L., Robertson, R.C.R., Clark, A.H. and Farrar, E. (1983). Magmatic and metallogenetic episodes in the northern tin belt, Cordillera Real, Bolivia, *Geol. Rundsch.* **72**, 685–714.

McCarthy, T.S. and Cawthorn, R.G. (1980). Changes in initial $^{87}Sr/^{86}Sr$ ratio during protracted fractionation in igneous complexes, *J. Petrol.*, **21**, 245–264.

McCourt, W.J. (1978). Geochemistry of the Coastal Batholith of Peru, *PhD Thesis, University of Liverpool* (unpubl.).

McCourt, W.J. (1981). The geochemistry and petrography of the Coastal batholith of Peru, Lima segment, *J. Geol. Soc. Lond.*, **138**, 407–420.

McCourt, W.J. and Taylor, W.P. (1979). La geoquímica de los magmas del Batolito de la Costa, *Bol. Soc. Geol. Peru,* **61**, 167–180.

McCulloch, M.T. and Perfit, M.R. (1981). $^{143}Nd/^{144}Nd$, $^{87}Sr/^{86}Sr$ and trace element constraints on the petrogenesis of Aleutian island arc magmas, *Earth Planet. Sci. Lett.*, **56**, 167–179.

McCulloch, M.T., Jaques, A.L., Nelson, D.R. and Lewis, J.D. (1983). Nd and Sr isotopes in kimberlites and lamproites from Western Australia: an enriched mantle origin, *Nature,* **302**, 400–403.

McKee, E.H. and Noble, D.C. (1982). Miocene volcanism and deformation in the western Cordillera and high plateaux of south-central Peru, *Bull. Geol. Soc. Am.*, **93**, 657–662.

McNutt, R.H., Clark, A.H. and Zentilli, M. (1979). Lead isotopic compositions of Andean igneous rocks, latitudes 26° to 29°S: petrogenetic and metallogenic implications, *Econ. Geol.*, **74**, 827–837.

McNutt, R.H., Crocket, J.H., Clark, A.H., Caelles, J.C., Farrar, E., Haynes, S.J. and Zentilli, M. (1975). Initial $^{87}Sr/^{86}Sr$ ratios of plutonic and volcanic rocks of the Central Andes between latitudes 26° and 29° south.

Mégard, F. (1973). Étude géologique d'une transversale des Andes au niveau de Pérou Central, *Thesis Univ. Montpellier II.*

Mégard, F. (1978). Etude géologique des Andes, du Perou central, *Mém. ORSTOM,* **86**.

Mégard, F., Dalmayrac, B., Laubacher, G., Marocco, R., Martinez, C.L., Paredes, J. and Tomasi, P. (1971). La Chaine hercynienne au Perou et en Bolivie; premiers resultats, *Cahiers ORSTOM, Ser. Geol.*, **3**, 5–43.

Meijer, A. (1976). Pb and Sr isotopic data bearing on the origin of volcanic rocks from the Mariana Island arc system, *Bull. Geol. Soc. Am.*, **87**, 1358–1369.

Menzies, M.A. (1983). Mantle ultramafic xenoliths in alkaline magmas: evidence for mantle heterogeneity modified by magmatic activity. In: *Continental Basalts and Mantle Xenoliths* (Eds. C.J. Hawkesworth and M.J. Norry), pp. 92–110; Shiva Publishing, Nantwich.

Menzies, M.A., Leeman, W.P. and Hawkesworth, C.J. (1983). Isotope geochemistry of Cenozoic volcanic rocks reveals mantle heterogeneity below western USA, *Nature,* **303**, 205–209.

Minster, J.B., Jordan, T.H., Molnar, P. and Haines, E. (1974). Numerical modelling of instantaneous plate tectonics, *Geophys. J. R. Astr. Soc.,* **36**, 541.

Moorbath, S., Thorpe, R.S. and Gibson, I.L. (1978). Strontium isotope evidence for petrogenesis of Mexican andesites, *Nature,* **271**, 437–439.

Moore, N.D. (1979). The geology and geochronology of the Arequipa segment of the Coastal Batholith of Peru, *PhD Thesis, University of Liverpool* (unpubl.).

Moreno, H. (1974). Airplane flight over active volcanoes of central-south Chile. *IAVCEI Guidebook excursion D-3;* Int. Symposium Volcanology: Andean and Antarctic Volcanology Problems, Santiago.

Moreno, H. (1975). Caracteristicas Petrológicas del volcanismo Cenozoico Superior en los Andes del Sur de Chile (39°00′–41°30′S), *Actas VI Congreso Geológico Argentino,* **2**, 131–147.

Moreno, H. (1976). The Upper Cenozoic volcanism in the Andes of Southern Chile, In: *IAVCEI Proc. Symp. on Andean and Antarctic Volcanology Problems* (Ed. O. González F.), pp. 143–171; Santiago.

Morioka, M. and Kigoshi, K. (1978). Lead isotopes in mantle derived xenoliths from Japan and South Africa, *Geoch. J.,* **12**, 223–228.

Morris, J.D. and Hart, S.R. (1983). Isotopic and incompatible element constraints on the genesis of island arc volcanics, Cold Bay and Amak Island, Aleutians, *Geochim. et Cosmochim. Acta,* in press.

Morris, R.C. and Aleman, A. (1975). Sedimentation and tectonics of middle Cretaceous Copa Sombrero formation in Northwest Peru, *Bol. Soc. Geol. Peru,* **48**, 49–64.

Muehlenbachs, K. and Clayton, R.N. (1972). Oxygen isotope studies of fresh and weathered submarine basalts, *Can. J. Earth Sciences,* **9**, 172–184.

Muehlenbachs, K. and Kushiro, I. (1974). Oxygen isotope exchange and equilibrium of silicates with CO_2 or O_2, *Yb. Carnegie Inst. Wash.,* **73**, 232–236.

Muehlenbachs, K. and Stern, C.R. (1980). $\delta^{18}O$ variations among calc-alkaline volcanic rocks from southern Chile; implications for the effects of ridge subduction on the source region of orogenic volcanism, *Trans. Am. Geophys. Union,* **61**, 401.

Muehlenbachs, K., Anderson, A.T. and Sigvaldason, G.E. (1974). Low-^{18}O basalts from Iceland, *Geochim. et Cosmochim. Acta,* **38**, 577–588.

Mullan, H.S. and Bussell, M.A. (1977). The basic rock series in batholithic associations, *Geol. Mag.,* **114**, 265–280.

Munizaga, F. (1978). Geologia del complejo volcánico Laguna del Maule, *Thesis Dep. Geol. Univ. Chile.*

Muñoz, J. and Niemeyer, H. (1983). Volcanismo andesitico miocénico de la cordillera de los Andes de Chile Central (35–38°S), *Ser. Geol. Min.,* Santiago, in press.

Myers, J.S. (1974). Cretaceous stratigraphy and structure, Western Andes of Peru between latitudes 10–10°30′, *Bull. Am. Ass. Petrol. Geol.,* **58**, 474–487.

Myers, J.S. (1975a). Cauldron subsidence and fluidisation: mechanisms of intrusion of the Coastal Batholith of Peru into its own volcanic ejecta, *Bull. Geol. Soc. Am.,* **86**, 1209–1220.

Myers, J.S. (1975b). Vertical crustal movements in the Andes of Peru, *Nature,* **254**, 672–674.

Mysen, B.O. (1979). Trace-element partitioning between garnet peridotite minerals and water-rich vapor: experimental data from 5 to 30 kbar, *Am. Mineral.,* **64**, 274–287.

Mysen, B.O. (1983). Rare earth element partitioning between (H_2O + CO_2) vapor and upper mantle minerals: experimental data bearing on the conditions of formation of alkali basalt and kimberlite, *N. J. Miner. Abh.,* **146**, 41–65.

Nelson, E. and Elthon, D. Geochemistry and field relations of granitic rocks from the Cordillera Darwin, Chile, *Contr. Mineral. Petrol.,* in press.

Neumann, E.R. (1978). Petrology of plutonic rocks. In: *The Oslo Paleorift* (Eds. J.A. Dons and B.T. Larsen), pp. 25–34; Norges Geologiske Undersøkelse Bulletin 45.

Newell, N.D. (1949). Geology of the Lake Titacaca region, Peru and Bolivia, *Mem. Geol. Soc. Am.,* **36**.

Newell, N.D., Chronic, J. and Roberts, T. (1953). Upper Paleozoic of Perú, *Mem. Geol. Soc. Am.*, **58**.

Newhall, C.G. (1981). Geology of the Atitlan area, Guatemala: A study of subduction zone volcanism and caldera formation, *PhD Dissertation, Dartmouth College* (unpubl.).

Nicholls, I.A. (1971). Petrology of Santorini volcano, cyclades, Greece, *J. Petrol.*, **12**, 67–119.

Nicholls, I.A. and Harris, K.L. (1980). Experimental rare earth element partition coefficients for garnet, clinopyroxene and amphibole coexisting with andesitic and basaltic liquids, *Geochim. et Cosmochim. Acta*, **44**, 287–308.

Nicholls, I.A. and Ringwood, A.E. (1973). Effect of water on olivine stability in tholeiites and the production of silica-saturated magmas in the island-arc environment, *J. Geol.*, **81**, 285–300.

Nicholls, J. and Carmichael, I.S.E. (1969). A commentary on the absarokite-shoshonite-banakite series of Wyoming, U.S.A., *Schweizerische Mineralogische und Petrographische Mitteilungen*, **49**, 47–64.

Niemeyer, H. (1978). Nódulos máficos y ultramáficos en basaltos alcalinos de la Meseta Buenos Aires, Lago General Carrera, Provincia de Aysén, Chile, *Revista Asociación Geológica Argentina*, **33**, 63–75.

Noble, D.C., Drake, J.C. and Whallen, M.K. (1969). Some preliminary observations on compositional variations within the pumice and scain flow deposits of Mount Mazama, *Proc. The Andesite Conference, Oregon Dept. Geol. Mineral Ind. Bull.*, **65**, 157–164.

Noble, D.C., Peterson, P.S., Vogel, T.A., Landis, G.P., Grant, N.K. and Jezek, P.A. (1982). Rare-element-enriched ilmenite-series ash flow tuffs containing phenocrystic muscovite, andalusite, and sillimanite, Macusani Region, southeastern Peru (Abstr.), *Geol. Soc. Am. Ann. Meeting, New Orleans*, Program with Abstracts, **14**.

Noble, D.C., Silbermann, M.L., Mégard, F. and Bowman, H.R. (1978). Comendite (peralkaline rhyolite) in the Mitu Group, Central Peru: Evidence of Permian-Triassic crustal extension in the Central Andes, *J. Res. US Geol. Surv.*, **6**, 453–457.

Norton, D. (1977). Fluid circulation in the earth's crust, *Am. Geophys. Union Monograph*, **20**, 693–703.

Nur, A. and Ben-Avraham, Z. (1981). Volcanic gaps and the consumption of aseismic ridges in South America, *Mem. Geol. Soc. Am.*, **154**, 729–740.

Offler, R., Aguirre, L., Levi, B. and Child, S. (1980). Burial metamorphism in rocks of the Western Andes of Peru, *Lithos*, **13**, 31–42.

O'Hara, M.J. (1977). Geochemical evolution during fractional crystallization of a periodically filled magma chamber, *Nature*, **266**, 503–507.

O'Hara, M.J. and Mathews, R.E. (1981). Geochemical evolution in an advancing, periodically replenished, periodically tapped, continually fractionated magma chamber, *J. Geol. Soc.*, **138**, 237–278.

O'Neil, J.R. and Taylor, H.P. (1967). The oxygen isotope and cation exchange chemistry of feldspars, *Am. Mineral.*, **52**, 1414–1437.

O'Nions, R.K. Hamilton, P.J. and Evensen, N.M. (1977). Variations in $^{143}Nd/^{144}Nd$ and $^{87}Sr/^{86}Sr$ ratios in oceanic basalts, *Earth Planet. Sci. Lett.*, **34**, 13–22.

Oversby, V.M. (1971). Lead in oceanic islands: Faial, Azores and Trinidade, *Earth Planet. Sci. Lett.*, **11**, 401–406.

Oversby, V.M. (1972). Genetic relations among volcanic rocks of Reunion: chemical and lead isotopic evidence, *Geochim. et Cosmochim. Acta*, **36**, 1167–1179.

Oyarzún, J. (1979). Depósitos porfiricos y metalogénesis Andina, *Actas II Congreso Geológico Chileno*, **2**, C161–C179.

Oyarzún, J. and Frutos, J. (1980). Metallogénesis and porphyry deposits of the Andes (Southeastern Pacific Region), *Publ. 26 Congres Geologique International*, **C1**, 50–62.

Palacios, C. and López-Escobar, L. (1979). Geoquímica y petrología de andesitas cuaternarias de los Andes Centrales (18°57′–19°28′S) *Actas II Congreso Geológico Chileno*, **3**, E73–E88.

Palma, V.V. (1981). The San Rafael Tin-Copper Deposit, Puno, S.E. Peru. *MSc Thesis, Queen's University, Kingston, Ontario* (unpubl.).

Peacock, M.A. (1931). Classification of igneous rock series, *J. Geol.* **39**, 54–67.

Pearce, J.A. (1982). Trace element characteristics of lavas from destructive plate boundaries. In: *Andesites: Orogenic Andesites and Related Rocks* (Ed. R.S. Thorpe), pp. 525–548; Wiley, New York.

Peccerillo, A. and Taylor, S.R. (1975). Geochemistry of upper Cretaceous volcanic rocks from the Pontic Chain, northern Turkey, *Bull. Volc.*, **39**, 557–569.

Peccerillo, A. and Taylor, S.R. (1976). Geochemistry of Eocene calc-alkaline volcanic rocks from the Kastamann area, Turkey, *Contr. Mineral. Petrol.*, **58**, 63–69.

Perfit, M.R., Gust, D.A., Bence, A.E., Arculus, R.J. and Taylor, S.R. (1980). Chemical characteristics of island-arc basalts: implications for mantle sources, *Chem. Geol.*, **30**, 227–256.

Petersen, G. and Vidal, C. Tres épocas metalogenéticas evidenciadas en el Cenozoico del Perú, *Bol. Soc. Geol. Peru*, in press.

Pineau, F. and Javoy, M. (1969). Déterminations des rapports $^{18}O/^{16}O$ et $^{13}C/^{12}C$ dans diverses carbonatites: implications génétiques, *C. R. Acad. Paris*, **269**, 1930–1933.

Pineau, F., Javoy, M., Hawkins, J.W. and Craig, H. (1976). Oxygen isotope variations in marginal basin and ocean ridge basalts, *Earth Planet. Sci. Lett.*, **28**, 299–307.

Pitcher, W.S. (1972). The Coastal Batholith of Peru: some structural aspects, *24th IGC sec. 2*, 156–163.

Pitcher, W.S. (1974). The Mesozoic and Cenozoic Batholiths of Peru, *Pacific Geology*, **8**, 51–63.

Pitcher, W.S. (1978). The anatomy of a batholith, *J. Geol. Soc. Lond.*, **135**, 157–182. Reprinted in Spanish, *Bol. Soc. Geol. Peru*, **60**, 13–60 (1979).

Pitcher, W.S. (1983). Granite type and tectonic environment. In: *Mountain Building Processes* (Ed. K. Hsu), pp. 19–40; Academic Press, London.

Pitcher, W.S. and Bussell, M.A. (1977). Structural control of batholith emplacement in Peru: a review, *J. Geol. Soc. Lond.*, **133**, 249–56.

Polanski, J. (1964). Descripcion geologica de la hoja 26c La Tosca, Provincia de Mendosa, *Direcion Geologica y Mineria Boletin*, **101**.

Powell, J.L. and Bell, K. (1974). Isotopic composition of strontium in alkaline rocks. In: *The Alkaline Rocks* (Ed. H. Sorensen), pp. 412–421; Wiley and Sons, London.

Pushkar, P. (1968). Strontium isotope ratios in volcanic rocks of three island arc systems, *J. geophys. Res.*, **73**, 2701–2714.

Quensel, P.D. (1911). Geologisch-petrographische studien in den Patagonischen Cordillera, *Bull. Geol. Inst. Upsala*, 1–114.

Quirt, S., Clark, A.H., Farrar, E. and Sillitoe, R.H. (1971). Potassium-Argon ages of porphyry copper deposits in northern and central Chile, *Geol. Soc. Am. Ann. Meetings*, **3**, 676–677.

Regan, P.F. (1976). The genesis and emplacement of mafic plutonic rocks of the coastal Andean batholith, Lima Province, Peru, *PhD Thesis, University of Liverpool.* (Unpubl., quoted in Cobbing *et al.*, 1981).

Reynolds, P.H. and Dasch, E.J. (1971). Lead isotopes in marine manganese nodules and the ore-lead growth curve, *J. Geophys. Res.*, **76**, 5124–5129.

Richardson, S.H., Erlank, A.J., Duncan, A.R. and Reid, D.L. (1982). Correlated Nd, Sr and Pb isotopic variation in Walvis Ridge basalts and implication for the evolution of their mantle source, *Earth Planet. Sci. Lett.*, **59**, 327–342.

Ritchey, J.L. (1980). Divergent magmas at Crater Lake, Oregon: Products of fractional crystallization and vertical zoning in a shallow, water-undersaturated chamber, *J. Volc. Geoth. Res.*, **7**, 373–386.

Rivera, R., Petersen, G. and Rivera, M. (1975). Estratigrafia de la Costa de Lima, *Bol. Soc. Geol. Peru*, **45**, 159–186.

Roden, M.F., Frey, F.A. and Francis, D.M. (1984). An example of consequent mantle metasomatism in peridotite inclusions from Nunivak Island, Alaska *J. Petrol.*, in press.

Roobol, M.J. and Francis, P.W. (1976). Physico-chemical characteristics of the Andean volcanic basic between latitudes 21 and 22°S, *Proc. Symposium Andean and Antarctic Vulcanology Problems*, Santiago, Chile, 1974, IAVCET.

Rose, Jr., W.I. (1972). Santiaguito volcanic dome, Guatemala, *Bull. Geol. Soc. Am.*, **83**, 1413–1434.

Rose, Jr., W.I., Anderson, A.T., Bonis, S. and Woodruff, L.G. (1978). The October 1974 basaltic tephra, Fuego volcano, Gautemalam description and history of the magma body, *J. Volc. Geoth. Res.*, **4**, 3–53.

Rose, Jr., W.I., Grant, N.K., Hahn, G.A., Lange, I.M., Powell, J.L., Easter, J. and DeGraff, J.M. (1977). The evolution of Santa Maria volcano, Guatemala, *J. Geol.*, **85**, 63–87.

Rose, Jr, W.I., Newhall, C.G., Bornhorst, T.J. and Self, S. Quaternary rhyolitic pyroclastic deposits of the Atitlan caldera, Guatemala, in prep.

Russo, A. and Rodrigo, G. (1965). Estratigrafiá y paleogeografiá del Grupo Puca en Bolivia, *Bolivian IBP*, **5**, 5–51.

Rutland, R.W.R. (1971). Andean orogeny and ocean floor spreading, *Nature*, **233**, 252–255.

Sacks, I.S. (1977). Interrelationships between volcanism, seismicity, and anelasticity in western Southern America, *Tectonophysics*, **37**, 131–139.

Sanchez, A.W. (1982). Edades Rb–Sr en los segmentos Arequipa-Toquepala del batolito de la costa del Perú, *Quinto Congreso Latinamericano de Geologîa*, Argentina, Actas, **111**, 487–504.

Sanderson, L.M. and Atherton, M.P. The geochemistry of the Coastal Batholith of Peru, in press.

Santana, R. (1971). Les Cendres Volcaiques de la vallee du Cachapoal-Rapel, Chile, *Cahiers de Geographie de Quebec*, **35**, 315–332.

Saunders, A.D. and Tarney, J. (1979). The geochemistry of basalts from a back-arc spreading centre in the East Scotia Sea, *Geochim. et Cosmochim. Acta*, **43**, 555–572.

Saunders, A.D. and Tarney, J. The geochemical characteristics of basaltic volcanism within back-arc basins. In: *Volcanic Processes in Marginal Basins* (Eds. B.P. Kokelaar, M.F. Howells and R.A. Roach); Spec. Publ. Geol. Soc. Lond., in press.

Saunders, A.D., Tarney, J., Stern, C.R. and Dalziel, I.W.D. (1979). Geochemistry of Mesozoic marginal basic floor igneous rocks from southern Chile, *Bull. Geol. Soc. Am.*, **90**, 237–258.

Saunders, A.D., Tarney, J. and Weaver, D. (1980). Transverse geochemical variations across the Antarctic Peninsula: implications for the genesis of calc-alkali magmas, *Earth Planet. Sci. Lett.*, **46**, 344–366.

Savin, S. and Epstein, S. (1970). The oxygen and hydrogen isotope geochemistry of ocean-sediments and shales, *Geochim. et Cosmochim. Acta*, **34**, 43–63.

Scholl, D.W., Marlow, M.S. and Cooper, A.K. (1977). Sediment subduction and offscraping at Pacific margins. In: *Island Arcs, Back-Arc Basins and Deep Sea Trenches* (Eds. M. Talwani and W.C. Pitman III), pp. 199–210; American Geophysical Union, Washington.

Schroeder, B., Thompson, G., Sulanowska, M. and Ludden, J.N. (1980). Analysis of geological materials using an automated X-ray fluorescence system, *X-ray Spectrom*, **9**, 198–205.

Shackleton, R.M., Ries, A.C., Coward, M.P., and Cobbold, P.R. (1979). Structure, metamorphism and geochronology of the Arequipa Massif of coastal Peru, *J. Geol. Soc. Lond.*, **136**, 195–214.

Sigurdsson, H. and Sparks, R.S.J. (1981). Petrology of rhyolitic and mixed magma ejected from the 1875 eruption of Askja, Iceland, *J. Petrol.*, **22**, 41–84.

Skewes, M.A. and Stern, C.R. (1979). Petrology and geochemistry of alkali basalts and ultramafic inclusions from the Palei-Aike volcanic field in southern Chile and the origin of the Patagonian plateau basalts, *J. Volc. Geoth. Res.*, **6**, 3–25.

Smith, R.L. and Shaw, H.R. (1975). Igneous related geothermal systems. In: *Assessment of geothermal resources of the United States*, pp. 58–84; US Geol. Survey, Circular 726.

SNGM (Servicio Nacional Geologia y Mineria) (1982). Geological Map of Chile, Toconao sheet and memoir; Santiago.

Sorensen, H. (Ed.) (1974). *The Alkaline Rocks*; J. Wiley and Sons, London.

Sparks, R.S.J., Meyer, P. and Sigurdsson, H. (1980). Density variation amongst mid-ocean ridge basalts: Implications for magma mixing and the scarcity of primitive lavas, *Earth Planet. Sci. Lett.*, **46**, 419–430.

Sparks, S.R.J., Sigurdsson, H. and Wilson, L. (1977). Magma mixing: A mechanism for triggering acid explosive eruptions, *Nature*, **267**, 315–318.

Spera, F.J. and Crisp, J.A. (1981). Eruption volume, periodicity, and caldera area: Relationships and inferences on development of compositional zonation in silicic magma chambers, *J. Volc. Geoth. Res.*, **11**, 169–187.

Stacey, J.S. and Kramers, J.D. (1975). Approximation of terrestrial lead isotope evolution by a two-stage model, *Earth Planet. Sci. Lett.*, **26**, 207–221.

Stauder, W. (1973). Mechanism and spatial distribution of Chilean earthquakes with relation to subduction of the oceanic plate, *J. Geophys. Res.*, **78**, 5033–5061.

Stern, C.R. (1974). Melting products of olivine tholeiite basalt in subduction zones, *Geology*, **2**, 227–230.

Stern, C.R. (1979). Recent calc-alkaline volcanism in southern South America: distribution and petrochemistry in relation to the subduction of the Chile Rise, *Trans. Am. Geophys. Union*, **60**, 414.

Stern, C.R. (1980) Geochemistry of Chilean ophiolites: evidence for the compositional evolution of the mantle source of back-arc basin basalts, *J. Geophys. Res.*, **85**, 955–966.

Stern, C.R. (1981). $^{87}Sr/^{86}Sr$ variations among calc-alkaline volcanic rocks from the southernmost Andes; implications for the effects of ridge subduction on orogenic volcanism, *Geol. Soc. Am.*, Abstracts with Programs, **13**, 560.

Stern, C.R. (1982). Mt. Burney, the southernmost active stratovolcano in the Andes: its history, petrology and origin, *Abs. Tercero Congreso Chileno*, 52–53.

Stern, C.R. and Futa, K. (1982a). Nd and Sr isotope ratios of calc-alkaline volcanic rocks from the southern Andes, *Trans. Am. Geophys. Union*, **63**, 456–457.

Stern, C.R. and Futa, K. (1982b). An Andean andesite derived directly from subducted MORB or from LIL-depleted subcontinental mantle, *Trans. Am. Geophys. Union*, **63**, 1148.

Stern, C.R. and Wyllie, P.J. (1978). Phase compositions through the crystallization intervals of basalt-andesite-H_2O at 30 kb with implications for subduction zone magmatism, *Am. Mineral.*, **63**, 641–663.

Stern, C.R., Charrier, R., Varela, J., Godoy, E., Hervé, F. and Skewes, M.A. Voluminous late-Pleistocene high-silica rhyolite ash-flow derived from the Maipo volcano, central Chilean Andes, *Revista Geologica de Chile*, in press.

Stern, C.R., Futa, K. and Peng, R.E. (1983). Sr, Nd and Pb isotopic compositions of alkali basalts from the Patagonian plateau lavas, South America, *Geol. Soc. Am.*, Abstracts with Programs.

Stern, C.R., Herve, M., Puig, A., Piraces, R., Suarez, M. and Cepeda, A. (1979). Petrology of post-Pleistocene volcanic domes on Cook Island in the southernmost Andes, *Abstracts, Segundo Congreso Geologico Chileno*, p. 78.

Stern, C.R., Skewes, M.A. and Duran, M. (1976). Volcanismo calc-alkaline en Chile austral, *Actas Primer Congreso Geologico Chileno*, **2**, 195–212.

Stern, R.J. (1979). On the origin of andesite in the Northern Mariana Island Arc: Implication from Agrigan, *Cont. Mineral. Petrol.*, **68**, 207–219.

Stern, R.J. (1981). A common mantle source for Mariana and Volcano arc and Caroline and Kilauea "hotspot" magmas? A consideration of the trace element and isotopic evidence, *Trans. Am. Geophys. Union*, **62**, 408.

Stern, R.J. and Ito, E. (1983). Trace element and isotopic constraints on the source of magmas in the active Volcano and Mariana Island Arcs, Western Pacific.

Stewart, J.W., Evernden, J.F. and Snelling, N.J. (1974). Age determinations from Andean Peru: A reconnaissance survey, *Bull. Geol. Soc. Am.*, **85**, 1107–1116.

Stille, P., Unruth, D.M. and Tatsumoto, M. (1983). Pb, Sr, Nd and Hf isotopic evidence of multiple sources for Oahu, Hawaii basalts, *Nature*, **304**, 25–29.

Strong, D.F. and Hanmer, S.F. (1981). The leucogranites of southern Britanny: origin by faulting, frictional history, fluid flux and fractional melting, *Can. Mineral.*, **19**, 163–176.

Sun, S.S. (1980). Lead isotopic study of young volcanic rocks from mid-ocean ridges, oceanic islands and island arcs, *Phil. Trans. R. Soc. Lond.*, **A297**, 409–445.

Sun, S.S. and Jahn, B.M. (1975). Lead and strontium isotopes from post-glacial basalts from Iceland, *Nature*, **255**, 527–530.

Sun, S.S., Nesbitt, R.W. and Sharaskin, A.Y. (1979). Geochemical characteristics of mid-ocean ridge basalts, *Earth Planet. Sci. Lett.*, **44**, 119–138.

Sun, S.S., Tatsumoto, M. and Schilling, J.G. (1975). Mantle plume mixing along the Reykjanes Ridge axis: lead isotope evidence, *Science*, **30**, 143–147.

Suwa, K., Oana, S., Wada, H. and Osaki, S. (1975). Isotope geochemistry and petrology of African carbonatites. In: *Physics and Chemistry of the Earth* (Eds. L.H. Ahrens, J.B. Dawson, A.R. Duncan and A.J. Grlark), **9**, 735–745; Pergamon Press, Oxford.

Swift, S. and Carr, M. (1975). The segmented nature of the Chilean seismic zone, *Phys. Earth Plan. Int.*, **9**, 183–191.

Takahashi, M., Aramaki, S. and Ishihara, S. (1980). Magnetite-series/ilmenite-series vs. I-type/S-type granitoids. In: *Granitic Magmatism and Related Mineralization* (Eds. S. Ishihara and S. Takenouchi), Mining Geology, Japan, Special Issue, **8**, 13–28.

Tarney, J. and Saunders, A.D. (1979). Trace element constraints on the origin of Cordilleran batholiths. In: *Origin of Granitic Batholiths: Geochemical Evidence* (Eds. M.P. Atherton and J. Tarney), pp. 90–105; Shiva Publishing, Nantwich.

Tarney, J., Saunders, A.D., Mattey, D.P., Wood, D.A. and Marsh, N.G. (1981). Geochemical aspects of back-arc spreading in the Scotia Sea and Western Pacific, *Phil. Trans. R. Soc. Lond.*, **A300**, 263–285.

Tatsumoto, M. (1978). Isotopic composition of lead in oceanic basalts and its implication to mantle evolution, *Earth Planet. Sci. Lett.*, **38**, 63–87.

Taylor, H.P. (1968). The oxygen isotope geochemistry of igneous rocks, *Contr. Mineral. Petrol.*, **19**, 1–71.

Taylor Jr, H.P. (1980). The effects of assimilation of country rocks by magmas on $^{18}O/^{16}O$ and $^{87}Sr/^{86}Sr$ systematics in igneous rocks, *Earth Planet. Sci. Lett.*, **47**, 243–254.

Taylor, H.P. and Epstein, S. (1962). Relationship between $^{18}O/^{16}O$ ratios in coexisting minerals of igneous and metamorphic rocks. Part 1. Principles and experimental results, *Bull. Geol. Soc. Am.*, **73**, 461–480; Part 2. Application to petrological problems, *Bull. Geol. Soc. Am.*, **73**, 675–694.

Taylor, H.P. and Forester, R.W. (1971). Low-^{18}O igneous rocks from the intrusive complexes of Skye, Mull, and Ardnamurchan, Western Scotland, *J. Petrol.*, **12**, 465–497.

Taylor, J.P., Giannetti, B. and Turi, B. (1979). Oxygen isotope geochemistry of potassic igneous rocks from the Roccamontina volcano, Roman Comagmatic Region, Italy, *Earth Planet. Sci. Lett.*, **46**, 81–106.

Taylor, R.P., Strong, D.F. and Fryer, B.J. (1981). Volatile control of contrasting trace element distributions in peralkaline granitic and volcanic rocks, *Contr. Mineral. Petrol*, **77**, 267–271.

Taylor, R.P., Strong, D.F. and Kean, B.F. (1980). The Topsails igneous complex: Silurian-Devonian peralkaline magmatism in western Newfoundland, *Can. J. Earth Sci.*, **17**, 425–439.

Taylor, S.R. and Gorton, M.P. (1977). Applications of spark source spectrometry III. Element sensitivity, precision and accuracy, *Geochim. et Cosmochim. Acta*, **41**, 1375–1380.

Taylor, S.R. and McLennan, S.M. (1981). The composition and evolution of the continental crust: rare earth evidence from sedimentary rocks, *Phil. Trans. R. Soc. Lond.*, **A301**, 381–399.

Taylor, W.P. (1976). Intrusion and differentiation of granitic magma at a high level in the crust: the Puscao pluton, Lima Province, Peru, *J. Petrol*, **17**, 194–218.

Tera, F. (1982). On the episodic evolution of the earth—the four sources of magma, *Trans. Am. Geophys. Union*, **63**, 460.

Thiele, R. and Katsui, Y. (1969). Contribucion al conocimiento del volcanismo post-Miocenico de los Andes en la provincia de Santiago, Chile, *Publicacion del Departmento de Geologia, Universidad de Chile*, **35**, 3–23.

Thornburg, T. and Kulm, L.O. (1981). Sedimentary basins of the Peru Continental margin: Structure, stratigraphy and Cenozoic tectonics from 6°S to 16°S latitude, *Mem. Geol. Soc. Am.*, **154**, 393–422.

Thorpe, R.S. and Francis, P.W. (1979). Variations in Andean andesite compositions and their petrogenetic significance, *Tectonophysics*, **57**, 53–70.

Thorpe, R.S., Francis, P.W., Hammill, M. and Baker, M.C.W. (1982). The Andes. In: *Andesites: Orogenic Andesites and Related Rocks* (Ed. R.S. Thorpe), pp. 187–206; Wiley, London.

Thorpe, R.S., Francis, P.W. and Harmon, R.S. (1981). Andean andesites and crustal growth, *Phil. Trans. R. Soc. Lond.*, **A301**, 305–320.

Thorpe, R.S., Francis, P.W. and Moorbath, S. (1979a). Rare earth and strontium isotope evidence concerning the petrogenesis of North Chilean ignimbrites, *Earth Planet. Sci. Lett.*, **42**, 359–367.

Thorpe, R.S., Francis, P.W. and Moorbath, S. (1979b). Strontium isotope evidence for petrogenesis of Central American andesites, *Nature,* **277**, 44–45.

Thorpe, R.S., Francis, P.W. and O'Callaghan. Relative roles of source composition, fractional crystallization and crustal contamination in the petrogenesis of Andean volcanic rocks, *Phil. Trans. R. Soc. Lond.,* **A**, in press.

Thorpe, R.S., Potts, P.J. and Francis, P.W. (1976). Rare earth data and petrogenesis of andesite from the North Chilean Andes, *Contr. Mineral. Petrol.,* **54**, 65–78.

Tilton, G.R. (1979). Isotopic studies of Cenozoic Andean calc-alkaline rocks, *Yb. Carnegie Inst. Wash.,* **78**, 298–304.

Tilton, G.R. (1983). Evolution of depleted mantle: the lead perspective, *Geochim. et Cosmochim. Acta,* **47**, 1191–1197.

Tilton, G.R. and Barreiro, B.A. (1980). Origin of lead in Andean calcalkaline lavas, southern Peru, *Science,* **210**, 1245–1247.

Tilton, G.R., Pollak, R.J., Clark, A.H. and Robertson, R.C. (1981). Isotopic compositions of Pb in Central Andean ore deposits. In: *Nazca Plate: Crustal Formation and Andean Convergence* (Eds. Kulm *et al*), Geol. Soc. Am. Mem. 154, 791–816.

Turi, B. and Taylor, H.P. (1976). Oxygen isotope studies of potassic volcanics of the Roman Province, central Italy, *Contr. Mineral. Petrol.,* **55**, 1–31.

Unruh, D.M. and Tatsumoto, M. (1976). Lead isotopic composition and uranium, thorium and lead concentrations in sediments and basalts from the Nazca plate. In: Initial Reports of the Deep Sea Drilling Project 34 (U.S. Government Printing Office, Washington, D.C.), 341–347.

Vargas, L. (1970). Geología del cuadrángulo de Arequipa, *Bol. Ser. Geol. Min. Peru.,* 24.

Vergara, M. (1970). Note on the Zonation of the Upper Cenozoic volcanism of the Andean area of central-south Chile and Argentina, *Conf. Solid Earth Problems, Buenos Aires, Internat. Upper Mantle Project,* **2**, 381–397.

Vergara, M. and Katsui, Y. (1969). Contribución a la geología y petrología del volcán Antuco, cordillera de los Andes, Chile central, *Publ. Dep. Geol. Univ. Chile,* **35**, 25–47.

Vergara, M. and López-Escobar, L. (1980). Geología y petrología de los intrusivos subvolcánicos de la pre-codillera andina entre Santiago y Colina, *Comunicaciones Dep. Geol. Univ. Chile,* **29**, 1–21.

Vergara, M. and Munizaga, F. (1974). Age and evolution of the Upper Cenozoic andesitic volcanism in central-south Chile, *Bull. Geol. Soc. Am.,* **85**, 603–606.

Vidal, C. (1980). Mineral deposits associated with the Peruvian Coastal Batholith, *PhD Thesis, University of Liverpool* (unpubl.).

Volfinger, M. (1976). Effet de la temperature sur les distributions de Na, Rb et Cs entre la sanidine, la muscovite, la phlogopite et une solution hydrothermale sous une pression de 1 kbar, *Geochim. et Cosmochim. Acta,* **40**, 267–282.

Wager, L.R. (1962). Igneous cumulates from the 1902 eruption of Soufriere, St. Vincent, *Bull. Volc.,* **24**, 93–99.

Wagner, L.R. (1982). Igneous cumulates from the 1902 eruption of Soufriere, St. Vincent, *Bull. Volc.,* **24**, 93.

Watson, E.B. (1982). Basalt contamination by continental crust: some experiments and models, *Contr. Mineral. Petrol.,* **80**, 73–87.

Watson, E.B. and Jurewicz, S.R. (1983). Behavior of alkalis during diffusive interaction of granitic xenoliths with basaltic magma, *Trans. Am. Geophys. Union,* **64**, 336.

Weaver, B.L. and Tarney, J. (1980). Rare-earth geochemistry of Lewisian granulite-facies gneisses, NW Scotland: implications for the petrogenesis of the lower continental crust, *Earth Planet. Sci. Lett.,* **51**, 279–296.

Weaver, B.L. and Tarney, J. (1981). Lewisian gneiss geochemistry and Archaean crustal development models, *Earth Planet. Sci. Lett.,* **55**, 171–180.

Weaver, B.L. and Tarney, J. (1983). Chemistry of the sub-continental mantle: inferences from Archaean and Proterozoic dykes and continental flood basalts. In: *Continental Basalts and Mantle Xenoliths* (Eds. C.J. Hawkesworth and M.J. Norry), pp. 209–229; Shiva Publishing, Nantwich.

Weaver, S.D., Saunders, A.D., Pankhurst, R.J. and Tarney, J. (1979). A geochemical study of magmatism associated with the initial stages of back-arc spreading: the Quaternary volcanics of Bransfield Strait, South Shetland Islands, *Contr. Mineral. Petrol.*, **68**, 151–169.

Webb, S. (1976). The volcanic envelope of the Coastal Batholith in Lima and Ancash, Peru, *PhD Thesis, University of Liverpool* (unpubl.).

White, A.J.R. and Chappell, B.W. (1977). Ultrametamorphism and granitoid genesis, *Tectonophysics*, **43**, 7–22.

White, W. (1979). Pb isotope geochemistry of the Galapagos Islands, *Yb. Carnegie Inst. Wash.*, **78**, 331–335.

Whitford, D.J. (1975). Geochemistry and petrology of volcanic rocks from the Sunda arc, Indonesia, *PhD Thesis, Australian National University, Canberra.*

Whitford, D.J. and Bloomfield, K. (1976). Geochemistry of late-Cenozoic volcanic rocks from the Nevado de Toluca area, Mexico, *Yb. Carnegie Inst. Wash.*, **75**, 207–213.

Whitford, D.J., Nicholls, I.A. and Taylor, S.R. (1979). Spatial variations in the geochemistry of quaternary lavas across the Sunda arc in Java and Bali, *Contr. Mineral. Petrol.*, **70**, 341–346.

Whitford, D.J., White, W.M. and Jezek, P.A. (1981). Neodymium isotopic composition of quaternary island arc lavas from Indonesia, *Geochim. et Cosmochim. Acta*, **45**, 989–995.

Wilson, J.J. (1963). Cretaceous stratigraphy of the central Andes of Peru, *Bull. Am. Ass. Pet. Geol.*, **47**, 1–54.

Wilson, J.J., Reyes, L. and Garayar, J. (1967). Geologia de los cuadrángulos de Mollebamba, Tayabamba, Huaylas, Pomabamba Carhuazy Huari. Bol. Ser. Geol. Min. Peru, **16**.

Wilson, P.A. (1975). K–Ar age studies in Peru with special reference to the emplacement of the Coastal Batholith, *PhD Thesis, University of Liverpool* (unpubl.).

Wortel, M.J.R. and Vlaar, N.J. (1978). Age-dependent subduction of oceanic lithosphere beneath western South America, *Phys. Earth Planet. Int.*, **17**, 201–208.

Wyllie, P.J. (1977), Crustal anatexis, an experimental review, *Tectonophysics*, **43**, 41–71.

Wyllie, P.J. (1982). Subduction products according to experimental prediction, *Bull. Geol. Soc. Am.*, **93**, 468–476.

Zartman, R.E. (1974). Lead isotopic provinces in the Cordillera of the western United States and their geologic significance, *Econ. Geol.*, **69**, 792–805.

Zartman, R.E. and Tera, F. (1973). Lead concentration and isotopic composition in five peridotite inclusions of probable mantle origin, *Earth Planet. Sci. Lett.*, **20**, 54–66.

Zeil, W. (1979). *The Andes, a Geological Review;* Gebruder Born traeger, Berlin.

Zindler, A. (1982). Nd and Sr isotopic studies of komatiites and related rocks. In: *Komatiites*, (Eds. N. Arndt and E.G. Nisbet), pp. 399–420; Allen and Unwin, Boston.

Zindler, A.Z., Jagoutz, E. and Goldstein, S. (1982). Nd, Sr and Pb isotopic systematics in a three component mantle: a new perspective, *Nature*, **298**, 519–523.

Other titles of interest

Origin of Granite Batholiths: Geochemical Evidence
M.P. Atherton & J. Tarney (Editors)

Migmatites, Melting and Metamorphism
M.P. Atherton & C.D. Gribble (Editors)

Continental Basalts and Mantle Xenoliths
C.J. Hawkesworth & M.J. Norry (Editors)

Applied Geochemistry in the 1980s
I. Thornton *et al.* (Editors)

European Dinantian Environments
A.E. Adams, J. Miller & V.P. Wright (Editors)

Geochemistry of Ocean Basalts
J. Tarney (Editor)

